Elementary Statics and Strength of Materials

Elementary Statics and Strength of Materials

Ladislav Cerny
Professor of Civil Engineering
University of Minnesota

McGraw-Hill Book Company
New York St. Louis San Francisco Auckland Bogotá
Hamburg Johannesburg London Madrid Mexico
Montreal New Delhi Panama Paris São Paulo
Singapore Sydney Tokyo Toronto

ELEMENTARY STATICS AND STRENGTH OF MATERIALS

2 3 4 5 6 7 8 9 0 DODO 8 9 8 7 6 5 4 3 2 1

This book was set in Press Roman by Automated Composition Service, Inc.
The editors were Diane D. Heiberg and Susan Hazlett; the design was done by Caliber Design Planning; the production supervisor was Diane Renda.
The drawings were done by J & R Services, Inc.
R. R. Donnelley & Sons Company was printer and binder.

Library of Congress Cataloging in Publication Data

Cerny, Ladislav, date
Elementary statics and strength of materials.

Bibliography: p.
Includes index.
1. Statics. 2. Structures, Theory of.
3. Strength of materials. I. Title.
TA351.C49 624.1'71 80-20442
ISBN 0-07-010339-9

To Mario Salvadori, the innovative,
enthusiastic, and unsurpassable
teacher of all between 6 and 96
who are willing to learn.

Contents

Preface

We are all interested in buildings. We use them, we like them or disapprove of them, and we rent them or pay for their construction. Buildings are no longer the exclusive interest of engineers and architects. City and community planners, environmentalists, sociologists, politicians, and citizen groups raise their voices in favor of or against the construction of housing, stadiums, airports, industrial parks, or factories. Thus engineering activity has acquired an interdisciplinary character, and more and more people become members of decision-making teams that influence construction projects.

This interest has penetrated the press as well as the educational systems of our society. Daily newspapers have columns on architecture and city planning; liberal arts students are exposed to courses in transportation and housing.

Yet very few of us understand how a building is built and why or how it is made to stand up under the weight of its occupants and furniture, the buffeting of wind, or the shaking of an earthquake.

This book has been written for those of you who are puzzled by such questions and would like to find how a building is designed and checked for safety.

The prerequisite for this book is knowledge of the basic algebraic operations and some elementary geometry; no special engineering background is necessary. Those of you who are interested in buildings and their construction and who have these prerequisites can read this book and solve its quantitative problems and exercises. The result of such a reading, followed by working through the problems, will give you a much deeper understanding of construction and possibly encourage you to continue studying this fascinating branch of engineering. But even if it does not, it will broaden your understanding of what your house is or could be, of why skyscrapers can be built in our cities, and of how we may see a solution in the future to the problem of decent housing for each one of us. Thus the book can do a service to a very broad spectrum of readers such as high school students, apprentices, masons, carpenters, draftspersons, accountants, real estate workers, and many others.

It goes without saying that students of architecture as well as students in vocational schools and technical institutes will find in this book those fundamentals of statics that will constitute their daily bread in school and in their professional lives.

It is in particular to these students that the author recommends more than just "reading the book." They should complete the quizzes at the end of each chapter and evaluate their test results by means of the given key, going through a chapter once again if their score is below the suggested minimum. After passing the quiz, the student should work out assigned problems and check the results with the answers given at the end of the book.

Obviously the material in this book is an introduction to a complex and important subject. The method of this book is based on the simplest educational process:

> Seeing an airplane for the first time, a child asks: "What is it?" Answers the parent, "It is an airplane . . . , it is flying," "It is an airplane, airplane . . . , it is flying," repeats the child, "Why is it that it flies?" "It is the engine that makes it to fly," goes the answer.

That is the (temporary) end to the sequence of questions and answers. Later, when the child is older, and interested in airplanes, he or she will receive the answers of any desired depth.

After reading this book, you will realize that there remain many unanswered questions. Some of the technical terms which you might have met for the first time might not be sufficiently clear to you. You will notice that only simplest structures and structural elements are discussed. To answer all your questions adequately, you would probably have to read many more books requiring much additional knowledge. It was not the intent of the author to write such an extensive and complete treatise on structures, but he hopes that your interest in structures will be awakened. If you will be motivated to continue studying this subject and realize why the subject is both essential and fascinating, the author will be totally rewarded for his effort.

ACKNOWLEDGMENT

I gratefully acknowledge the help I have received from Dr. Mario Salvadori, who extensively edited the manuscript. I am also indebted to several "anonymous" reviewers for their valuable comments which helped considerably to improve it. Last but not least, my thanks are directed to Kenneth G. Powell and Louis Ho, both now practicing architects, for the preparation of numerous drawings. Without all this help, the publishing of the book would not be possible.

Ladislav Cerny

Think Metric!

Any practical field of endeavor involving quantities to be measured uses a system of units. In the English-speaking countries the so-called British gravitational system has been used exclusively until some time ago, and we in the United States still talk and measure in inches, feet, or miles and in ounces or pounds. This is particularly true in the construction industry and structural engineering.

But the British system presents many inconvenient features. For example, you have to multiply by 12 to convert from feet to inches and by 16 to go from pounds to ounces. For this and many other reasons almost the entire world has adopted a system called the modern metric, or SI system of units (for Système International d'Unités, in French). This system is so convenient and universal that the United States has set in motion machinery for the conversion from the British to the SI system of units. It is therefore imperative that you start thinking metric, that is, in the SI system of units.

In this book we are going to be interested in measuring lengths, areas, volumes, and forces, so that the units to be considered will be few and easy to learn. Here they are.

The basic unit of length used in the metric system is one meter (1 m). This measure is divided into 10 decimeters (dm). Each decimeter is divided into 10 centimeters (cm), and each centimeter into 10 millimeters (mm)† so that the sequence of units of length is

$$1 \text{ m} = 10 \text{ dm} = 100 \text{ cm} = 1\ 000 \text{ mm}$$

$$1 \text{ dm} = 10 \text{ cm} = 100 \text{ mm}$$

$$1 \text{ cm} = 10 \text{ mm}$$

One square decimeter (1 dm^2) may be divided into 10×10, or 100 square centimeters:

$$1 \text{ dm}^2 = 100 \text{ cm}^2$$

†Deci = 1/10; centi = 1/100; milli = 1/1 000.

Each square centimeter can be further divided into 10 × 10, or 100 square millimeters

$$1\ \text{cm}^2 = 100\ \text{mm}^2$$

Moreover, since one square meter has 10 × 10, or 100 square decimeters

$$1\ \text{m}^2 = 100\ \text{dm}^2$$

The sequence of units of area is

$$\begin{aligned} 1\ \text{m}^2 = 100\ \text{dm}^2 &= 10\ 000\ \text{cm}^2 = 1\ 000\ 000\ \text{mm}^2 \\ 1\ \text{dm}^2 &= 100\ \text{cm}^2 = 10\ 000\ \text{mm}^2 \\ 1\ \text{cm}^2 &= 100\ \text{mm}^2 \end{aligned}$$

If we construct a cube from six plates, each with an area of 1 dm^2, we obtain an enclosed volume of one cubic decimeter (dm^3) [which measures one liter (1 L) of fluid content]. Again, one cubic meter (m^3) contains 10 × 10 × 10, or 1 000 such cubes, that is, 1 000 dm^3. Further, each cubic centimeter has 1 000 cubic millimeters (mm^3), so that the sequence of units of volume is

$$\begin{aligned} 1\ \text{m}^3 = 1\ 000\ \text{dm}^3 &= 1\ 000\ 000\ \text{cm}^3 = 1\ 000\ 000\ 000\ \text{mm}^3 \\ 1\ \text{dm}^3 &= 1\ 000\ \text{cm}^3 = 1\ 000\ 000\ \text{mm}^3 \\ 1\ \text{cm}^3 &= 1\ 000\ \text{mm}^3 \end{aligned}$$

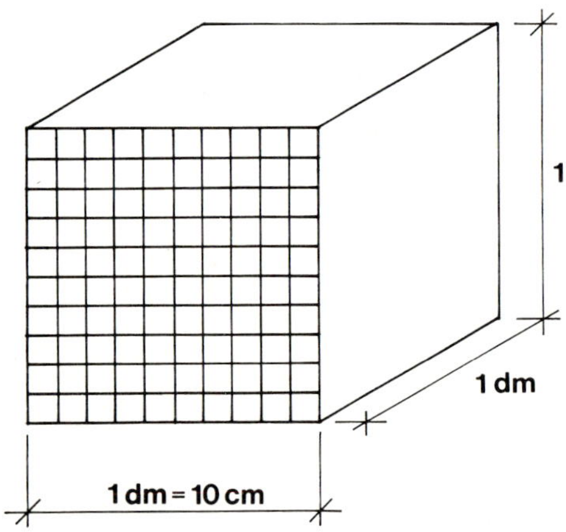

The basic unit of force is one newton (1 N) and

1 000 newtons = one kilonewton (1 kN)

1 000 000 newtons = one meganewton (1 MN)†

†Deca = 10; hecta = 100; kilo = 1 000; mega = 1 000 000.

The preceding units are basic ones from which we can derive the units of other mechanical quantities such as a unit of ***pressure*** or ***stress***, that is, of force per unit area

$$\frac{\text{N}}{\text{mm}^2} = \frac{100\ \text{N}}{\text{cm}^2} = \frac{10\,000\ \text{N}}{\text{dm}^2} = \frac{1\,000\,000\ \text{N}}{\text{m}^2}$$

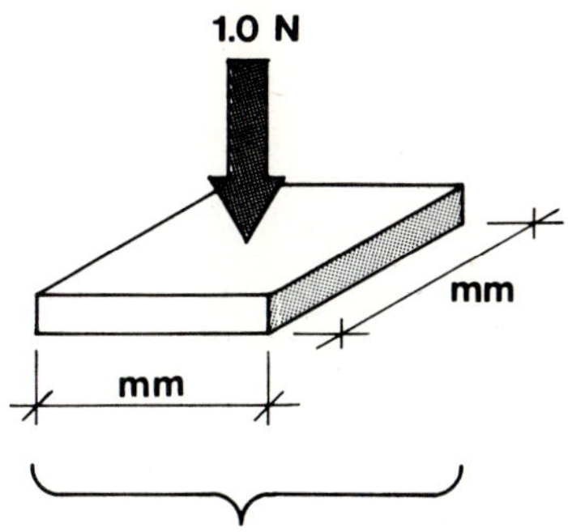

$$\text{PRESSURE} = \frac{1.0\ \text{N}}{\text{mm}^2}$$

or, in abbreviated notation,

$$\frac{\text{N}}{\text{mm}^2} = \frac{100\ \text{N}}{\text{cm}^2} = \frac{10\ \text{kN}}{\text{dm}^2} = \frac{1\ \text{MN}}{\text{m}^2}$$

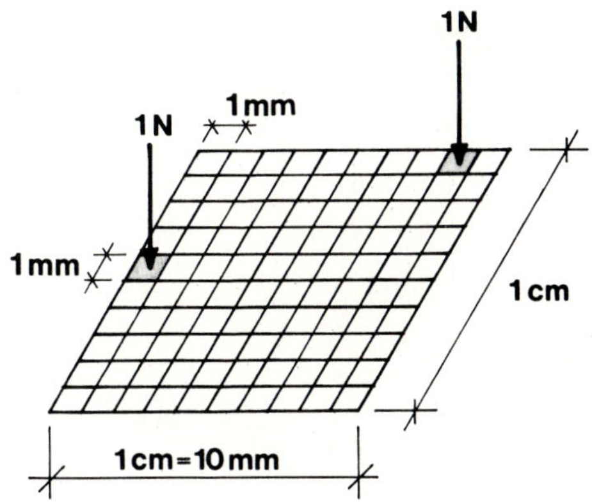

$$\frac{1\ \text{N}}{\text{mm}^2} = \frac{100\ \text{N}}{\text{cm}^2}$$

For your convenience a conversion table with basic and some derived units can be found on the front and back inside cover.

Since now is the time to start thinking and feeling "metric," to help you develop this feeling, we give metric SI equivalents in parentheses next to the British units you are familiar with, wherever this can be done without distracting you from the topic being presented or the sequence of calculations.

Elementary Statics and Strength of Materials

Franklin Avenue concrete arch bridge over the Mississippi River in Minneapolis (*Photo by L. Cerny*).

Statics and Strength of Materials

1

In our daily life, we come into contact with a variety of architectural or mechanical structures. We use them and enjoy them, but we rarely think about how they behave, how they carry loads, or how they transfer *safely* the loads they carry to the foundations and to the soil. This, of course, is not surprising; how many drivers know how their cars run?

Did you ever wonder how the floor in your living room can safely support a piano, bookcases, and other pieces of furniture, in addition to a group of dancing guests? You pass some highway bridge daily, but you rarely notice how it was constructed. Is it built of steel or concrete? What parts is it composed of? How is it supported? Why does an old reinforced concrete building have so many massive columns, while a modern building has just a few slender ones? Why did an old steel bridge collapse after years of carrying much larger loads than the one which caused the collapse?

All structures deform when they are subjected to loads. But when a train passes over a bridge, why are the deflections of the bridge so small that we cannot see them with the naked eye?

The top of a tall building sways laterally under the pressure of wind gusts. Does it move by a fraction of an inch or by a few feet?

Statics and *strength of materials* are the subjects that try to give answers to these questions: they are parts of *mechanics*, which is the oldest branch of *physics*.

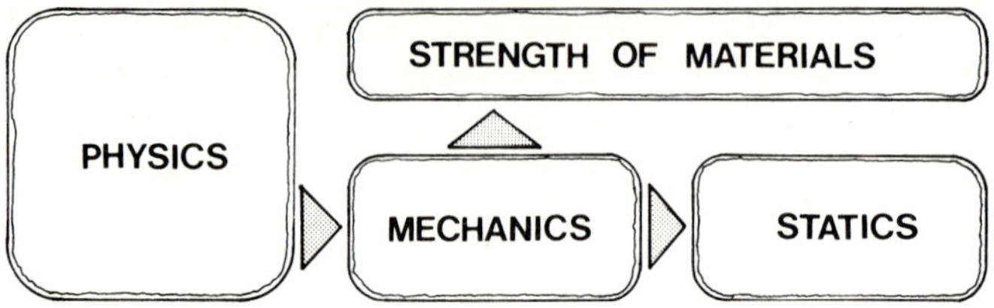

Structures are exposed to the action of a variety of loads, and these loads are transmitted through the structure to its *supports*, most often its foundations and the soil on which it rests. For example, a human arm can be thought of as a structure which transfers a weight in the hand to its supporting shoulder. The supports resist the loads coming from the structure; they *react* to these loads. The genius of Newton proved that the *reactions* of the supports acting on a stable structure must be *in equilibrium*, that is, in balance with the applied loads, since when the loads and the reactions are in equilibrium, the structure does not move; it is *at rest.*

Statics is concerned with the conditions to be satisfied by all the *forces* that act on a structure (applied loads and resulting reactions) to guarantee its equilibrium. (You will see, in fact, that most of the time these conditions allow us to evaluate the reactions, once we know what loads act on a structure.)

When a structure transmits loads to its supports, say, to the ground, *internal forces* develop through it. For example, when a hemp rope is used to lift the weight of a pail of water, an internal tensile force develops in the rope, thus causing the rope to stretch. A structure can be thought of as a mechanism that *balances* the applied loads by means of internal forces. Hence statics, which is concerned with equilibrium, allows us also to determine the *internal* forces in a structure submitted to *external* loads.

Once we have determined, by statics, the reactions and the internal forces in the various parts of a structure, we must design it properly. We must choose the most suitable materials to build it with and decide the shape and the dimensions of its parts. This job is the concern of strength of materials, which deals with the study of the internal forces holding elements of structures together and the laws governing these internal forces. Strength of materials teaches us how to size the parts of the structure so that they will develop safely the internal forces due to the external loads and also how to guarantee that the deformations of the structure under load will not impair its use. For example, if the sway due to wind gusts at the top of a tall building were to be too great, people in the top floors of the building would feel airsick or unsafe; these floors could not be profitably rented.

Moreover, strength of materials teaches us how to make sure that a structure will not fail or break. If the weight of a pail of water is too large for the rope used to lift it, the hemp rope will break. But if you use a steel cable instead, the cable probably will not break even if it is much thinner than the rope. Strength of materials allows us to evaluate the deformations of the structure, and hence its *serviceability*, as well as its strength, and hence its *safety*.

Statics and strength of materials are interesting subjects to learn, as you will realize in going through this book. They will allow you to understand, and design, most of the structures that you will be interested in. If in years to come you decide to become a *professional structural designer*, you will need to acquire additional knowledge of structures, but even then you will be amazed to realize that a clear understanding of statics and strength of materials is the basis for a sound grasp of *structural design principles.*

First National Pillsbury towers in downtown Minneapolis (*Photo by L. Cerny*).

Loads on Structures

2

Houses, apartment buildings, factories, theaters, stadiums, warehouses, bridges, and dams are some of a wide variety of structures. What loads act on these structures during their lifetime?

2.1 DEAD LOAD

Vertical load due to the weight of a structure itself (such as walls, floors, and roofs) and other bodies *permanently* attached to it during the entire life of a structure is called the *dead load.*

We calculate the dead load of a structural element by multiplying its volume by the weight per unit volume, or the *unit weight* of its material. Let the volume of the element be V and the unit weight of its material γ;† then the dead load W is given by

$$W = V\gamma$$

Volume V is measured in cubic feet (ft^3) or cubic meters (m^3); the units of unit weight γ are pounds per cubic foot (lb/ft^3) or newtons per cubic meter (N/m^3). Thus the dimension of W is pounds or newtons:

$$ft^3 \times \frac{lb}{ft^3} = lb \qquad \left(m^3 \, \frac{N}{m^3} = N\right)$$

Unit weights of some basic construction materials are given in Table A-1 (see Appendix).

†Greek letter gamma.

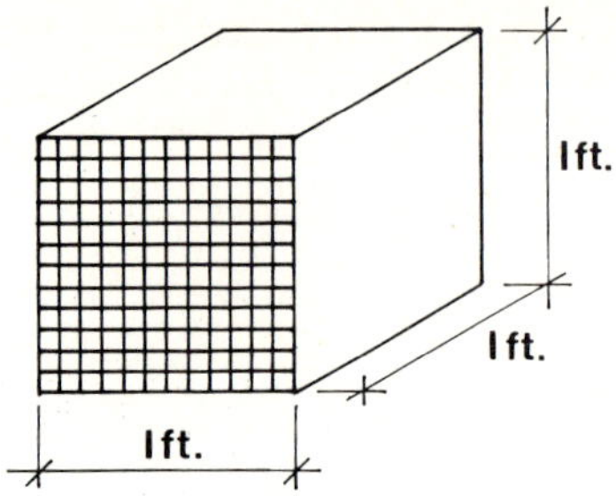

2.2 LIVE LOAD

In addition to their dead load, structures support the loads due to their use and occupancy. These include the weight of furniture and people in houses, apartments, schools, and offices; the weight of goods in warehouses; the weight of vehicles on bridges; water flowing in the aqueduct; and so on. Such load is called the *live load.* Live loads on buildings according to their use or occupancy are given in *building codes.* Data found in the *Uniform Building Code*† represent observations, measurements, and experience accumulated through many decades, justified by and checked against theoretical calculations.

Table A-2 shows the distributed and concentrated live loads on structures given by the *Uniform Building Code* which are to be used in design. For example, the weight of furniture and people in a room of a residential building is represented by a continuous load uniformly distributed over the entire area of the room and equal to 40 pounds per square foot (40 lb/ft^2) or 1.92 kilonewtons per square meter (1.92 kN/m^2). This means that whatever the random accumulation of furniture, appliances, and people may be, its effect on the structure need not be larger than that of 40 lb on each square foot of the floor area.

2.3 FORCES OF NATURE

A third important group of loads on structures are forces of nature, such as snow, wind, and earthquake. For example, the magnitude of the *snow load* on roofs depends on the geographic location of the building and the pitch and the shape of the roof. The snow load given in the codes is always considered to act on the horizontal projection of the roof surface rather than on the inclined surface itself. For any pitch smaller than 20°, it can be taken as a constant, basic snow load. However, for a pitch larger than 20°, the snow load can be proportionally reduced, since the steeper the slope of the roof is, the harder it is for snow to accumulate on it. For roofs with a pitch larger than 60°, the effect of the snow load on the roof is almost negligible. The basic snow load in the United States varies from 5 lb/ft^2 (0.24 kN/m^2) in some southern states to 80 lb/ft^2 (3.85 kN/m^2) in the northeastern corner of New England. The values of the

†*Uniform Building Code*, 1976 Edition, International Conference of Building Officials, Whittier, Calif.

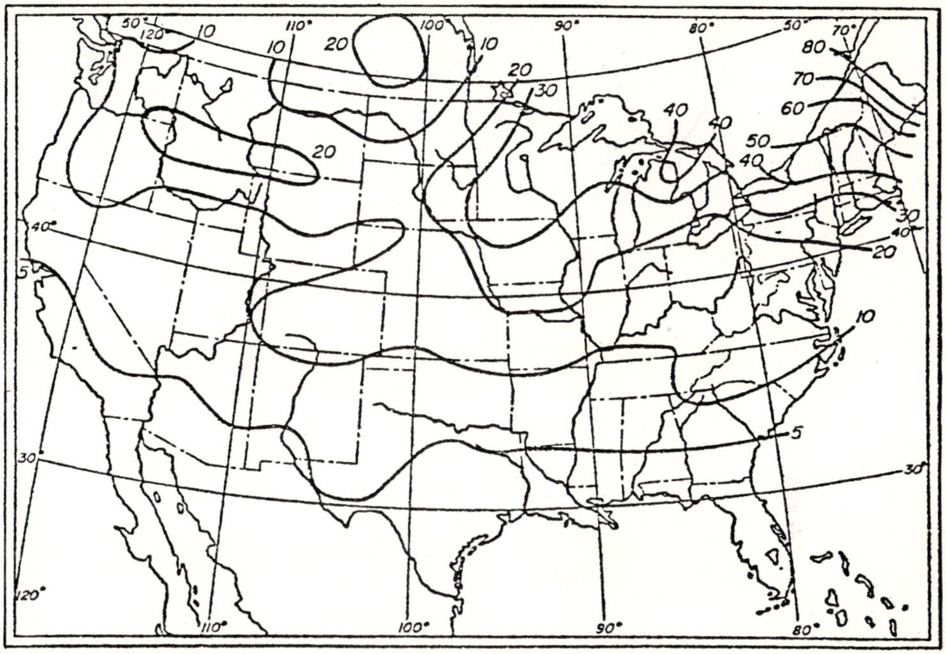

FIGURE 2-1 Minimum basic snow loads (lb/ft^2).

basic snow load may be found in snow maps such as that in Fig. 2-1. For each degree of pitch over 20°, basic snow loads in excess of 20 lb/ft^2 (0.96 kN/m^2) may be reduced by a value R_s determined by the following formula (taken from the *Uniform Building Code*):

$$R_s = \frac{S}{40} - \frac{1}{2}$$

where R_s is the snow load reduction in pounds per square foot per degree of pitch over 20° and S is the basic snow load in pounds per square foot.

Example

Compute the snow load on a 40°-pitch roof located in northern Michigan.

(*a*) From Fig. 2-1, the basic snow load is

$$S = 40\ \frac{\text{lb}}{\text{ft}^2} \qquad \left(1.92\ \frac{\text{kN}}{\text{m}^2}\right)$$

(*b*) The reduction for each degree of pitch over 20° is

$$R_s = \tfrac{40}{40} - \tfrac{1}{2} = 1.0 - 0.5 = 0.5\ \frac{\text{lb}}{\text{ft}^2} \qquad \left(0.024\ \frac{\text{kN}}{\text{m}^2}\right)$$

(*c*) The total reduction for a difference of 20°, between 40° and 20°, is

$$R_s = 20(0.5) = 10\ \frac{\text{lb}}{\text{ft}^2} \qquad \left(0.48\ \frac{\text{kN}}{\text{m}^2}\right)$$

(*d*) The final design snow load is

$$p_s = 40 - 10 = 30 \frac{\text{lb}}{\text{ft}^2} \qquad \text{(pounds per square foot of horizontal projection of the roof)}$$

$$= 1.44 \frac{\text{kN}}{\text{m}^2}$$

Humans, observing nature, learn how to fight natural odds. In mountain regions with heavy snowfalls, roofs are built with a steep pitch to prevent snow accumulation.

A moving car, earthquake shocks, or repeated gusts of wind may produce vibrations or oscillations in a structure. Such types of rapidly varying loads are called *dynamic* loads, while those which do not change with time are *static* loads. In this book we will be concerned mostly with static loads, even though dynamic loads are at times of fundamental importance.

The problem of finding the effect of *wind* on structures is complex. It depends on the geographic location, the velocity of wind (which increases with the altitude of a building above the ground), and the shape of the building. The surface of an enclosed building is exposed to wind pressure on the windward side, while suction develops on its leeward side. Glass panes tend to be pushed in on the windward side and sucked out on the leeward side. The action of the wind on low and stiff structures is usually considered as a static load, because the wind gusts have a long duration in comparison to the duration of the building oscillations. If, however, structures are tall and flexible —as those of high-rise buildings, cable roofs, television towers, or suspension bridges— their oscillations have long duration, and the effect of wind has to be considered as an oscillating, dynamic load. The magnitude and the distribution of the wind load (pressure and suction) on more complicated structural shapes must be determined experimentally by testing a scaled model of the structure in a *wind tunnel.*

Total wind loads (i.e., the sum of pressure and suction) on square or rectangular buildings, and for different heights above ground, are given in Table A-3. Enter the column in Table A-3 with a value corresponding to the basic wind pressure indicated on the wind map for the particular locality. For example, if the basic wind load from the wind map is 40 lb/ft^2 (1.92 kN/m^2), the portion of a building between 50 and 100 ft (15 and 30 m) above ground is exposed to a total (pressure plus suction) unit load of 50 lb/ft^2 (2.4 kN/m^2). The values given in Table A-3 are recommended as minimum and do not take into account possible tornadoes. Figures 2-2 and 2-3 show the distribution of total wind pressure on short and tall buildings, respectively.

When we investigate all the possible loads which may act on a retaining seawall, we find that a variety of loads of different nature must be considered. In addition to the vertical weight of the wall itself, lateral soil pressure acts on one side and water

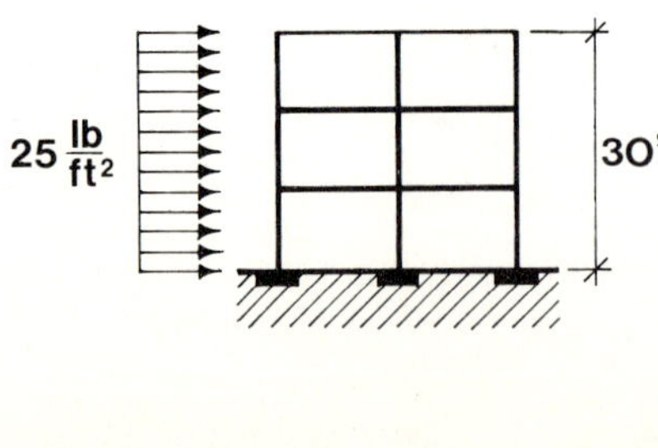

FIGURE 2-2 Wind pressure on a low building.

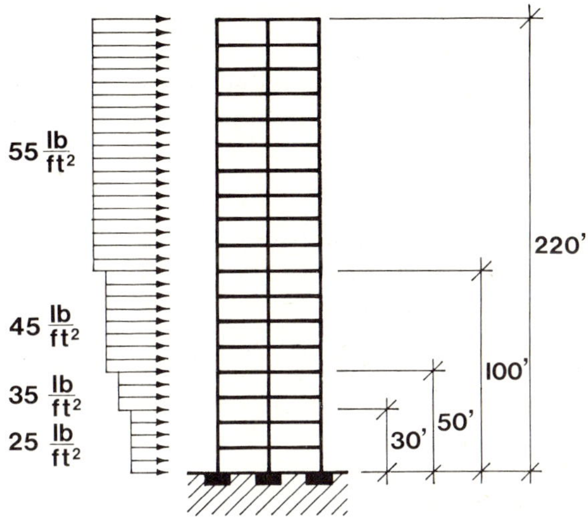

FIGURE 2-3 Wind pressure on a tall building.

pressure acts on the other. Wind, wave, and current forces acting on ships tied to the wall may cause tensile or compressive forces on the wall. Loads from cranes, trains, or other traffic near the wall may also be transmitted to the wall by the adjoining soil.

The identification of all the kinds of loads which may act on a structure during its construction and length of service may be a fairly complicated matter, but the designer must determine all these loads before starting the design, in order to select the proper materials for the structure and the shapes and the dimensions of its members or elements.

As shown in this chapter, the effects of loads on structures must be understood and calculated as accurately as possible. Incorrect estimation of loads may cause structural failure, with loss of human lives and property.

Illustrative Problem 2-1

Find the weight per foot w_D of a concrete beam (Fig. 2-4) which is 20 ft (5.10 m[†]) long and has a rectangular cross section 8 × 12 in (20 × 30 cm). The unit weight of concrete $\gamma = 144$ lb/ft^3 (22.6 kN/m^3).

[†]The metric equivalents are rounded to give a better feeling for conversion.

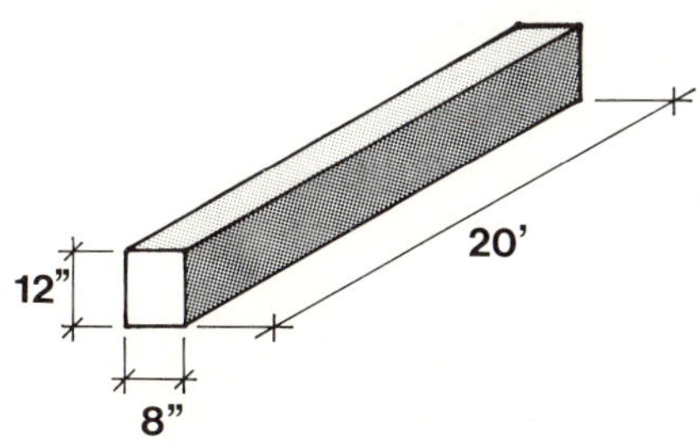

FIGURE 2-4 A concrete beam 20 ft long, 8 in wide, and 12 in deep.

SOLUTION In changing the cross-sectional dimensions of the beam from inches into feet, notice that

$$1 \text{ ft} = 12 \text{ in} \quad \text{and} \quad \frac{12 \text{ in}}{1 \text{ ft}} = 1 \quad \text{or} \quad \frac{1 \text{ ft}}{12 \text{ in}} = 1$$

Then the volume of the beam is

$$V = \text{width} \times \text{depth} \times \text{length} = 8 \text{ in} \times \frac{1 \text{ ft}}{12 \text{ in}} \times 12 \text{ in} \times \frac{1 \text{ ft}}{12 \text{ in}} \times 20 \text{ ft} = 13.33 \text{ ft}^3$$

The weight of the beam W is obtained by multiplying its volume by the unit weight of the concrete,

$$W = V\gamma = 13.33 \text{ ft}^3 \times 144 \frac{\text{lb}}{\text{ft}^3} = 1\,920 \text{ lb} = 1.92 \text{ kips}^\dagger$$

The weight of one lineal foot of beam w_D is computed by dividing its total weight W by its length

$$w_D = \frac{\text{total weight}}{\text{total length}} = \frac{1\,920 \text{ lb}}{20 \text{ ft}} = 96 \frac{\text{lb}}{\text{ft}} \text{(pounds per lineal foot)}$$

In metric this is

$$V = 0.2 \times 0.3 \times 6.10 \text{ m} = 0.366 \text{ m}^3$$

(*Note:* 20 cm = 0.2 m; 30 cm = 0.3 m)

Then the weight of the beam is ($\gamma = 22.6 \text{ kN/m}^3$)

$$W = 0.366 \text{ m}^3 \times 22.6 \frac{\text{kN}}{\text{m}^3} = 8.27 \text{ kN} = 8\,270 \text{ N}^\ddagger$$

†k is used as abbreviation for kip (kilopound) in figures, where 1 k = 1 000 lb.

‡If we used unrounded metric values for the dimensions of the beam, its volume would be

$$V = 0.2032 \times 0.3048 \times 6.096 \text{ m} = 0.37756 \text{ m}^3$$

and its weight would be

$$W = 0.37756 \text{ m}^3 \times 22.6224 \frac{\text{kN}}{\text{m}^3} = 8.54 \text{ kN} = 8\,540 \text{ N}$$

which is the value we would get by multiplying $W = 1\,920$ lb by 4.448, the conversion factor for changing pounds to newtons (1 lb = 4.448 N). Then

$$W = 1\,920 \times 4.448 = 8\,540 \text{ N}$$

and

$$w_D = \frac{8\,540}{6.096} = 1\,401 \frac{\text{N}}{\text{m}} = 1.401 \frac{\text{kN}}{\text{m}}$$

(You may see from these calculations why we suggest the use of rounded values in converting quantities from one unit system to the other.)

and

$$w_D = \frac{8\,270\text{ N}}{6.10\text{ m}} = 1\,356\,\frac{\text{N}}{\text{m}} = 1.356\,\frac{\text{kN}}{\text{m}}$$

The weight of the beam can be represented by a continuous *uniformly distributed load* with an intensity of 96 lb/ft of length of the beam (1.4 kN/m).

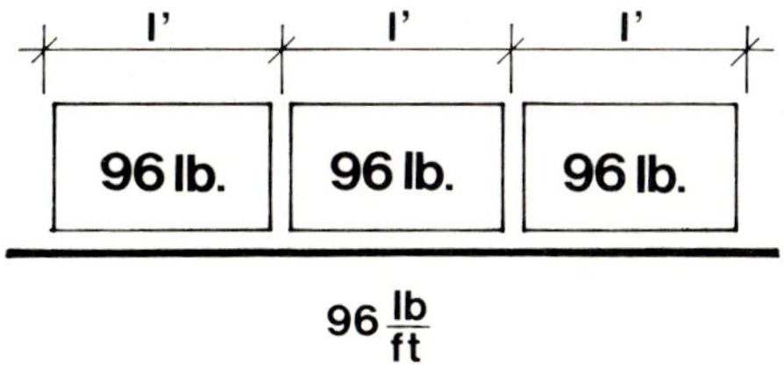

At this time it should suffice to explain that a beam is a structural element whose cross-sectional dimensions are small in comparison with its length. It is loaded mostly by forces that act at right angles to its longitudinal axis. It is supported in many different ways.

If the beam is supported at its ends on two walls, the force transmitted to each wall will be equal to one-half the total weight of the beam (concluded by intuitively observing the rules of symmetry),

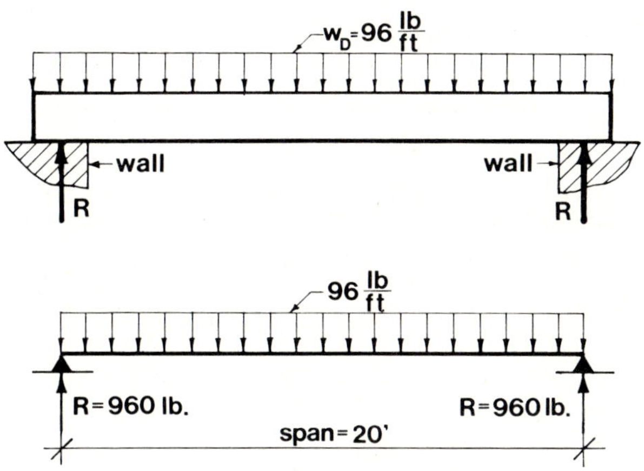

$$R = \frac{1\,920\text{ lb}}{2} = 960\text{ lb} \qquad (4\,270\text{ N} = 4.27\text{ kN})$$

The force R is concentrated on a small area of the wall called the *bearing area.* Such a force is called a *concentrated force.* A typical example of a concentrated force is the force exerted by a wheel on the ground.

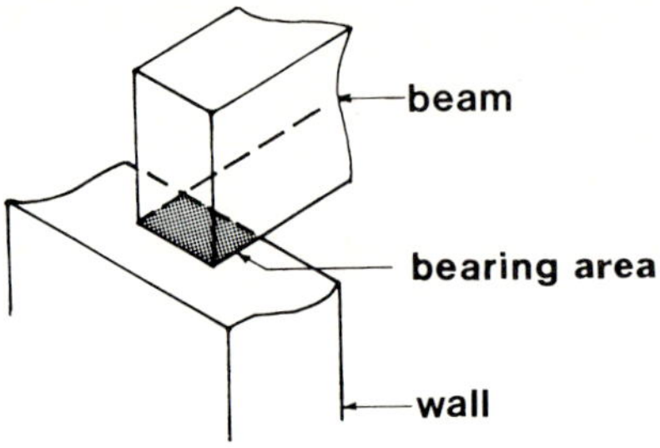

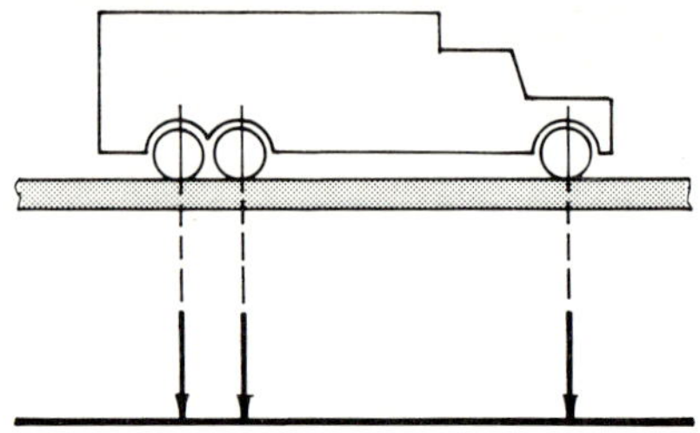

Illustrative Problem 2-2

Assume that the beam of Illustrative Prob. 2-1 resting on two walls at its ends supports a lightweight concrete slab 36 in (90 cm) wide and 4 in (10 cm) thick, in addition to its own weight (see Fig. 2-5). What is the value of the force R_T transmitted to each wall, if the slab has the same length as the beam (20 ft)?

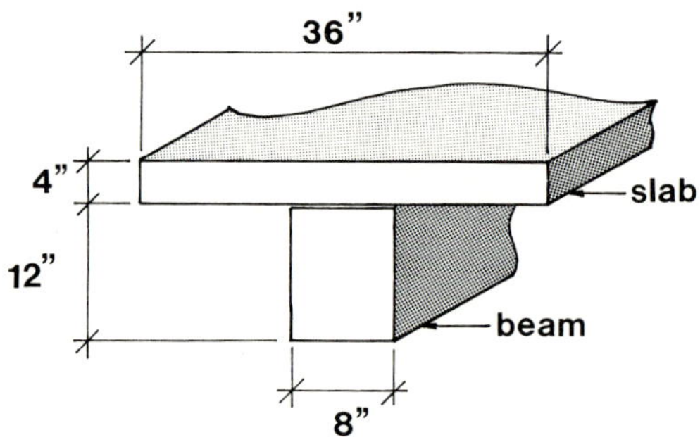

FIGURE 2-5 A beam supports a 36-in-wide and 4-in-thick slab.

SOLUTION Assume the unit weight of the lightweight structural concrete to be

$$\gamma = 80\,\frac{\text{lb}}{\text{ft}^3} \qquad \left(12.57\,\frac{\text{kN}}{\text{m}^3}\right)$$

The weight of the slab is called a *superimposed dead load* since it is added to the weight of the beam. The weight of the slab 20 ft long is

$$\begin{aligned} W_{SD} &= \text{volume} \times \text{unit weight} = V \times \gamma \\ &= \text{width} \times \text{thickness} \times \text{length} \times \text{unit weight} \\ &= 36 \text{ in} \times \frac{1 \text{ ft}}{12 \text{ in}} \times 4 \text{ in} \times \frac{1 \text{ ft}}{12 \text{ in}} \times 20 \text{ ft} \times 80\,\frac{\text{lb}}{\text{ft}^3} \\ &= 1\,600 \text{ lb} \qquad (7.12 \text{ kN}) \end{aligned}$$

The corresponding force transmitted to the walls is

$$R_{SD} = \frac{1\,600 \text{ lb}}{2} = 800 \text{ lb} \qquad (3.56 \text{ kN})$$

One lineal foot of the slab weighs

$$w_{SD} = \frac{\text{total weight of the slab}}{\text{total length}} = \frac{1\,600 \text{ lb}}{20 \text{ ft}} = 80 \frac{\text{lb}}{\text{ft}} \qquad \left(1.2 \frac{\text{kN}}{\text{m}}\right)$$

The total force on each wall due to the weight of the beam and the slab is

$$R_T = R + R_{SD} = 960 + 800 = 1\,760 \text{ lb} = 1.76 \text{ kip} \qquad (7.83 \text{ kN})$$

where R = 960 lb as obtained in Illustrative Prob. 2-1.

By putting such beams side by side and sealing the joints between them, we may build a *T-beam* floor (Fig. 2-6). The forces R_T which the walls exert on the beam ends are called the beam end *reactions.*

A common steel element used for spanning the space between two walls, and thus building floors or roofs, is composed of two *flanges* connected by a *web* (Fig. 2-7). Such elements are manufactured in steel mills by forming or rolling and are called *American standard beams* (S), or, if they have wider flanges, *wide-flange sections* (W).

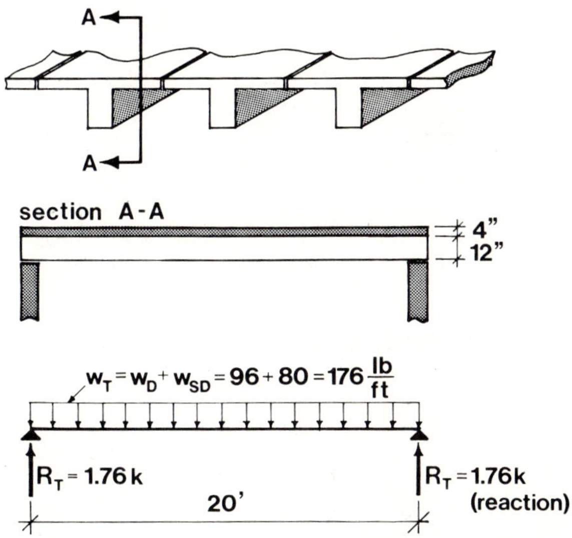

FIGURE 2-6 (*a*) Axonometric view of a T-beam floor; (*b*) section *A-A* through the floor; (*c*) static scheme of the T beam, with its loads and reactions.

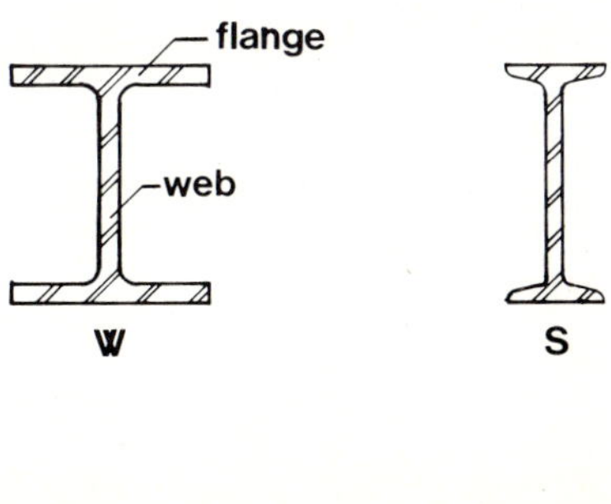

FIGURE 2-7 A wide-flange section (W) and an American standard section (S).

Wide-flange sections are used as columns as well as beams. American standard beams, which have interior flange surfaces sloped toward the web, are used only as beams. At this time it should suffice to explain that a column is a structural element whose transverse dimensions are much smaller than its length. In this aspect it is similar to a beam. However, a column primarily transfers compressive axial loads (acting in the direction of its longitudinal axis).

Commonly manufactured wide-flange sections and their dimensions and design properties are shown in Table A-4. (At this point, do not worry if you do not understand what some of the quantities shown in Table A-4 mean.) The equivalent values in the metric system are shown in Table A-5.

If, for example, we select from Table A-4 a wide-flange beam **W** 24 X 100 as an element of a structure, we find that **W** stands for wide flange, 24 stands for the depth of the section in inches, and 100 denotes the weight in pounds of one lineal foot of this beam (100 lb/ft). The metric equivalent to the wide-flange **W** 24 X 100 would be **W** 610 X 1 460 (see Table A-5), which has a depth of 610 mm, while each meter of its length weighs 1 460 N (1 460 N/m).

Illustrative Problem 2-3

A wide-flange beam **W** 14 X 103 (**W** 360 X 1 500) is 36 ft (11 m) long and is supported on two columns at its ends. The superimposed dead load per lineal foot of the beam is w_{SD} = 200 lb/ft (not including weight of the beam) and the live load per lineal foot of the beam is w_L = 1 000 lb/ft (Fig. 2-8). Calculate the force transmitted by the beam to each column.

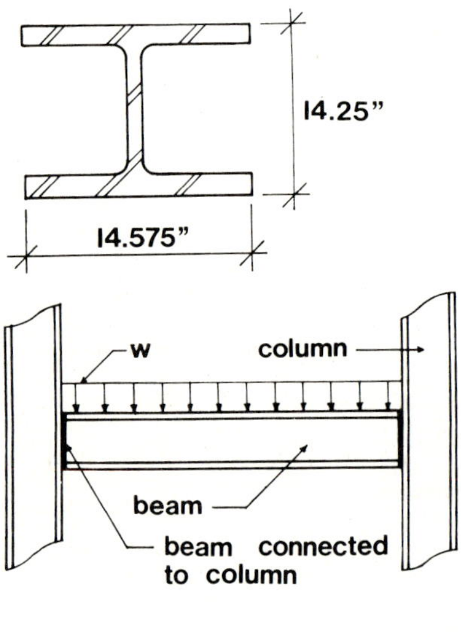

FIGURE 2-8 A beam transmits load to the columns.

SOLUTION The weight per lineal foot of the beam **W** 14 X 103 is

$$w_D = 103 \frac{\text{lb}}{\text{ft}} \quad \left(1.5 \frac{\text{kN}}{\text{m}}\right)$$

The total uniformly distributed load per lineal foot of the beam is

$$w_T = w_D + w_{SD} + w_L = 103 + 200 + 1\,000 = 1\,303\,\frac{\text{lb}}{\text{ft}} \qquad \left(19\,\frac{\text{kN}}{\text{m}}\right)$$

The total load on the beam is

$$W_T = \text{total load per lineal foot} \times \text{total length of beam}$$

$$= w_T \times 36\text{ ft} = 1\,303\,\frac{\text{lb}}{\text{ft}} \times 36\text{ ft} = 46\,908\text{ lb} \qquad (208.6\text{ kN})$$

Then the force on each column is

$$R = \frac{W_T}{2} = \frac{46\,908}{2} = 23\,454\text{ lb} \qquad (104.3\text{ kN})$$

We could also find the weight of *one lineal foot* of **W** 14 × 103 beam by finding the volume V of one lineal foot of that beam and multiplying it by the unit weight of steel, $\gamma = 490\text{ lb/ft}^3$. The area of this section is given in Table A-4 to be $A = 30.3\text{ in}^2$. Changing the units of the area from square inches to square feet, we have

$$V \times \gamma = \text{area of the cross section} \times \text{length} \times \text{unit weight}$$

$$= 30.3\text{ in}^2 \times \frac{\text{ft}}{12\text{ in}} \times \frac{\text{ft}}{12\text{ in}} \times 1\text{ ft} \times 490\,\frac{\text{lb}}{\text{ft}^3}$$

$$= 103\text{ lb}$$

Then the weight of the beam per lineal foot is

$$w_D = 103\,\frac{\text{lb}}{\text{ft}}$$

In metric units the area of the cross section is $A = 195\text{ cm}^2$ (see Table A-5) and the unit weight of steel $\gamma = 77\text{ kN/m}^3$ [see Table A-1(A)]. Changing square centimeters into square meters, we have

$$w_D = 195\text{ cm}^2 \times \frac{\text{m}}{100\text{ cm}} \times \frac{\text{m}}{100\text{ cm}} \times 77\,\frac{\text{kN}}{\text{m}^3} \times \frac{1}{\text{m}} = 1.5\,\frac{\text{kN}}{\text{m}}$$

$$= 1500\,\frac{\text{N}}{\text{m}}$$

which is the value shown in Table A-5.

There are many other steel and aluminum shapes used as structural elements. Some of these are channels, angles, **Z** shapes, **T** shapes, and tubes (see Fig. 2-9). Tables A-6 and A-7 give the design properties of channels (**C**) and angles (**L**) in U.S. units, while Tables A-8 and A-9 list their metric equivalents.

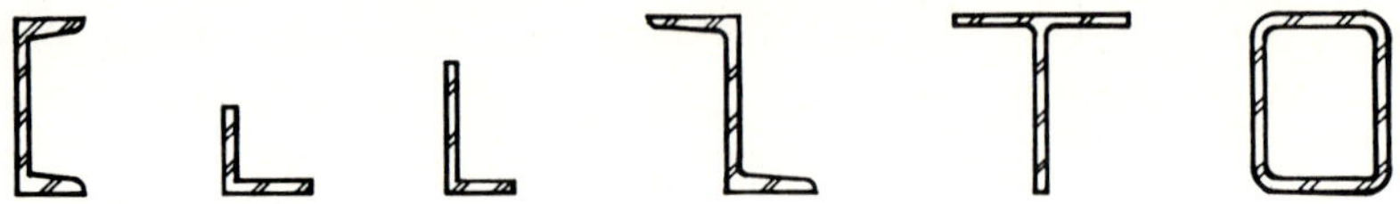

FIGURE 2-9 From left to right: channel section, equal-legs angle, unequal-legs angle, Z-shape, T-shape, and a tube.

Illustrative Problem 2-4

Wide-flange beams **W** 18 X 105 (**W** 460 X 1 530), 36 ft long (11 m), are placed across a long, rectangular hall from wall to wall at 12-ft (3.66-m) intervals. In structural terms, we say that the beams are spaced 12 ft on centers, as shown in Fig. 2-10. Across the beams and parallel to the walls, a *steel deck* spans from beam to beam; its weight is 5 lb/ft^2 (0.24 kN/m^2). A lightweight concrete slab is poured on top of the steel deck (Fig. 2-11). The weight of the slab is 23 lb/ft^2 (1.1 kN/m^2). A live load of 100 lb/ft^2 (4.8 kN/m^2) acts on the floor. Compute the total load per lineal foot of each beam.

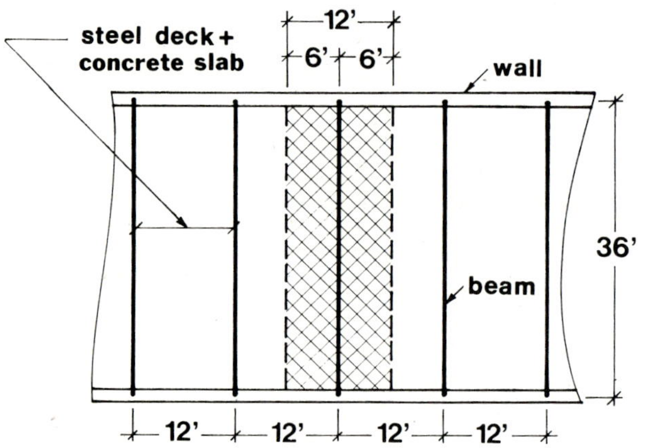

PLAN

FIGURE 2-10 Plan of the floor system in Illustrative Prob. 2-4.

CROSS SECTION

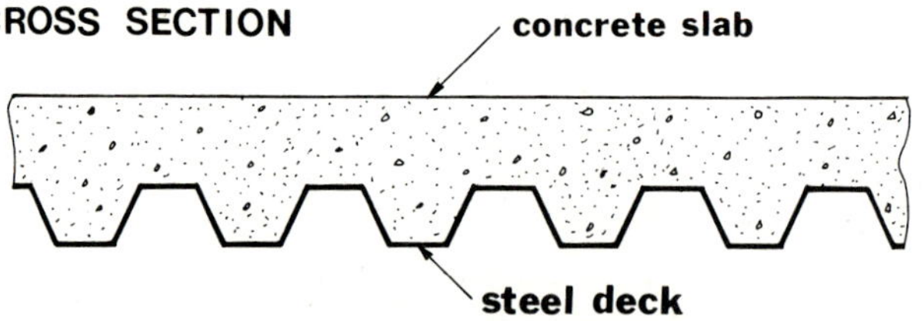

FIGURE 2-11 Section through the steel deck topped by concrete.

SOLUTION The total load p_T on each square foot of floor area consists of

Steel deck:

$$5\ \frac{\text{lb}}{\text{ft}^2} \qquad \left(0.24\ \frac{\text{kN}}{\text{m}^2}\right)$$

Concrete slab:

$$23\ \frac{\text{lb}}{\text{ft}^2} \qquad \left(1.10\ \frac{\text{kN}}{\text{m}^2}\right)$$

Live load:

$$100\ \frac{\text{lb}}{\text{ft}^2} \qquad \left(4.80\ \frac{\text{kN}}{\text{m}^2}\right)$$

$$\text{Total} \quad p_T = 128\ \frac{\text{lb}}{\text{ft}^2} \qquad \left(6.14\ \frac{\text{kN}}{\text{m}^2}\right)$$

Each beam carries the load from the tributary area shown crosshatched in Fig. 2-10. The dimensions of the tributary area are 12 X 36 ft (3.66 X 11 m). The total load on the beam from the tributary area is

$$W = p_T \times \text{tributary area} = 128\ \frac{\text{lb}}{\text{ft}^2} \times 12 \times 36\ \text{ft}$$

$$= 55\ 296\ \text{lb} \qquad (247\ \text{kN})$$

The load per lineal foot of the beam is

$$w = \frac{W}{\text{total length of beam}} = \frac{55\ 296\ \text{lb}}{36\ \text{ft}} = 1\ 536\ \frac{\text{lb}}{\text{ft}} \qquad \left(22.4\ \frac{\text{kN}}{\text{m}}\right)$$

Add the weight of the beam itself, 105 lb/ft (1 530 N/m), to get the total load per lineal foot of beam (Fig. 2-12)

$$w_T = 1\ 536 + 105 = 1\ 641\ \frac{\text{lb}}{\text{ft}} = 1.641\ \frac{\text{kips}}{\text{ft}}$$

or in metric

$$w_T = 22.4 + 1.53 = 23.93\ \frac{\text{kN}}{\text{m}}$$

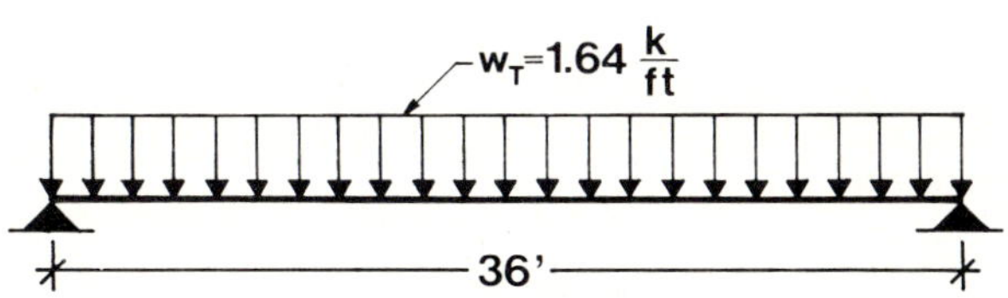

FIGURE 2-12 Total uniformly distributed load per lineal foot of the beam in Illustrative Prob. 2-4. [*Note:* In all figures k is used as an abbreviation for kips (kilopounds), where 1 k = 1 000 lb.]

Illustrative Problem 2-5

A concrete pedestal with square cross section 15 X 15 in (38 X 38 cm) is 10 ft (3 m) tall. It supports a sculpture weighing 4 000 lb, or 4 kips (17.8 kN). What force W_T does the column transmit to the foundation? What pressure p is exerted on the foundation?

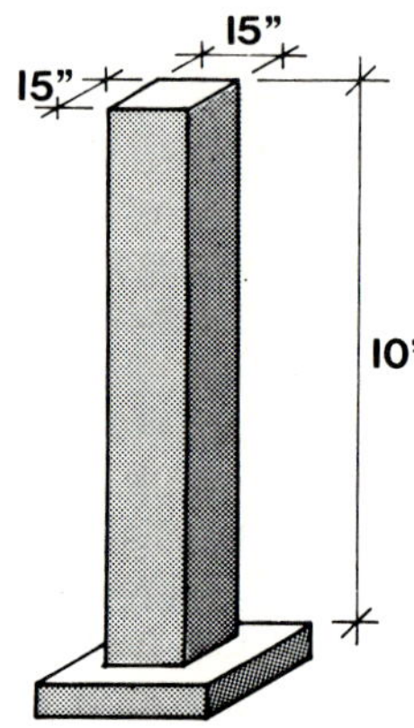

SOLUTION If the unit weight of concrete $\gamma = 150$ lb/ft³ (23.5 kN/m³), the weight of the pedestal is

$$W = 15 \times 15 \text{ in} \times \frac{\text{ft}}{12 \text{ in}} \times \frac{\text{ft}}{12 \text{ in}} \times 10 \text{ ft} \times 150 \frac{\text{lb}}{\text{ft}^3}$$

$$= \frac{15 \times 15}{144} \times 10 \times 150 = 2\,344 \text{ lb} \qquad (10.4 \text{ kN})$$

Adding the weight of the sculpture,

$$W_T = 2\,344 + 4\,000 = 6\,344 \text{ lb} = 6.344 \text{ kips} \qquad (28.2 \text{ kN})$$

Now divide the area on which the pedestal rests into small squares 1 X 1 in or 1 in². We will have 15 X 15, or 225 such squares. Divide the total force of

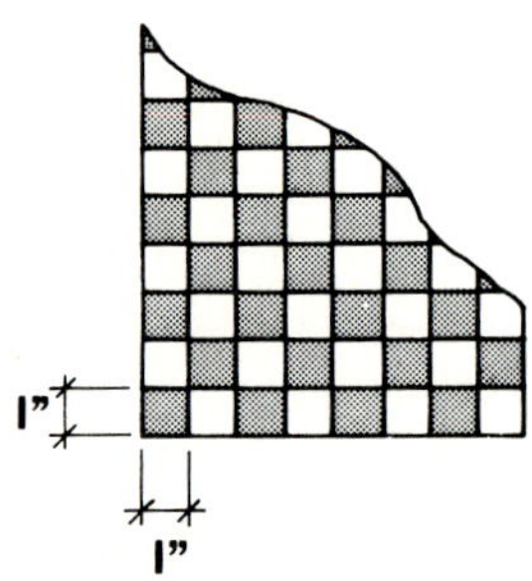

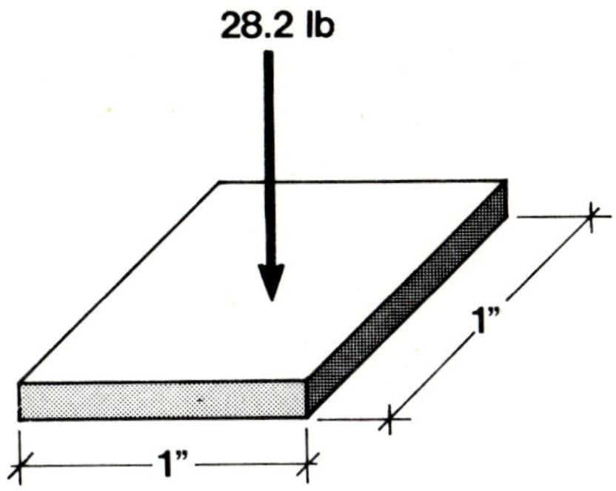

FIGURE 2-13

6 344 lb equally among the 225 squares,

$$p = \frac{6\,344}{225} = 28.2 \text{ pounds per square inch} = 28.2 \frac{\text{lb}}{\text{in}^2} \qquad \left(195 \frac{\text{kN}}{\text{m}^2}\right)$$

The quantity p, which is given in units of force per unit area, is called ***pressure.*** The pressure exerted by the pedestal on the footing is 28.2 lb/in^2. We have assumed in this example that the total force of 6.344 kips is distributed uniformly to all squares (Fig. 2-13).

Illustrative Problem 2-6

Compute the weight of a floor composed of a 6-in concrete slab, a 1-in plaster ceiling underneath it, a 2-in lightweight concrete fill (for electrical distribution ducts), and a $\frac{7}{8}$-in hardwood parquet, as shown in Fig. 2-14. If the floor is used for a classroom and is supported on columns on a grid 20 X 24 ft on centers, determine the column reactions.

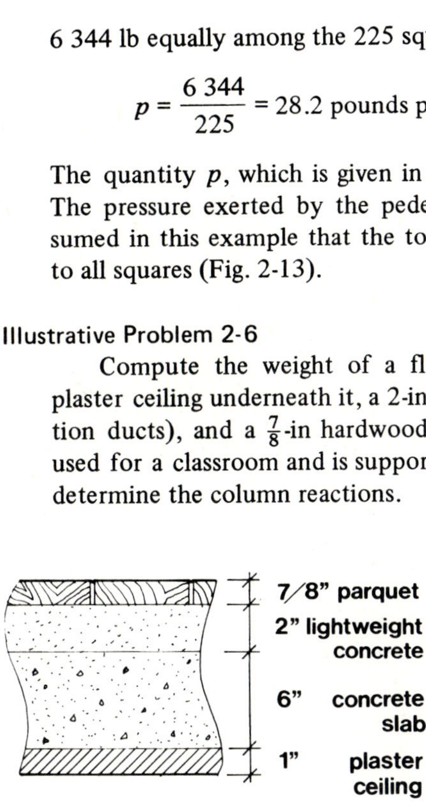

FIGURE 2-14 Section through the floor in Illustrative Prob. 2-6.

SOLUTION From Table A-1(*B*), select the weights of the materials used:

Hardwood parquet:

$$4.0 \frac{\text{lb}}{\text{ft}^2} \qquad \text{(1 in thick)}$$

Lightweight concrete:

$$6.0\ \frac{\text{lb}}{\text{ft}^2} \qquad \text{(1 in thick)}$$

Reinforced concrete:

$$12.5\ \frac{\text{lb}}{\text{ft}^2} \qquad \text{(1 in thick)}$$

Plaster:

$$10.0\ \frac{\text{lb}}{\text{ft}^2} \qquad \text{(1 in thick)}$$

For the thicknesses given in our problem:
Hardwood parquet:

$$\tfrac{7}{8} \times 4.0 = \ 3.5\ \frac{\text{lb}}{\text{ft}^2}$$

Lightweight concrete:

$$2 \times 6.0 = 12.0\ \frac{\text{lb}}{\text{ft}^2}$$

Reinforced concrete:

$$6 \times 12.5 = 75.0\ \frac{\text{lb}}{\text{ft}^2}$$

Plaster:

$$1 \times 10.0 = 10.0\ \frac{\text{lb}}{\text{ft}^2}$$

$$\text{Total weight of the floor } p_D = 100.5\ \frac{\text{lb}}{\text{ft}^2} \qquad \left(4.8\ \frac{\text{kN}}{\text{m}^2}\right)$$

The total dead load of the floor is 100.5 lb/ft^2 of floor.

For classrooms, Table A-2 gives a live load

$$p_L = 40\ \frac{\text{lb}}{\text{ft}^2} \qquad \left(1.95\ \frac{\text{kN}}{\text{m}^2}\right)$$

Thus the total dead and live load acting on each square foot of the classroom floor is

$$p_T = p_D + p_L = 100.5 + 40 = 140.5\ \frac{\text{lb}}{\text{ft}^2} \qquad \left(6.72\ \frac{\text{kN}}{\text{m}^2}\right)$$

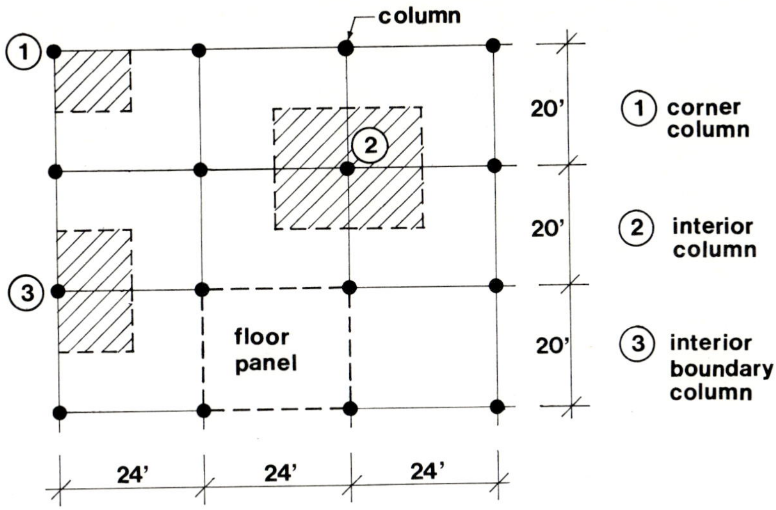

FIGURE 2-15 Plan of a slab floor system supported on circular columns.

Assume further that the floor slab is supported on columns arranged in a rectangular grid 20 X 24 ft (6.1 X 7.32 m) on centers (Fig. 2-15). The total area of one rectangular floor panel limited by four column lines is

$$A = 20 \times 24 = 480 \text{ ft}^2 \qquad (44.7 \text{ m}^2)$$

The total load on the floor panel is

$$W = \text{area of floor panel} \times \text{total load on 1 ft}^2 \text{ of floor}$$

$$= A \times p_T = 480 \text{ ft}^2 \times 140.5 \frac{\text{lb}}{\text{ft}^2}$$

$$= 67\,440 \text{ lb} \qquad (300 \text{ kN})$$

The tributary areas for individual columns are crosshatched in Fig. 2-15. Thus the total force transmitted to the corner column 1 (tributary area is $\frac{1}{4}$ the area of the full floor panel) is

$$F_1 = \frac{W}{4} = \frac{67\,440}{4} = 16\,860 \text{ lb} \qquad (75 \text{ kN})$$

An interior column 2 transfers the force from the full panel.

$$F_2 = W = 67\,440 \text{ lb} \qquad (300 \text{ kN})$$

An interior boundary column 3 transfers the force from half a panel

$$F_3 = \frac{W}{2} = 33\,720 \text{ lb} \qquad (150 \text{ kN})$$

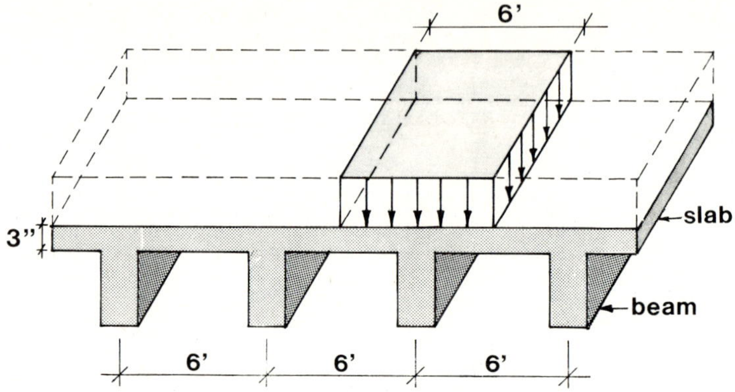

FIGURE 2-16 Reinforced concrete floor system in Illustrative Prob. 2-7.

Illustrative Problem 2-7

Compute the dead load carried by the reinforced concrete beams of the floor shown in Fig. 2-16. The floor consists of beams 6 ft on centers. A reinforced concrete slab 3 in (7.6 cm) thick spans between the beams. A 1-in (2.5-cm) layer of cement mortar and $\frac{3}{4}$ in (2.0 cm) of terrazzo tiles lie on top of the slab. The slab and the beams are paneled with softwood paneling $\frac{3}{4}$ in (2.0 cm) thick (Fig. 2-17).

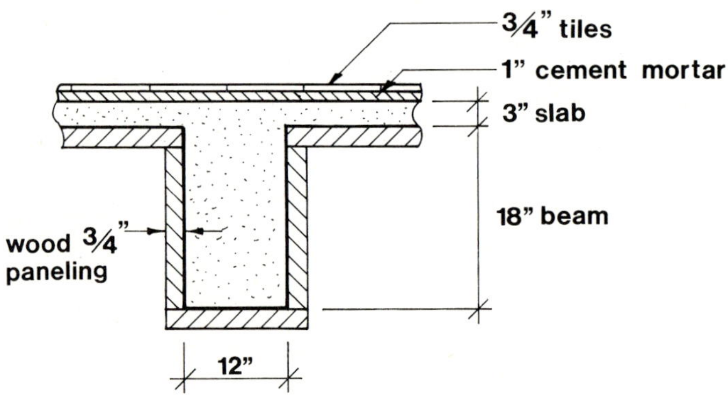

FIGURE 2-17 Section through the floor system in Illustrative Prob. 2-7.

SOLUTION The weights of the materials used [from Table A-1(*B*)] are

Concrete slab:

$$12.5\ \frac{\text{lb}}{\text{ft}^2} \qquad \text{(1 in thick)}$$

Cement finish:

$$12.0\ \frac{\text{lb}}{\text{ft}^2} \qquad \text{(1 in thick)}$$

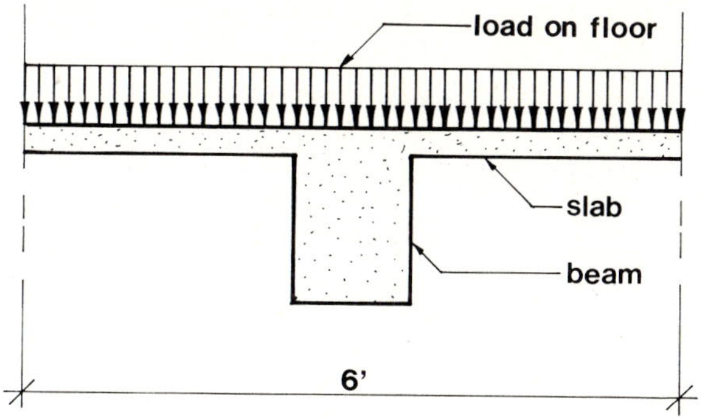

FIGURE 2-18 Tributary width for computation of load on the beam in Illustrative Prob. 2-7.

Terrazzo:

$$13.0 \frac{\text{lb}}{\text{ft}^2} \quad \text{(1 in thick)}$$

Softwood paneling:

$$2.2 \frac{\text{lb}}{\text{ft}^2} \quad \text{(1 in thick)}$$

Since the beams are 6 ft on centers, each will support the load from a tributary width of 6 ft as seen in Figs. 2-16 and 2-18:

3-in concrete slab:

$$3.0 \times 12.5 \frac{\text{lb}}{\text{ft}^2} \times 6 \text{ ft} = 225 \frac{\text{lb}}{\text{ft}} \quad \text{(pounds per lineal foot of beam)}$$

1-in cement mortar:

$$1.0 \times 12.0 \frac{\text{lb}}{\text{ft}^2} \times 6 \text{ ft} = 72 \frac{\text{lb}}{\text{ft}}$$

$\frac{3}{4}$-in terrazzo tiles:

$$0.75 \times 13.0 \frac{\text{lb}}{\text{ft}^2} \times 6 \text{ ft} = 59 \frac{\text{lb}}{\text{ft}}$$

Softwood paneling (Fig. 2-17):

On the horizontal surface 6 ft wide,

$$0.75 \times 2.2 \frac{\text{lb}}{\text{ft}^2} \times 6 \text{ ft} = 10 \frac{\text{lb}}{\text{ft}}$$

On the two vertical surfaces of the beam,

$$0.75 \times 2.2 \frac{\text{lb}}{\text{ft}^2} \times 2 \times 18 \text{ in} \times \frac{\text{ft}}{12 \text{ in}} = 5 \frac{\text{lb}}{\text{ft}}$$

Weight of the 12- X 18-in beam with $\gamma = 150\ \text{lb/ft}^3$

$$12\ \text{in} \times \frac{\text{ft}}{12\ \text{in}} \times 18\ \text{in} \times \frac{\text{ft}}{12\ \text{in}} \times 150\ \frac{\text{lb}}{\text{ft}^3} = 225\ \frac{\text{lb}}{\text{ft}}$$

Total load acting on each lineal foot of the beam $\qquad w = 596\ \frac{\text{lb}}{\text{ft}} \quad \left(8.7\ \frac{\text{kN}}{\text{m}}\right)$

Thus the beam carries 596 lb/ft of its length (8.7 kN/m). If the beam is 16 ft (4.88 m) long, the total load it carries is (Fig. 2-19)

$$W = w \times 16\ \text{ft} = 596\ \frac{\text{lb}}{\text{ft}} \times 16\ \text{ft} = 9\ 536\ \text{lb} \qquad (42.4\ \text{kN})$$

If it is supported at its ends, each support receives

$$R = \frac{9\ 536}{2} = 4\ 768\ \text{lb} \qquad (21.2\ \text{kN})$$

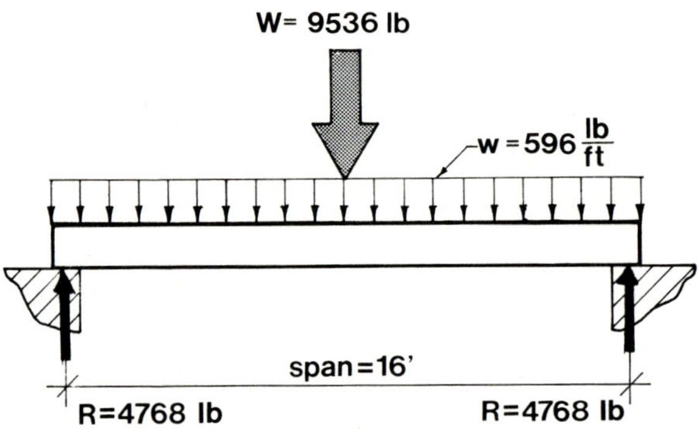

FIGURE 2-19 Load and reactions on the beam in Illustrative Prob. 2-7.

New Terms Appearing in Chapter 2

(Arranged in the order of their appearance in the text)

Loads on structures	Mechanics
Foundations	Physics
Soil	Supports
Floor	Equilibrium
Bridge	Forces
Reinforced concrete	Internal forces
Column	External forces
Deflections of structures	Serviceability
Statics	Strength
Strength of materials	Safety

Structural engineer
Structural design
Weight
Unit weight
Dead load
Live load
Uniform Building Code
Snow load
Dynamic loads
Static loads
Wind load
Wind pressure
Wind suction
Wind tunnel
Beam
Uniformly distributed load
Concentrated force
Bearing area
Wall
Slab
T-Beam floor
Reactions
American standard beam
Wide-flange beam
Flange
Web
Steel shapes
Channels
Angles
Z Shapes
T Shapes
Tubes
Steel deck
Pressure
Floor panel

Three Important Structural Elements Appearing in Chapter 2

Beam. This is a structural element whose cross-sectional dimensions are small in comparison with its length (span). It is loaded mostly by forces acting at right angles to its longitudinal axis. It is supported in many different ways (as we shall see later).

Slab (plate). This is a structural element whose thickness is small in comparison with its width and length. It covers large floor and roof areas. It is loaded mostly by forces acting at right angles to the plane of a slab.

Column. This is a structural element whose transverse dimensions are much smaller than its length. In this aspect it is similar to a beam. However, a column primarily transfers compressive **axial loads** (acting in the direction of its longitudinal axis).

Quiz Number 1 (The keys to quizzes and grading are at the end of the book.)

A. TRUE OR FALSE

1. In technological societies interest in engineering is limited to a small group of specialists called engineers.
2. 1 meter = 10 decimeters = 100 centimeters = 10 000 millimeters.
3. The volume of one liter is exactly equal to one cubic decimeter (1 L = 1 dm^3).
4. A prefix *deci* means 10 times, while *centi* means 100 times.
5. A newton is a unit of density.

6. Structures do not deform under the action of loads, since they are absolutely rigid.
7. Statics and strength of materials are part of physics.
8. In statics we study the equilibrium of all the forces acting on a structure.
9. No internal forces develop in a structure which transmits loads to its foundations.
10. The internal tensile force in a cable causes its elongation.
11. A serviceable structure must not have excessive deformations; however, it can sway considerably under the effect of wind, thus causing occupants to feel discomfort.
12. A well-designed structure possesses strength and serviceability.
13. The dead load on a structure acts only occasionally, while the live load is present permanently.
14. The snow load is to be found in the snow maps.
15. The wind load on tall, flexible structures may cause excessive vibrations or oscillations.
16. Static loads change rapidly with time.
17. Loads on structures are given in building codes.
18. The snow load given in the codes is always considered to act on the actual plane of the roof rather than on the horizontal projection of the roof plane.
19. The larger the pitch of the roof, the smaller the snow load.
20. There is suction on the windward side of a building and pressure on its leeward side.
21. The action of wind on low and stiff structures is usually considered as a static load.
22. The higher the portion of the building above ground, the smaller the wind pressure on it.
23. The intensity of a uniformly distributed load on a slab is measured in pounds per square foot, whereas that on a beam is measured in pounds per lineal foot.

B. MULTIPLE CHOICE

1. Which of the following loads is not considered to be a live load?
   (*a*) piano (*b*) train (*c*) people (*d*) permanent partitions
2. If a force F acts on an area A, the pressure on that area is defined as
   (*a*) $F \times A$ (*b*) F/A (*c*) $F + A$ (*d*) A/F
3. The unit of pressure in the metric system is
   (*a*) N/m^2 (*b*) Nm^2 (*c*) kN/m^3 (*d*) Nft^2
4. A structural element which has cross-sectional dimensions much smaller than its length and is loaded at right angles to its longitudinal axis is called a
   (*a*) slab (*b*) plate (*c*) column (*d*) beam
5. The unit weight of reinforced (stone) concrete in lb/ft^3 is
   (*a*) 15 (*b*) 150 (*c*) 1 500 (*d*) 1.5
6. If a beam supported at its ends on walls is loaded by a vertical concentrated load F equal to 10 kips, located at the midpoint of its span, the beam reactions R are equal to
   (*a*) 10 kips (*b*) 2.5 kips (*c*) 5 kips (*d*) 50 kips

7. A force of 1 lb is approximately equal to
   (*a*) 10 N (*b*) 4.5 N (*c*) 45 N (*d*) 2 kN
8. One foot is approximately equal to
   (*a*) 30 cm (*b*) 300 cm (*c*) 5 cm (*d*) 12 cm
9. A T-beam floor is defined as a floor:
   (*a*) covered with T-shirts
   (*b*) composed of beams connected by a slab
   (*c*) used as a bus station for public transportation
   (*d*) consisting of a series of steel wide-flange beams

C. SHORT ANSWERS

1. A majority of the loads acting on structures does not change rapidly with time. They are called ______________.
2. Loads which change rapidly with time are ______________.
3. Conditions of equilibrium for structures or their parts are established by means of the rules of ______________.
4. The forces exerted on loaded structures by their supports are called ______________.
5. By means of strength of materials we are able to find ______________ and ______________ of structural elements.
6. One of the requirements for serviceability of structures is to have small ______________.
7. When a structural element is under the action of loads, ______________ forces develop in its interior.
8. The ratio of force and the corresponding area on which it acts is called ______________.
9. 100 cm = 1.0 m = ________ in (approximately).
10. Ten newtons is a force approximately equal to a force of ________ lb.
11. One meter is approximately ________feet.
12. The intensity of the snow load in a given geographic location changes with the ________ of a roof.
13. The magnitude and the distribution of the wind load depend on the ________ of the structure.
14. Wind effects on complicated structures are investigated in a ______________.
15. The weight of permanent partitions is considered a ________ load.

EXERCISES (For answers see the end of the book.)

2-1 Find the dead load, the live load, and the total load in pounds per square foot on the floor slab of a classroom. The slab is 10 in thick and is made of stone reinforced concrete. On top of the slab there is a 2-in layer of lightweight concrete ($\gamma = 50$ lb/ft^3) and a $\frac{1}{8}$-in layer of vinyl tiles. Attached to the bottom of the slab is an acoustical fiber tile ceiling.

2-2 A masonry wall 8 in thick and 12 ft high supports a roof load equal to 2 000 lb/ft of wall. The wall rests on reinforced concrete footing 1.5 ft deep and 2 ft wide. The reinforced concrete is stone concrete, and the masonry is of common brick. Find the pressure between the footing and the soil (Fig. 2-20). (*Hint:* Investigate 1 ft of the wall.)

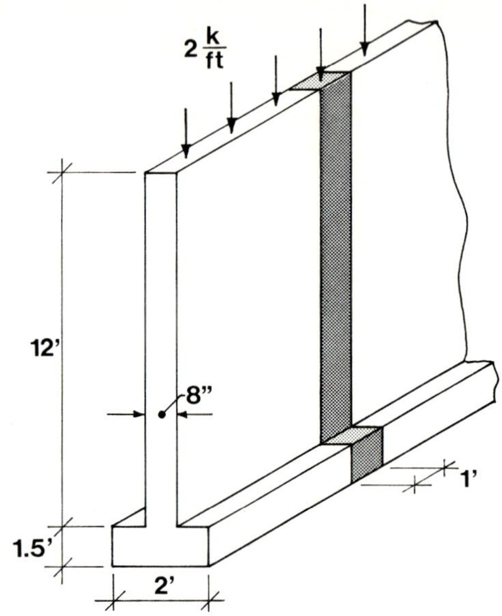

FIGURE 2-20 A masonry wall and footing in Exercise 2-2.

2-3 Find the pressure which a hollow cast-iron column exerts on its foundation (Fig. 2-21). The outside diameter (OD) of the column is 18 in, and the column is $\frac{1}{4}$ in thick and is loaded by a total axial load of 40 kips.

Hint: Divide the axial load on the column by its cross-sectional area equal to the area of the circle with the outside diameter of 18 in minus the area of the inside circle with the diameter

$$18 - 2 \times 0.25 = 17.5 \text{ in}$$

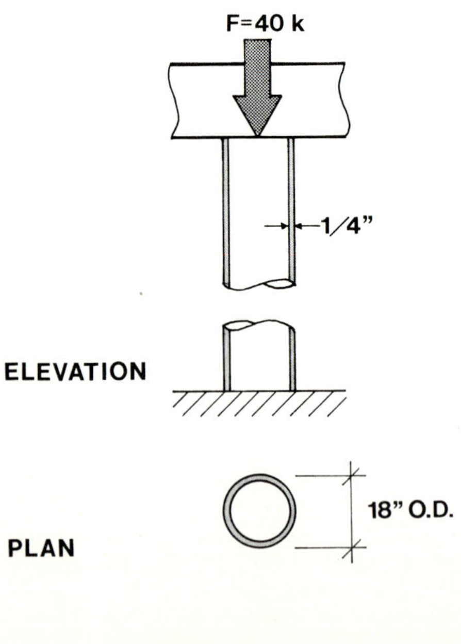

FIGURE 2-21 A hollow circular column in Exercise 2-3.
(OD = outside diameter.)

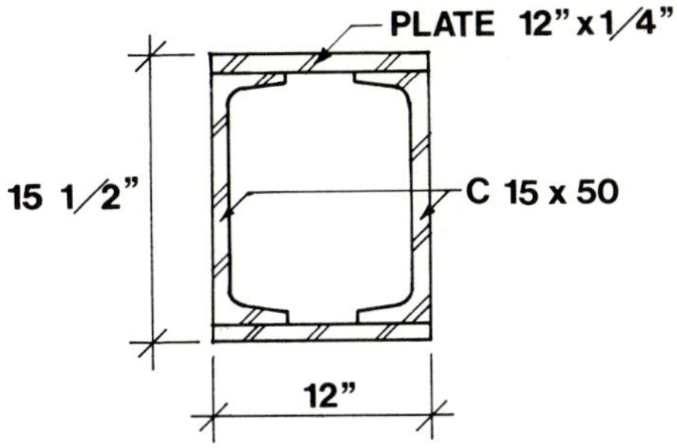

FIGURE 2-22 Cross section of a built-up steel beam in Exercise 2-4.

2-4 Find the weight per lineal foot of a built-up steel beam composed of two American standard channels **C** 15 × 50 connected by two cover plates 12 × $\frac{1}{4}$ in (Fig. 2-22).

Hint: Compute the volume of 1 ft of the section and multiply it by the γ of steel. The volume is equal to the cross-sectional area times the length (= 1.0 ft).

2-5 The roof of a square building consists of an 8-in-thick reinforced (stone) concrete slab. Roofing and insulation weighs 10 lb/ft^2, and a snow load of 30 lb/ft^2 must be considered. The roof is supported on nine columns as shown in Fig. 2-23. Compute the total force which will be transmitted to

(*a*) the central column
(*b*) a corner column
(*c*) a column at the center of a side

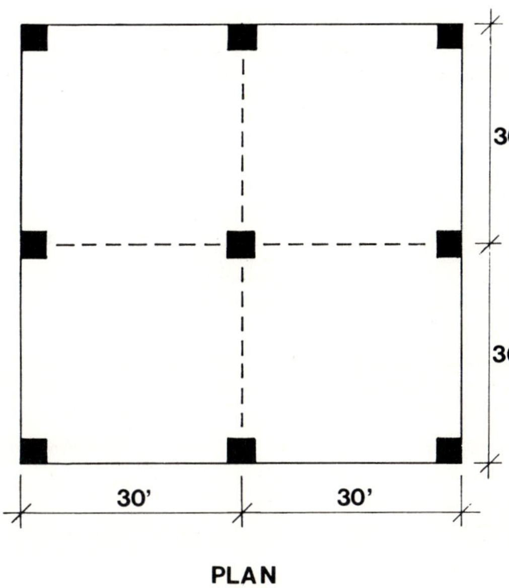

FIGURE 2-23 Plan of the roof in Exercise 2-5.

2-6 The timber floor of a house consists of timber *joists* 2 × 10 in, spaced 16 in on centers (Fig. 2-24). The span of the joists between the walls is 16 ft. Consider the total dead load of the floor to be 10 lb/ft^2 (including the weight of joists) and the live load to be 40 lb/ft^2. Compute the load on one joist in pounds per lineal foot of the joist length. (Joists are narrow, closely spaced beams.)

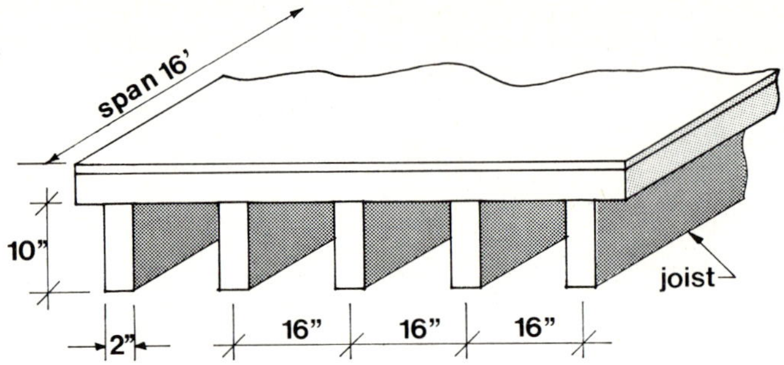

FIGURE 2-24 Timber joist floor system in Exercise 2-6.

2-7 An apartment tower to be built in southern Florida is 250 ft high and has plan dimensions of 70 × 140 ft. What will be the distribution of wind pressure along the height of the building, if the basic wind pressure (shown on the wind map) is 50 lb/ft^2? What will be the total wind forces acting on the narrow and on the wide sides of the building, respectively?

Hint: Divide the building into portions with the heights indicated in Table A-3. Find the pressures. The force on any partial area of the surface of the building is obtained by multiplying the corresponding pressure by the corresponding area. Then add all the partial forces acting on the side of the building. Do this for the narrow and the wide sides.

Exercises in Metric Units

2-8 Find the dead load, the live load, and the total load in kilonewtons per square meter on the floor slab of a classroom. The slab is 25 cm thick and is made of stone reinforced concrete. On top of the slab there is a 5-cm layer of lightweight concrete (γ = 7.8 kN/m^3) and a 3-mm layer of vinyl tiles. Attached to the bottom of the slab is an acoustical fiber tile ceiling.

2-9 A masonry wall of Exercise 2-2 is 20 cm thick and 366 cm high. It supports a roof load equal to 30 kN/m of wall. The wall rests on reinforced concrete footing 45 cm deep and 60 cm wide. The reinforced concrete is stone concrete, and the masonry is of common brick. Find the pressure between the footing and the soil. (*Hint:* consider one lineal meter of the wall.)

2-10 Find the pressure which a hollow cast-iron column of Exercise 2-3 exerts on its foundation. The outside diameter (OD) of the column is 45 cm, and the column is 6 mm thick. It is loaded by a total axial load of 180 kN.

2-11 Find the weight per lineal meter of a built-up steel beam of Exercise 2-4 composed of two American standard channels **C** 380 × 730 (see Table A-8) connected by two cover plates 300 × 6 mm.

2-12 The roof of a square building of Exercise 2-5 consists of a 20-cm-thick reinforced concrete slab. Roofing and insulation weighs 480 N/m^2, and a snow load of 1 400 N/m^2 must be considered. Compute the total force which will be transmitted to

(*a*) the central column

(*b*) a corner column

(*c*) a column at the center of a side

Columns are spaced 9 m apart.

2-13 The timber floor of a house of Exercise 2-6 consists of timber joists 5 × 25 cm, spaced 40 cm on centers. The span of the joists between the walls is 490 cm. Consider the total dead load of the floor to be 0.48 kN/m^2 (including the weight of joists) and the live load to be 1.9 kN/m^2. Compute the load on one joist in kilonewtons per lineal meter of the joist length.

2-14 An apartment tower of Exercise 2-7 to be built in southern Florida is 76 m high and has plan dimensions of 21 × 42 m. What will be the distribution of wind pressure along the height of the building, if the basic wind pressure (shown on the wind map) is 2.4 kN/m^2? What will be the total wind forces acting on the narrow and on the wide sides of the building, respectively? (See hint in Exercise 2-7.)

A portal crane in a storage yard of Fabcon, Minnesota, a producer of hollow prestressed concrete slabs. (*Courtesy of Fabcon, Minnesota.*)

Forces and Their Actions

3

3.1 FORCES ON STRUCTURES

We have seen that structures are subject to all kinds of forces and that one part of a structure transmits forces to all its adjoining parts until the loads are transferred to the foundations and to the subgrade soil.

For example, the roof structure of an industrial building (Fig. 3-1) transfers the loads from the roof to the roof trusses,† thus, in turn, bringing them to the columns and down to the footings. But in addition to carrying their own weight and the loads from the roof, in this case the columns are also loaded by forces from the crane-runway girders‡ spanning longitudinally from column to column. These girders support the transverse crane-bridge beam, along which the load (picked up anywhere inside the building) moves while the whole bridge runs along the girders to drop the load to any desired point.

To design such a complex structure, we must investigate the most severe position of the load for the design of each of its components: (1) the crane-bridge beam, (2) the crane-runway girder, and (3) the columns. It is intuitively clear that:

1. For the design of the crane-bridge beam, the load should be placed in the middle of its span, since in this position the concentrated force due to the load produces the maximum deflection of the beam. We can say that the load in this position is most effective for the design of the beam.
2. Similarly, for the maximum load effect on the crane-runway girder, the crane-bridge beam should be placed at midspan of the girder, but in this case the load should be as close to one of the beam ends as possible, so as to be almost on top of the girder (Fig. 3-2).
3. For the maximum load effect on a column, one of the crane-bridge beam ends

†*Truss* a structural member composed of straight elements in a triangular pattern. Used for large-span floors, roofs, bridges, and so on.

‡*Girder* a large main (or primary) beam usually supporting smaller, secondary beams.

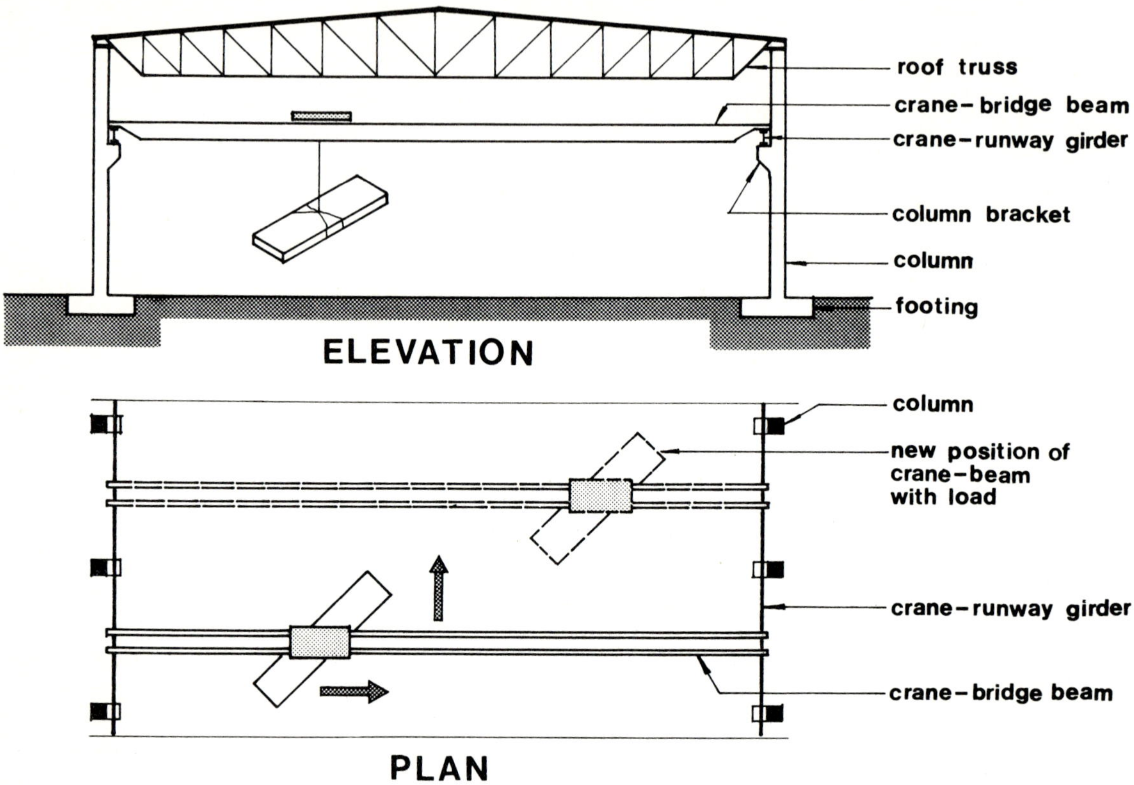

FIGURE 3-1 Plan and elevation of the structural system of an industrial building.

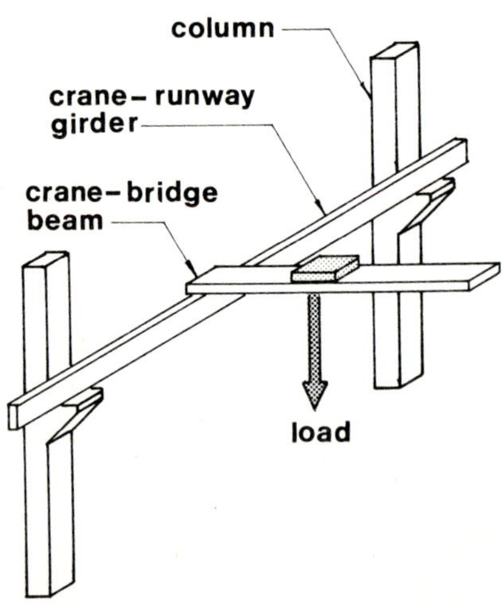

FIGURE 3-2 The crane-bridge beam is in the mid-span of the girder, and the load is as close to the girder as feasible.

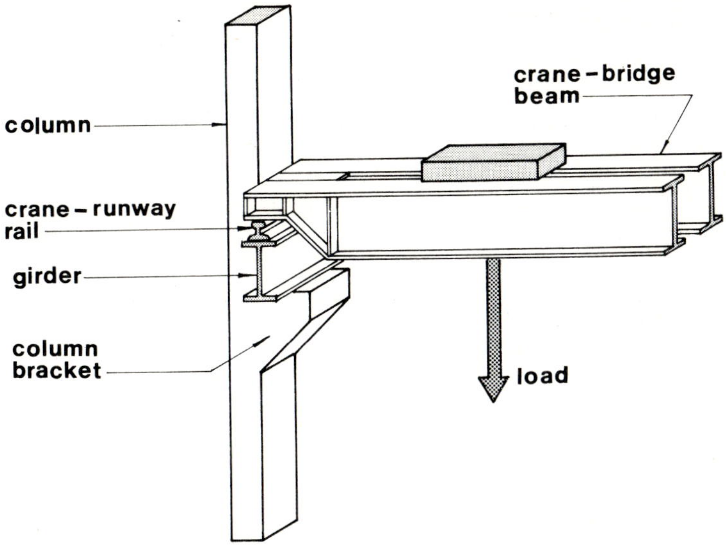

FIGURE 3-3 The crane-bridge beam is above the column bracket, and the load is as close to the column as feasible.

must be placed over the bracket of the column and the load brought as close to the column as feasible (Fig. 3-3).

At this point the above "intuitive" reasoning might be confusing. To alleviate some of this confusion, a suggestion is offered.

Suggestion: Buy a few rectangular pieces of balsa wood and use a knife and some glue to build the elements of a crane runway (columns, girders, and crane beam). Load the beam with a concentrated load hanging from a wire hook and change its position on the beam while moving the beam along the girders. Observe the deflections of the beam and girders while you increase the magnitude of the concentrated load and change its position. Your observations will confirm the above statements (1 through 3).

This example shows the importance of knowing the effects of forces on structures. To do so, we may have to add the effects of several forces on a beam to find their cumulative effect, or to apportion to the supports the forces acting at various points of the beam in order to find its reactions. In this chapter we will begin to study problems of this type.

3.2 FORCES ACTING ALONG THE SAME STRAIGHT LINE (COLLINEAR FORCES); RESULTANT OF FORCE SYSTEMS; EQUILIBRIUM OF BODIES

Let's try to pull a loaded cart. We can do it by applying to it a force of certain magnitude in the direction in which we want the cart to move. Obviously, this force must

be applied (somehow) to a point of the cart. Let us suppose that one person can develop the necessary force. The point of the cart where the force is applied is called the *point of application* of the force, and the line along which it acts is called its *line of action.* The *sense* of the force, that is, the direction in which it acts, is shown by an arrow.

A *force* is completely defined by two characteristics (Fig. 3-4): (1) its *magnitude* (measured in pounds, kilopounds, and tons or newtons, kilonewtons, etc.) and (2) its *sense* or *direction* (defined by the arrow). All quantities that possess magnitude and direction are called *vector* quantities; hence a force is a vector quantity.

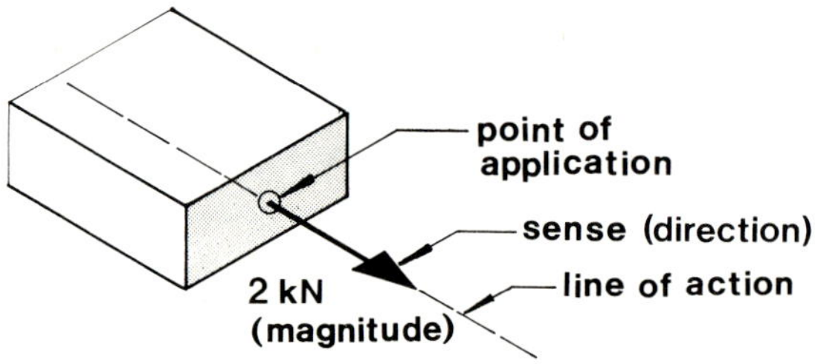

FIGURE 3-4 A force is completely defined by its magnitude and direction.

A force may be considered to act *at any point on its line of action* so long as its direction and magnitude are unchanged. The force on the cart will have the same effect whether we push it or pull it by means of a rope. And, if we pull, the effect will be the same whether the rope is long or short.

If the cart load is increased, two people may be needed to move the cart. Their action will be most efficient if both pull or push in the same direction. When two forces F_1 and F_2 act along the same straight line, they are called *collinear forces*,

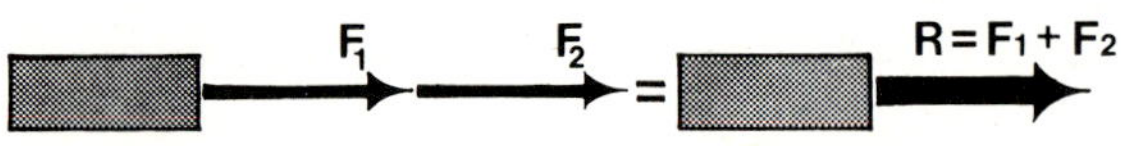

since they have a common line of action. If they also have the same direction, they are physically equivalent to a single force, their *resultant* R, whose magnitude is obtained by adding the magnitudes of the two forces

$$R = F_1 + F_2$$

If, for example, $F_1 = 140$ lb (623 N) and $F_2 = 180$ lb (800 N), then

$$R = 140 + 180 = 320 \text{ lb} \qquad (1\ 423 \text{ N})$$

The resultant has, of course, the same direction as the two forces.

Generally, two or more forces may be replaced by a single resultant force which will produce the same result (effect) as the forces it replaces.

As a matter of convenience, when a force acts in a certain direction, say, to the right, it is considered positive, and if it acts in the opposite sense, say, to the left, it is considered negative. The purpose of the sign is to indicate the direction of the force along its line of action.

The resultant of two forces can also be obtained graphically. The graphical representation of a force (as a vector quantity) with its proper magnitude and direction is called a *vector*.

Consider two horizontal collinear forces $F_1 = 20$ lb (89 N) and $F_2 = 30$ lb (133 N) having the same direction (to the right). First choose a force scale which allows us to express the magnitude of a force by the length of a line segment. For example, let a force of 10 lb be represented by a segment $\frac{1}{4}$ in long, so that

$$10 \text{ lb} = \tfrac{1}{4} \text{ in} = 0.25 \text{ in}$$

Force $F_1 = 20$ lb is then represented by a vector of length $2 \times 0.25 = 0.50$ in and force $F_2 = 30$ lb by a vector of length $3 \times 0.25 = 0.75$ in.

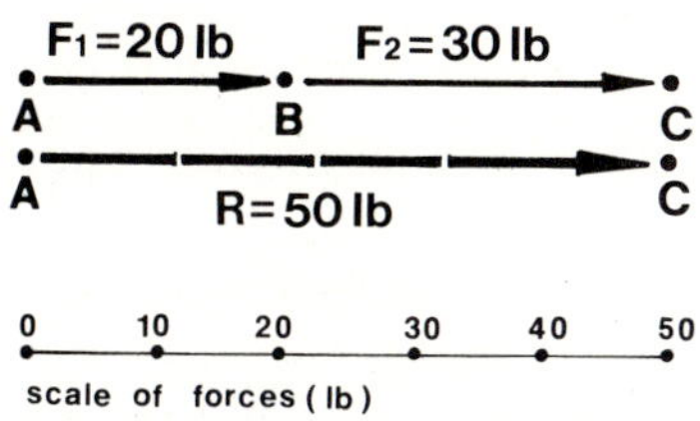

FIGURE 3-5 Resultant of two collinear forces.

Plot vector F_1 from point A to point B (Fig. 3-5) and add vector F_2 from point B to point C. Show the sense of both forces by arrows. We may conveniently call this arrangement for the summation of force vectors the ***tip-to-tail rule***, since we attach the tail of the vector of any subsequent force to the tip of the vector of the previous force. The length AC represents the magnitude of the resultant R of the two forces, which equals

$$R = 20 + 30 = 50 \text{ lb} \qquad (222 \text{ N})$$

The sense of the resultant is to the right, the same as the sense of the two forces. The arrow of the resultant points from the tail of the first force vector to the tip of the last force vector in the sequence.

According to the scale of forces, the vector of the resultant is 1.25 in long since

$$R = 50 \text{ lb} = \tfrac{50}{10} \times 0.25 = 1.25 \text{ in}$$

The same result is obtained by adding the lengths of the two vectors, 0.50 and 0.75 in, representing F_1 and F_2, respectively.

If two forces F_1 and F_2 acting on a body are of equal magnitude but opposite sense, they cancel each other. Their resultant is equal to zero (Fig. 3-6), and the body is said to be in ***equilibrium.*** If we increase one of the forces by even a small amount, the equilibrium is disturbed, and the body starts to move in the direction of the larger force. (Structures, in general, are supposed to remain at rest and not to move.

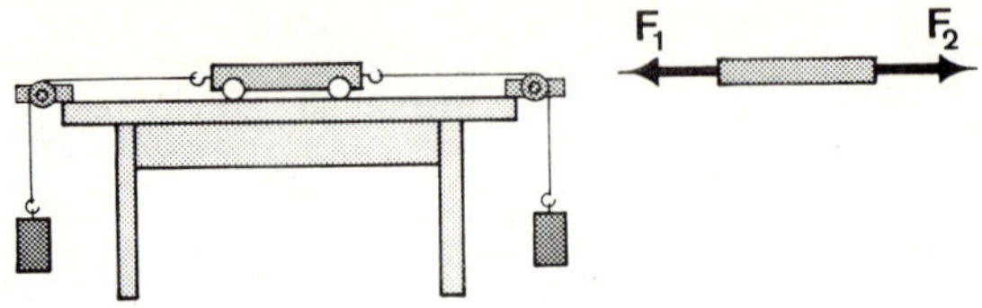

FIGURE 3-6 If $F_1 = F_2$, the body is said to be in equilibrium.

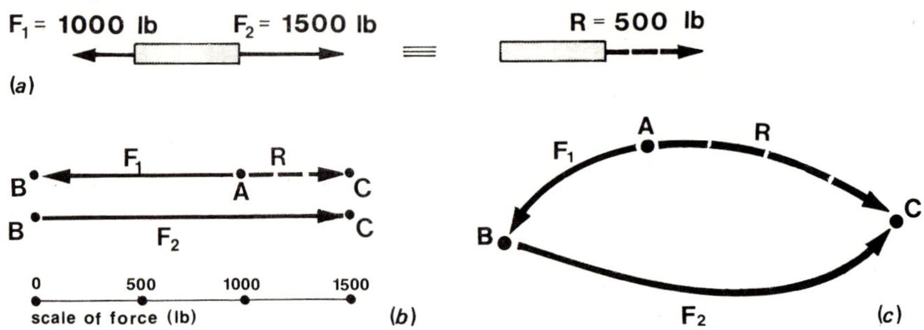

FIGURE 3-7 Result of two forces: (*b*) graphical solution; (*c*) the arrow of *R* points against the continuous flow of arrows of F_1 and F_2.

Most unpleasant experiences may be felt when equilibrium is disturbed—just think of tornadoes and earthquakes!)

At this point we introduced the term "equilibrium of a body." We will return to this principle and explain it more thoroughly in Sec. 3.4.

Let us assume that a body is acted on by two collinear forces F_1 and F_2 of opposite sense and different magnitudes (Fig. 3-7*a*). Let $F_1 = 1\,000$ lb (4 450 N) act to the left and $F_2 = 1\,500$ lb (6 670 N) act to the right. Considering the sense to the right as positive, the resultant of F_1 and F_2 is

$$R = +1\,500 - 1\,000 = +500 \text{ lb} \qquad (2\,220 \text{ N})$$

The positive sign of the resultant shows that its sense is to the right. In the graphic solution (Fig. 3-7*b*), we use the tip-to-tail rule: start plotting the vector $F_1 = 1\,000$ lb from point *A* to point *B* with the arrow to the left. Plot $F_2 = 1\,500$ lb from *B* to *C* with the arrow to the right. The length *AC* represents the magnitude of the resultant *R*. The sense of the resultant is indicated by the arrow, heading from *A* to *C*, that is, opposite to the continuous "flow" of the arrows of F_1 and F_2 (Fig. 3-7*c*).

3.3 FORCES THROUGH A POINT (CONCURRENT FORCES); RESULTANT OF CONCURRENT FORCE SYSTEMS; FORCE PARALLELOGRAM; FORCE TRIANGLE

Forces whose lines of action pass through a common point of a body are called *concurrent forces.* As an example, consider a motorboat crossing a stream (Fig. 3-8*a*). The force F_1 due to the stream current and the propulsive force F_2 of the engine act on a point *M* of the boat in different directions. Figure 3-8*a* is called a *force plan.*

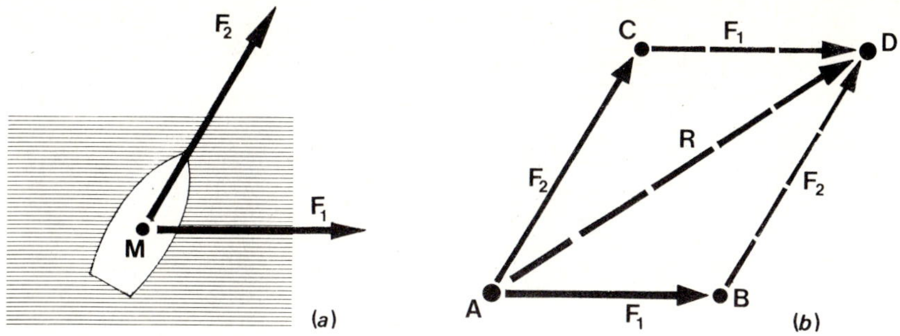

FIGURE 3-8 (*a*) Force plan; (*b*) force parallelogram.

Under the action of these forces, the boat will move in the direction of their resultant. But how can we find the magnitude and the direction of the resultant of two concurrent forces? The resultant can be found by the so-called parallelogram law, which is nothing more than an extension of the tip-to-tail rule.

Choose a force scale and plot the vector representing the force F_1 from A to B in the direction of F_1 (Fig. 3-8*b*). Similarly, draw the vector of F_2 from A to C. Remember that both vectors must have the magnitude and the direction of the original forces F_1 and F_2 as shown in the force plan. Now draw a parallel to F_2 through B and a parallel to F_1 through C. The two parallels intersect at D. The diagonal AD of the parallelogram $ABDC$ represents the vector of the resultant R of the forces F_1 and F_2 with its magnitude (found on the scale of the forces) and direction. Figure 3-8*b* is called a *force parallelogram.*

The parallelogram law cannot be proved mathematically. It is accepted because it corresponds to physical facts; the boat actually moves in the direction of the resultant obtained by the parallelogram law. Analyzing the resulting displacement of the boat, we may say that it moves first in the direction AB because of the action of the current only and then travels in the direction BD as a result of the push of the engine.

The force parallelogram is usually drawn separately, next to the force plan. Let us notice that we do not have to complete the whole parallelogram. Instead, a force triangle is sufficient to determine the resultant vector. Starting from point A (Fig 3.9*b*), draw the vector AB of F_1 with its given magnitude and in a direction parallel to the actual force F_1 (the arrow pointing to the right). From B draw the vector BD of the force F_2, parallel to the direction of the actual force F_2. The directed segment AD is the vector of the resultant R of the two forces F_1 and F_2. The arrow of the resultant is from A to D, that is, *opposite* to the continuous flow of arrows of the forces F_1 and F_2.

In constructing the force triangle, we can start with any of the forces (Fig. 3-10). If, for example, we start with the vector AC of F_2 and add the vector CD of F_1, we obtain the same point D, as previously, and the vector AD is the resultant vector.

Returning from the force triangle (Fig. 3-9*b*) to the force plan (Fig. 3-9*a*), draw the parallel to R through M in the force plan. Show its direction by an arrow, and show its magnitude (in the figure it is shown arbitrarily to be $R = 5$ kN), which can be

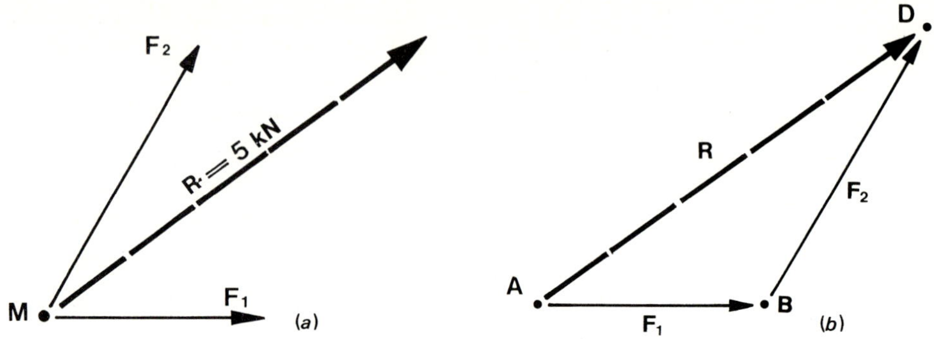

FIGURE 3-9 (*a*) Force plan (point *M* defines the boat). (*b*) Force triangle. The arrow of *R* points against the continuous flow of arrows of F_1 and F_2.

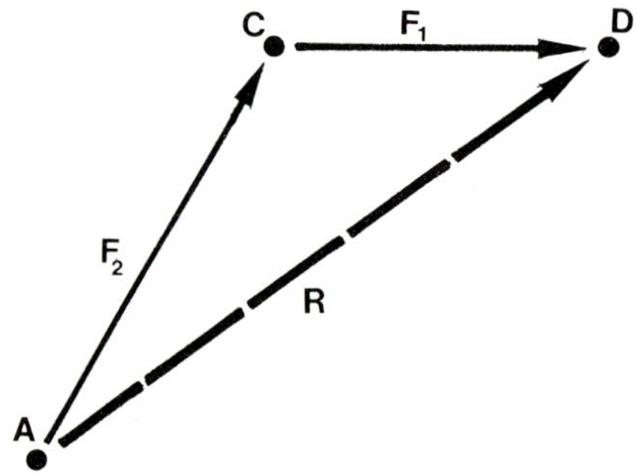

FIGURE 3-10 Alternate force triangle.

found by measuring the length *AD* in the force triangle and converting it by means of the force scale.

Illustrative Problem 3-1

Find the magnitude and the direction of the resultant *R* of the two forces acting on point *M* of the helicopter shown in Fig. 3-11*a*. Let F_1 = 2 kips (8.90 kN), F_2 = 5 kips (22.24 kN), and the angle between their lines of action = 60°.

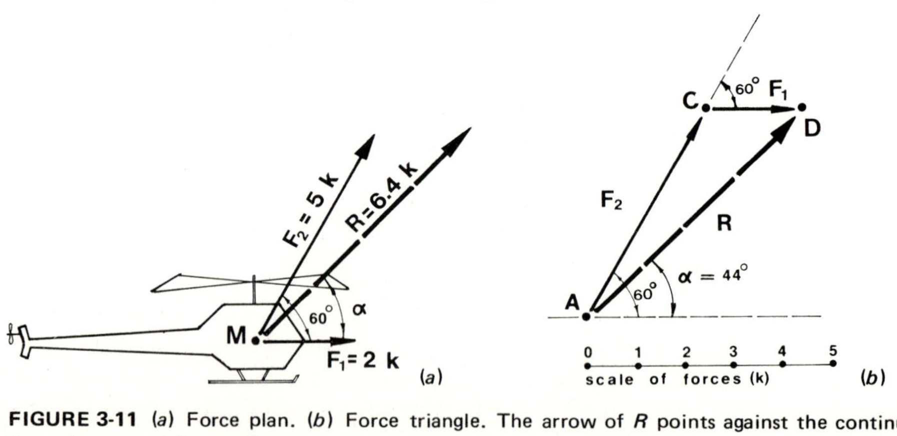

FIGURE 3-11 (*a*) Force plan. (*b*) Force triangle. The arrow of *R* points against the continuous flow of arrows of F_1 and F_2.

SOLUTION Choose the scale of forces (1 kip = $\frac{1}{2}$ in) and draw the force triangle (Fig. 3-11*b*). The length of the resultant R is 3.2 in, that is, $R = 6.4$ kips (28.5 kN), and its direction makes an angle of $\alpha = 44°$ with the horizontal. Transfer R to point M in the force plan and show its sense and magnitude. (Remember that the sense of the arrow of the *resultant* is always *opposite* to that of the continuous flow of the force arrows.)

Note: A graphical solution is seldom as accurate as an analytical one. However, the accuracy obtained graphically is sufficient for the preliminary design of structures, since the loads can be estimated only approximately and the actual behavior of the structure is often represented by a simplified model.

Illustrative Problem 3-2

A concrete wall 2 ft thick (61 cm) and 12 ft high (366 cm) is exposed to a wind pressure $p = 30$ lb/ft^2 (1.44 kN/m^2) (Fig. 3-12). If the unit weight of concrete γ is 150 lb/ft^3 (23.5 kN/m^3), find the resultant of the forces acting on the wall.

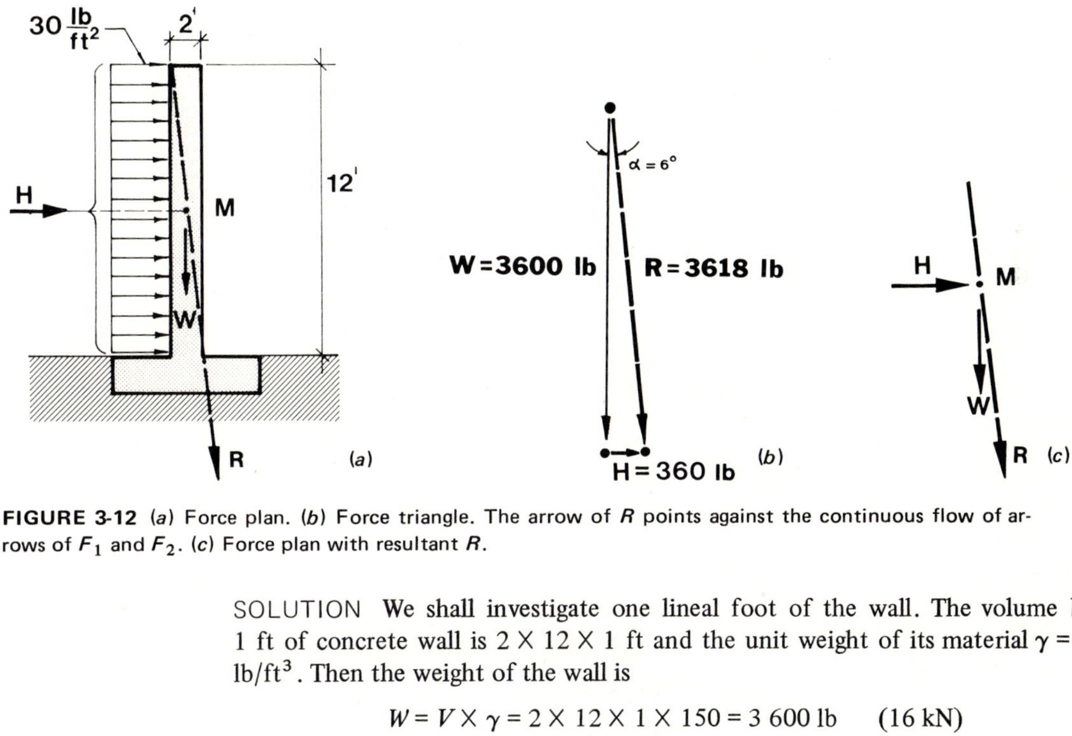

FIGURE 3-12 (*a*) Force plan. (*b*) Force triangle. The arrow of *R* points against the continuous flow of arrows of F_1 and F_2. (*c*) Force plan with resultant *R*.

SOLUTION We shall investigate one lineal foot of the wall. The volume V of 1 ft of concrete wall is $2 \times 12 \times 1$ ft and the unit weight of its material $\gamma = 150$ lb/ft^3. Then the weight of the wall is

$$W = V \times \gamma = 2 \times 12 \times 1 \times 150 = 3\,600 \text{ lb} \qquad (16 \text{ kN})$$

The total horizontal wind force on the wall surface of 12×1 ft is

$$H = (\text{area of the surface}) \times (\text{unit pressure of wind})$$

$$= (12 \times 1)(30) = 360 \text{ lb} \qquad (1.6 \text{ kN})$$

The weight of the wall W is assumed to be concentrated at point M, which is called the *center of gravity* of the wall. It is located halfway up the wall. The total horizontal wind force H represents a resultant due to the uniform wind pressure on the wall and is assumed to act on the surface of the wall in the middle of its height. It can be shifted to point M, since a force may be considered to act at any point on its line of action so long as its sense and magnitude are unchanged. Thus this force system is concurrent. The resultant R of the two forces is found by the force triangle (Fig. 3-12*b*). In this problem its magnitude can easily be computed numerically by using the pythagorean theorem:

$$R^2 = 3\,600^2 + 360^2 = 13\,089\,600$$

$$R = \sqrt{13\,089\,600} = 3\,618 \text{ lb} \qquad (16.1 \text{ kN})$$

As shown by the graphical solution, the resultant makes an angle $\alpha = 6^\circ$ with the vertical. We can now draw its line of action through the point M of the force plan (Fig. 3-12*a, c*).

Note: In Fig. 3-12*a* the resultant R appears as if it went through the diagonal corners of the wall. This is only accidental. It can assume any direction, depending on the magnitude of W and H.

Illustrative Problem 3-3

The beam shown in Fig. 3-13*a* is loaded by two forces acting in a vertical plane:

$$F_1 = 500 \text{ kN} \qquad (112.4 \text{ kips})$$

and

$$F_2 = 300 \text{ kN} \qquad (67.45 \text{ kips})$$

Find a single force equivalent to the two forces acting on the beam (i.e., their resultant).

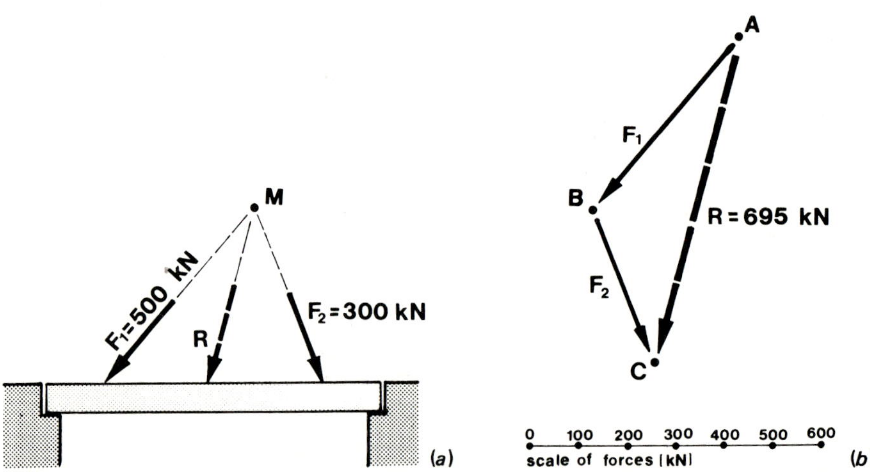

FIGURE 3-13 (*a*) Force plan. (*b*) Force triangle. The arrow of R points against the continuous flow of arrows of F_1 and F_2.

SOLUTION The force vectors shown in Fig. 3.13*a* do not intersect at one point, and you might think that the two forces are not concurrent. However, by extending the lines of action of the two forces, we can determine point M where they intersect and where both forces can be considered to act. Therefore, their resultant can be found by the *force-triangle* method (Fig. 3-13*b*). The resultant vector R turns out to be equal to 695 kN (156 kips), and its arrow is *opposite* to the continuous flow of the arrows of F_1 and F_2. Finally, draw the parallel to R through M in the force plan.

3.4 EQUILIBRIUM OF CONCURRENT FORCE SYSTEMS; EQUILIBRANT

In Sec. 3.2 we introduced the definition of a resultant and the principle of equilibrium of a body under the action of two equal forces of opposite sense.

Now we can extend this principle to a concurrent force system: a body under the action of a concurrent force system is in equilibrium if the resultant of the force system is equal to zero. In statics we presume that bodies are in static equilibrium. They do not move; they remain at rest.

When the resultant of a force system acting on a body is zero, the system is balanced, and we say that the *force system is in equilibrium.* Under the action of an *equilibrium force system*, a body is in equilibrium.

The cart shown in Fig. 3-14*a* is pulled by two people who exert on it equal horizontal forces $F_1 = F_2 = 6$ kips (26.7 kN). These forces are concurrent since their lines of action meet at point M. What is the direction and the magnitude of a horizontal

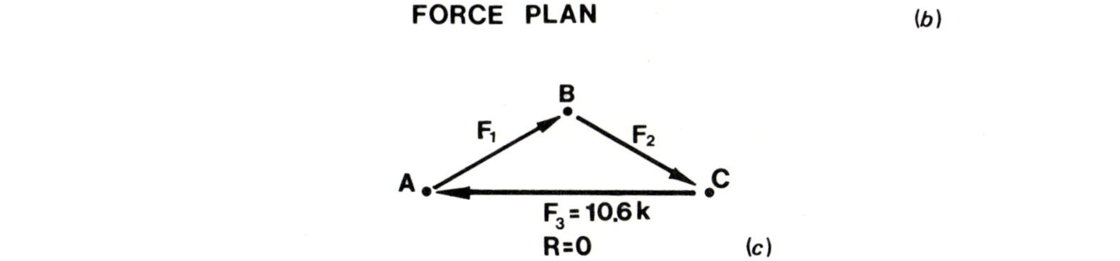

FIGURE 3-14 (*a*) Force plan. (*b*) Force triangle. The arrow of R points against the continuous flow of F_1 and F_2. (*c*) Equilibrium of forces F_1, F_2, and F_3. Their resultant R is equal to zero ($R = 0$). For equilibrium, all arrows follow the *same* direction (tip to tail).

force F_3 exerted at the same point M by a third man, such that F_1, F_2, and F_3 will be in equilibrium, that is, so that the cart will not move? In other words, in what direction and how hard must the third person pull to prevent the cart from moving?

To solve this problem, we first construct the force triangle of F_1 and F_2 (Fig. 3-14*b*). The vector AC defines the magnitude and the direction of the resultant R of these two forces, which would move the cart in the direction AC. To prevent this movement, a third force F_3, equal in magnitude to the resultant R but of *opposite* sense, must act at point M.

If we add all three forces F_1, F_2, and F_3 according to the tip-to-tail rule as in Fig. 3-14*c* (i.e., draw F_3 with its tail at the tip of F_2), we find that its tip is at A, the tail of F_1. The force triangle of the three forces is *closed*; it starts and ends at A, and hence the resultant of the three forces is zero. All arrows of the equilibrated force triangle follow the *same* direction in tip-to-tail fashion.

If a concurrent force system has a zero resultant, the forces are said to be in equilibrium. The body (the cart) remains at rest; it is in equilibrium.

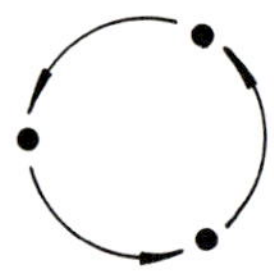

all arrows run tip - to - tail

We can now draw the parallel to CA through point M in the force plan to locate the direction of F_3 on the body. In this example its magnitude F_3 = 10.6 kips (47.15 kN) and its direction is to the left. The force F_3, needed to keep the body at rest, is called the *equilibrant* of the forces F_1 and F_2.

Since the lines of action of all three forces lie in the same plane, we call such a force system a *coplanar concurrent force system*.

In designing beams we must sometimes determine the equilibrium of a known

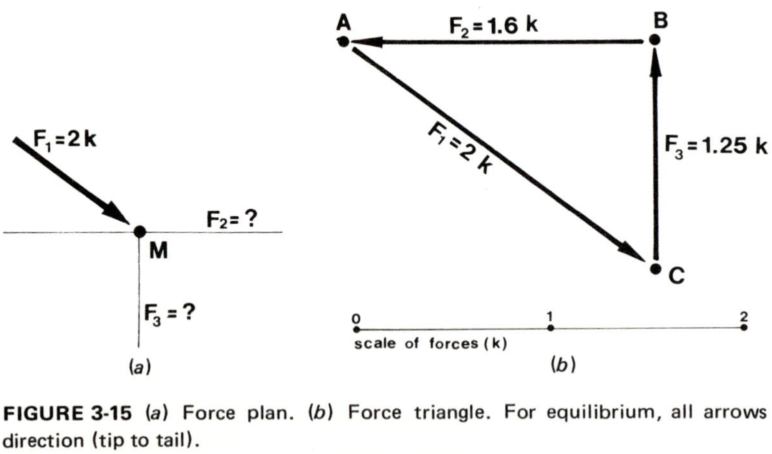

FIGURE 3-15 (*a*) Force plan. (*b*) Force triangle. For equilibrium, all arrows follow the *same* direction (tip to tail).

force F_1 and two additional forces, F_2 and F_3 of the known lines of action but unknown magnitudes and directions (Fig. 3-15*a*). If all three forces constitute an equilibrium concurrent force system at point M, the force-triangle method can be used to solve the problem as follows.

Consider (Fig. 3-15*b*) the given force F_1 to be an inclined 2-kip (8.9-kN) force, the line of action of F_2 to be horizontal, and that of F_3 vertical. Draw the vector AC of the known force F_1. Through point A draw a parallel to the line of action of F_2, and through point C draw a parallel to the line of action of F_3. These parallels intersect at point B, which defines the magnitude of F_2 and F_3. The vectors AC, CB, and BA, with all arrows flowing in the *same* direction (tip to tail), define a *closed* triangle with zero resultant, whose sides represent the given force F_1, a horizontal force F_2, and a vertical force F_3, as required by the problem.

3.5 PRINCIPLE OF ACTION AND REACTION; FREE BODY AND FREE-BODY DIAGRAM

A push against a wall causes an action to be exerted on the wall. The wall resists the push with an equal and opposite force called *reaction*.

To every action there is an equal and opposite reaction. This principle was first formulated by the physicist Isaac Newton (1642 to 1727) in his fundamental laws of mechanics and is referred to as the principle of action and reaction.

If a beam is supported by a wall, it exerts an action, while the wall reacts by exerting an equal and opposite reaction on the beam. In solving problems of statics, we can isolate a whole structure from its supports and replace their effects by reactions. When so isolated from its supports, a body is said to be "free" and is called a *free body*. A schematic diagram of a free body showing all forces acting on it is called a *free-body diagram*.

A structural system as a whole, as well as any of its parts, must be in equilibrium under the applied forces. Thus we can isolate any part of a structure as a free body. If we apply to it all forces (external forces and reactions of the removed parts of the structure), we obtain a free-body diagram of that part. If a free body of a structural part has practically negligible dimensions, we can represent it schematically by a point.

The following example illustrates practical use of these principles.

Illustrative Problem 3-4

Investigate the bridge structure shown in Fig. 3-16. The loads from the bridge floor are transferred to the longitudinal beams B_1 and B_2, which have one end supported on the piers and the other on the transverse beam T. The transverse beam is supported by the struts† S_1 and S_2. If the end reactions of B_1 and B_2 on the transverse beam add up to a force F of 10 kips (44.5 kN), what are the forces in the struts? What forces will the struts exert on the piers?

†A *strut* is a stiff columnlike element that transfers compressive forces.

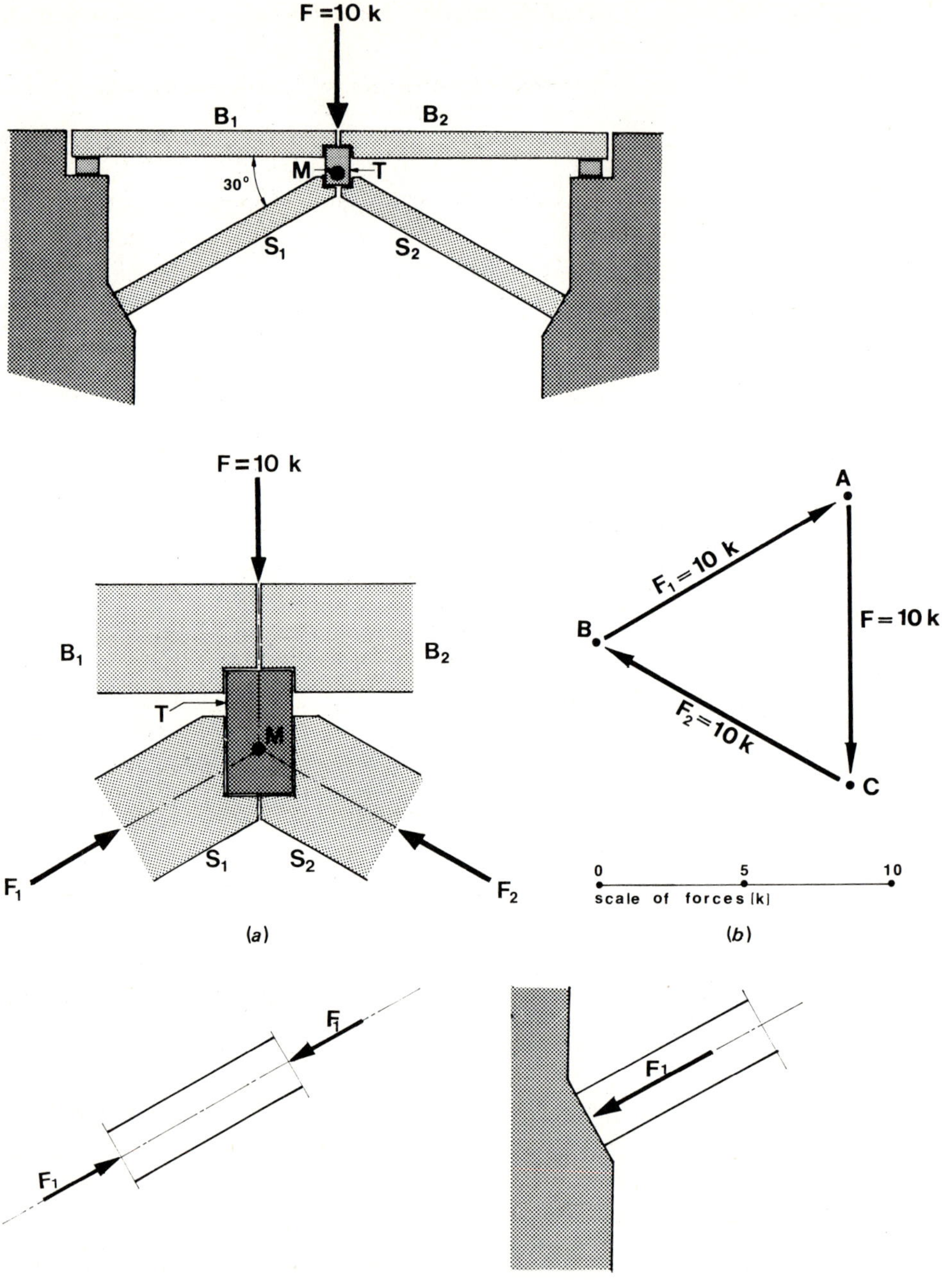

FIGURE 3-16 Schematic elevation of a bridge system: B_1, B_2—longitudinal beams; T—transverse beam; S_1, S_2—struts; M—joint. (*a*) Free-body diagram of joint M; (*b*) for equilibrium, all arrows follow the same direction (tip to tail); (*c*) the strut is compressed by equal and opposite forces applied to its ends; (*d*) compressive force F_1 acting on the pier.

SOLUTION The structural system as a whole, as well as any of its parts, must be in equilibrium under the applied forces. Isolate the part of the structure around point M, intersection of the struts and the transverse beam T, from the rest of the system, thus obtaining a free body of what we can call *joint M*. Figure 3-16*a* shows a free-body diagram of joint M. This free body must be in equilibrium under the action of the force F and the two unknown actions from the struts, F_1 and F_2. The lines of action of the strut forces are assumed to coincide with the longitudinal axes of the struts. All three forces intersect at point M, thus constituting a concurrent force system.

Let us draw the force triangle F, F_1, F_2 starting with the vector AC of the known force F (Fig. 3-16*b*). Its sense is downward. Through its endpoints draw parallels to the directions of F_1 and F_2. The intersection of the two parallels defines point B of the force triangle. For equilibrium, all arrows in the force triangle must follow the *same* direction, defined by the arrow of F. Transfer the arrows from the force triangle to the free-body diagram, and you will find that the forces F_1 and F_2 exerted by the struts on joint M are directed *toward M*; that is, they push on M. Hence they are *compressive* forces. From the force scale we have

$$F_1 = F_2 = 10 \text{ kips} \qquad (44.5 \text{ kN})$$

To determine the forces in the struts, we must appeal to the *principle of action and reaction*.

According to this principle, if the struts exert compressive forces on joint M, M exerts the same compressive forces on the struts. The struts are compressed. Now, for equilibrium of the struts, the piers must exert on their lower end a force equal and opposite to the forces F_1 and F_2 on their upper ends† (Fig. 3-16*c*). The struts are compressed by the equal and opposite forces applied to their ends.

Finally, by the principle of action and reaction, if the piers exert compressive forces on the struts, these must exert the same compressive forces on the piers (Fig. 3-16*d*). Thus the forces on the piers are also compressive and have a magnitude of 10 kips. The struts channel the force F to the piers and hence to the soil by compressive action (Fig. 3-17).

†We neglect the weight of the struts, which is usually small compared to F_1 and F_2.

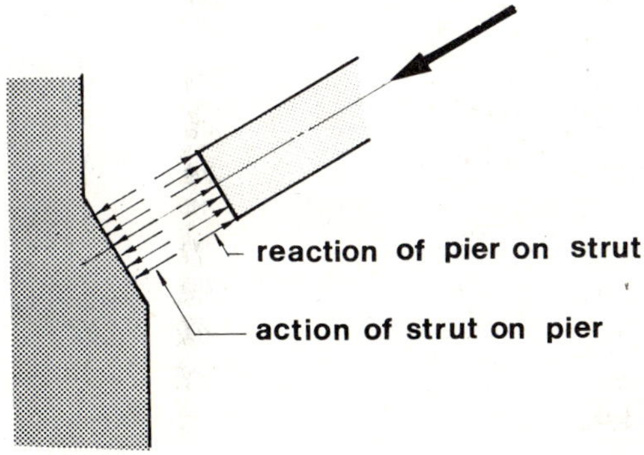

FIGURE 3-17 Principle of action and reaction (pier-strut).

There are numerous examples from our daily activity which remind us of the validity of the principle of action and reaction (Fig. 3-18). Can you mention some of them?

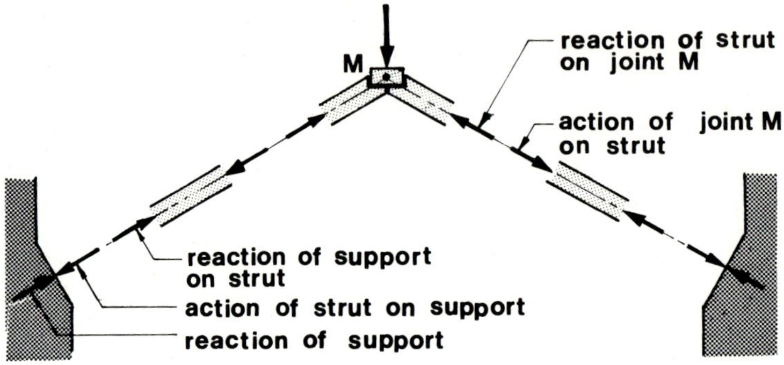

FIGURE 3-18 Principle of action and reaction (pier-strut-joint *M*).

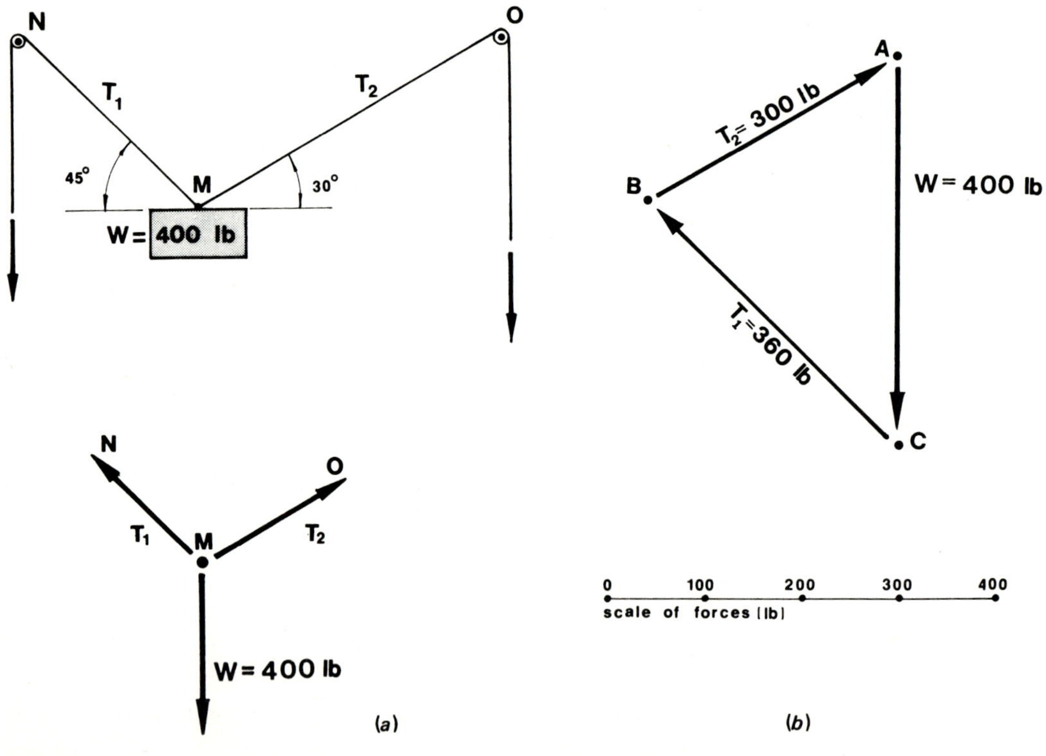

FIGURE 3-19 Force plan. (*a*) Free body of joint *M*. (*b*) Force triangle. For equilibrium, all arrows follow the same direction defined by the arrow of *W*.

Illustrative Problem 3-5

A weight $W = 400$ lb (1.78 kN) is to be lifted by a rope going over two pulleys, as shown in Fig. 3-19. Find the tensile forces T_1 and T_2 in the two inclined branches of the rope.

SOLUTION The force $W = 400$ lb at joint M and the lines of action of the forces T_1 and T_2 are known. We must find the magnitude and the sense of T_1 and T_2, the unknown forces in the two branches of the rope. Cut out the part of the system about joint M from the rest of the system. Joint M (Fig. 3-19*a*) is now changed into a free body. Joint M itself, as a free body, must be in equilibrium under the action of the forces W, T_1, and T_2. Let us draw the force triangle, starting with the fully defined vector AC of the known force, W (Fig. 3-19*b*). From its endpoints, draw the parallels to MN and MO, which are the lines of action of T_1 and T_2, thus obtaining point B on their intersection. For equilibrium, all arrows must follow the *same* direction given by the arrow of W. Transfer the arrows from the force triangle to the free-body diagram, and you will find that the forces T_1 and T_2 are pulling *away* from joint M—they are tensile forces. Rope part MN is in tension as a result of the force $T_1 = 360$ lb (1.6 kN), while the part MO is in tension as a result of the force $T_2 = 300$ lb (1.33 kN). Figure 3-20 shows the system separated into the free bodies of joint M and rope segments MN and MO.

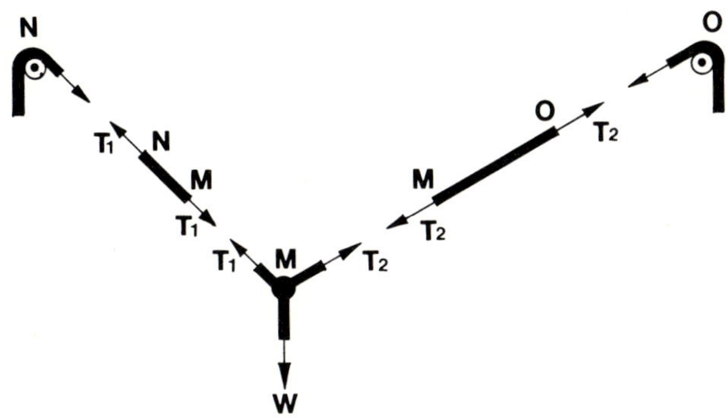

FIGURE 3-20 Principle of action and reaction (pulleys-rope-joint *M*).

Illustrative Problem 3-6

A hanger† consists of two rods connected by a pin (Fig. 3-21), making an angle of 60° with the horizontal. It is loaded at the pin by a vertical force $F = 120$ kN (27 kips). Find forces F_1 and F_2 acting on each rod of the hanger.

†A *hanger* is a long, slender element that transfers tensile forces.

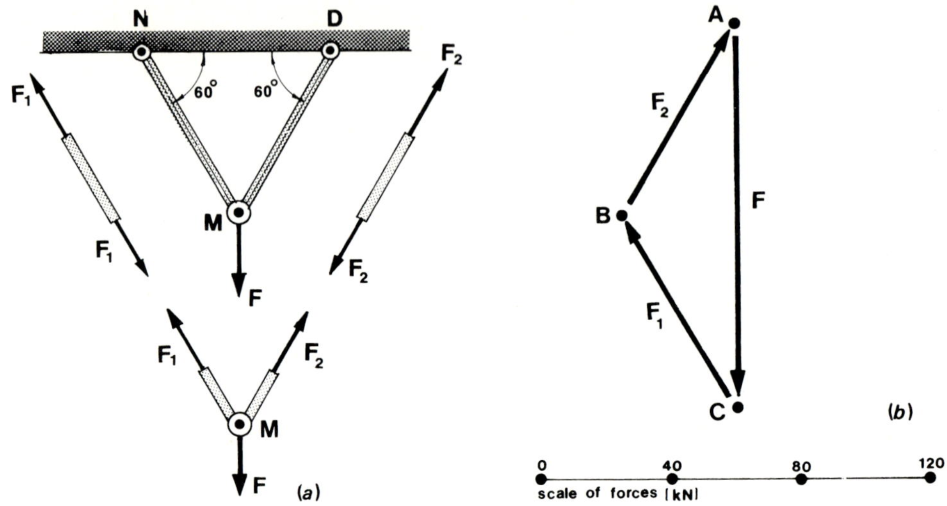

FIGURE 3-21 *(a)* A hanger and free-body diagrams of its parts. *(b)* Force triangle. For equilibrium, all arrows follow the same direction defined by the arrow of *F*.

SOLUTION Isolate a part of the structure about joint M from the rest of the structure and investigate its equilibrium as a free body (Fig. 3-21*a*). Start by drawing the vector AC of the known force F (Fig. 3-21*b*). Draw a parallel to the line of action of F_1 through point C and parallel to the line of action of F_2 through point A. The intersection of the two parallels defines point B of the force triangle and therefore the magnitude of the forces F_1 and F_2. Since all the forces acting on M must be in equilibrium, all arrows follow the *same* direction defined by the arrow of F.

Measuring the length of the vectors AB and BC, we find the magnitude of the two unknown forces to be

$$F_1 = 70 \text{ kN} \qquad F_2 = 70 \text{ kN} \qquad (15.74 \text{ kips})$$

Each rod of the hanger is stretched by a tensile force $F_1 = F_2 = 70$ kN. The cross-sectional dimensions of the rods must be sufficiently large to transfer the forces safely to the ceiling.

3.6 COMPONENTS OF A FORCE

The resultant of two forces was previously defined as a force replacing the effect of two forces. We can now reverse the problem and find two *components* of the resultant.

The resultant R is known; the magnitude and the direction of the two compo-

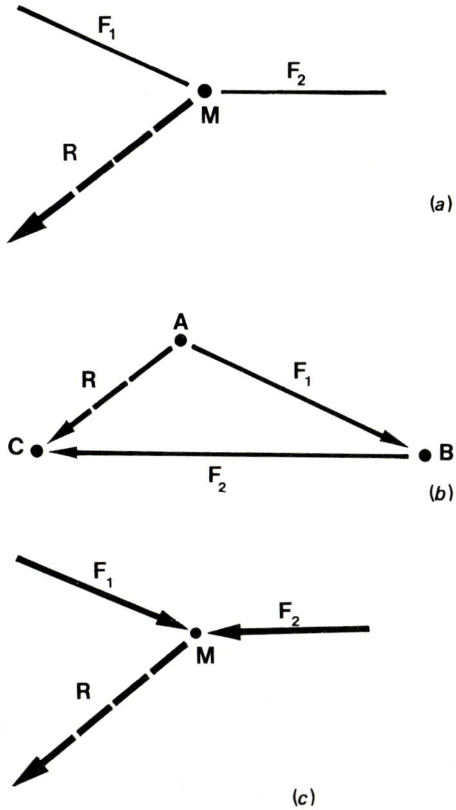

FIGURE 3-22 (*a*) Force plan; R is known; F_1 and F_2 are unknown. (*b*) Force triangle. Arrows of the components F_1 and F_2 run against the arrow of R. (*c*) The effect of forces F_1 and F_2 can be replaced by their resultant R.

nents F_1 and F_2 are unknown (Fig. 3-22*a*), but we know their lines of action. Again, three such forces form a triangle (Fig. 3-22*b*). The arrows of the components must run ***against*** the arrow of R. Once F_1 and F_2 have been found, transfer the arrows to the force plan to show their direction (Fig. 3-22*c*).

The resultant R can now be replaced by its two components, the forces F_1 and F_2, whose combined action is equal to that of R. Since R is any given force, we have learned now to split any force F into two components of given lines of action.

Note: Figure 3-15 (p. 46) shows the problem of finding two forces, F_2 and F_3, so that they are in equilibrium with the given (known) force F_1. For equilibrium force system ($R = 0$), all arrows follow the *same* direction defined by the arrow of the known force F_1 (see Fig. 3-15*b* and compare it with Fig. 3-22*b*).

Illustrative Problem 3-7

A column is loaded by the inclined force $F = 100$ kips (445 kN), making an angle of 60° with the horizontal (Fig. 3-23*a*). Replace the action of this force by two forces, one vertical (V) and the other horizontal (H). Find the magnitude of V and H.

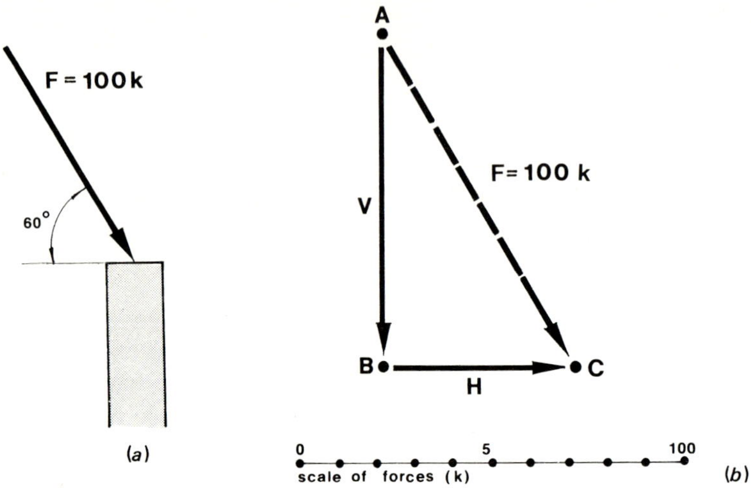

FIGURE 3-23 (*a*) What is the magnitude of horizontal component *H* and vertical component *V* of force *F*? (*b*) Force triangle. Arrows of components *V* and *H* run against the arrow of their resultant *F*.

SOLUTION Draw the vector *AC* of the force *F* (Fig. 3-23*b*). Through its end-points draw the parallels to the directions of *V* and *H*. They intersect at *B*, thus determining the magnitude of the forces *V* and *H*. The arrows of *V* and *H* run ***against*** the arrow of *F*. The magnitudes of *V* and *H* are found from the scale of forces to be $V = 86$ kips (383.5 kN) and $H = 50$ kips (222.4 kN). The effect of $F = 100$ kips can be replaced by that of the two forces, $V = 86$ kips and $H = 50$ kips (Fig. 3-24).

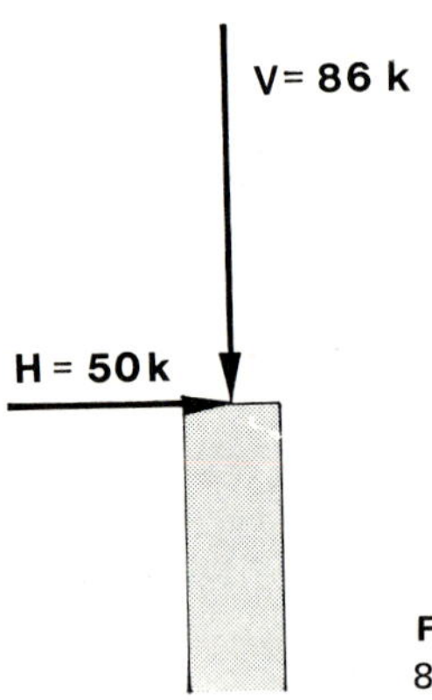

FIGURE 3-24 The effect of $F = 100$ kips can be replaced by that of $V =$ 86 kips and $H = 50$ kips.

Illustrative Problem 3-8

A body weighing 300 lb (1.33 kN) rests on an inclined plane, making an angle of 20° with the horizontal (Fig. 3-25). Find the magnitude of the force *N*, normal to the inclined plane, and the force *P*, parallel to the inclined plane. The force *N*

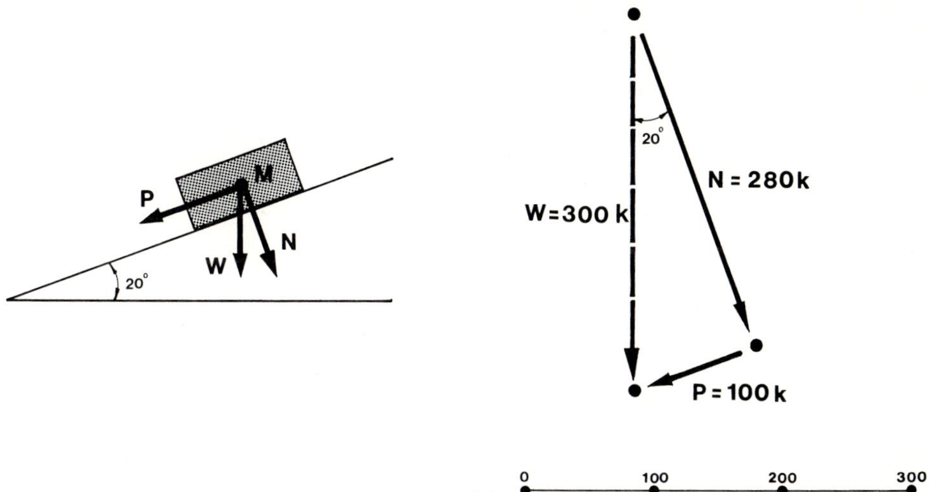

(a)

FIGURE 3-25 Force plan. What are the components *P* and *N* of the weight *W*? (*a*) Force triangle. The arrows of components *P* and *N* run against the arrow of their resultant *W*.

presses against the plane, while *H* is responsible for pulling the body down the plane. They both are the components of the weight *W*.

SOLUTION Replace the 300-lb force *W* by the two forces *N* and *P*. Using the force triangle (Fig. 3-25*a*), we find that $N = 280$ lb (1.25 kN) and $P = 100$ lb (0.45 kN). The arrows of the components run *against* the arrow of *W*. Transfer these arrows to point *M* of the body.

3.7 RESULTANT AND EQUILIBRIUM OF MORE THAN TWO CONCURRENT FORCES IN A PLANE

If more than two forces in the same plane act on a point *M*, the principles of the previous sections of this chapter can be used to obtain their resultant or to determine equilibrium.

Consider, for example, that in addition to the current force F_1 and the motor force F_2, a wind force F_3 acts on the boat in Fig. 3-26. We wish to find the direction and the magnitude of the resultant of the three forces, that is, a single force equivalent to F_1, F_2, and F_3.

To solve the problem, we draw a *force polygon* instead of a force triangle. From any convenient point *A* (Fig. 3-26*a*) draw the vector *AB*, parallel and equal to F_1. From *B* draw the vector *BC*, parallel and equal to F_2. Notice that the vector *AC* is the resultant R_{12} of the two forces F_1 and F_2. To complete the force polygon, from *C* draw the vector *CD*, parallel and equal to F_3. The arrows of all three forces run tip to

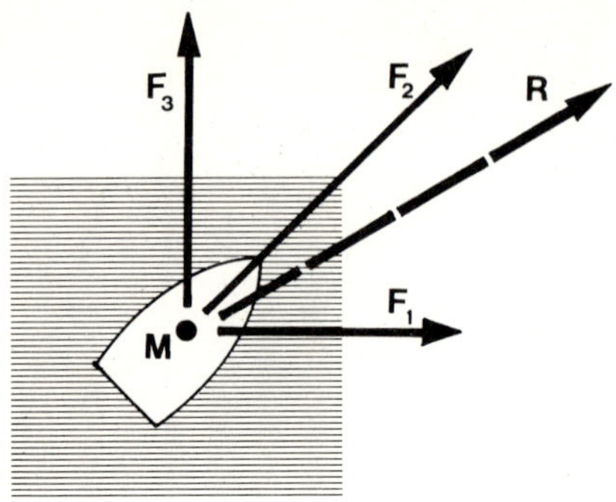

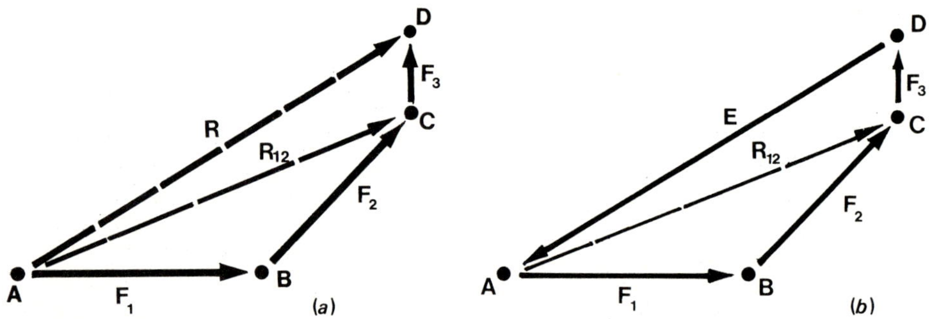

FIGURE 3-26 Resultant of three forces. (*a*) Force polygon. The arrows of forces F_1, F_2, and F_3 run against the arrow of R. (*b*) Equilibrant E of a force system. For equilibrium, all arrows follow the same direction.

tail. The resultant of the forces F_1, F_2, and F_3 is given by the vector AD. It connects the tail of the first force F_1 with the tip of the last force F_3 in the force polygon. The arrow of the resultant goes ***against*** the arrows of the forces F_1, F_2, and F_3.

Now transfer the resultant R of the three forces to point M on the boat and draw its arrow. Under the action of the three given forces the boat will move in the direction of R. Note that it does not matter what the sequence of the component forces is in the force polygon.

In the preceding example, the resultant R was found, thus replacing the effect of the three given forces on the body. The next question is: what force would cancel the effect of the three forces? As before, this force is the equilibrant E, equal and opposite to the resultant R. The equilibrant E (Fig. 3-26*b*) is needed to equilibrate the forces F_1, F_2, and F_3. If the equilibrant acts, the boat does not move. In the force polygon, all the arrows run in the *same* direction, showing that the system is in equilibrium ($R = 0$).

Illustrative Problem 3-9

Three known forces, F_1, F_2, and F_3, act on the body shown in Fig. 3-27. Find the force E which would ensure the equilibrium of the system. The magni-

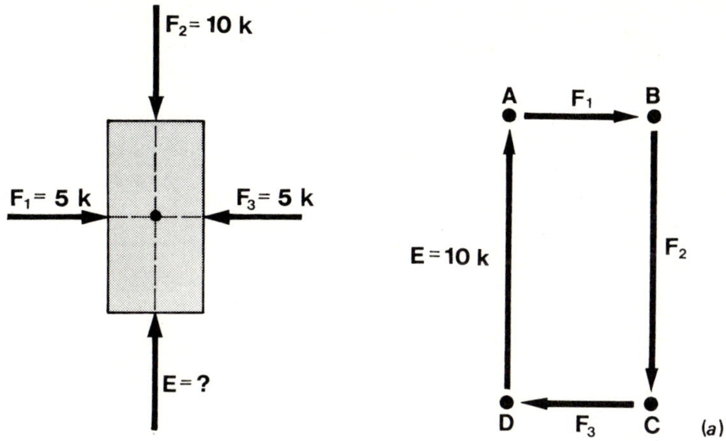

FIGURE 3-27 Force plan in Illustrative Prob. 3-9. (*a*) Force polygon. For equilibrium, all arrows run in the same direction.

tudes of the known forces are

$$F_1 = 5 \text{ kips} \quad (22.25 \text{ kN}) \qquad F_2 = 10 \text{ kips} \quad (44.5 \text{ kN}) \qquad F_3 = 5 \text{ kips}$$

SOLUTION Draw the force polygon *ABCD* as shown in Fig. 3.27*a*, starting with the three known forces, F_1, F_2, and F_3. The vectors *AB*, *BC*, and *CD* represent the magnitude and the direction of the forces F_1, F_2, and F_3, respectively. The vector of the equilibrant E is obtained by connecting points A and D of the force polygon. All the arrows run in the *same* direction. The magnitude of the equilibrant E is equal to 10 kips and points upward. (*Note:* The sequence of forces in the force polygon is arbitrary.)

Illustrative Problem 3-10

A roof truss is composed of bars connected at the joints as shown in Fig. 3-28. Joint J, shown as a free body in Fig. 3-28*a*, consists of four such bars. Let us assume that the magnitude and the sense of the forces in bars 1 and 2 have been found previously. They are the actions F_1 and F_2 on joint J. Let F_1 be a compressive force equal to 2.5 kips (11.12 kN) and F_2 another compressive force equal to 1.5 kips (6.67 kN). Find the magnitude and the sense of the remaining two unknown forces, F_3 and F_4, acting on the joint, such that equilibrium of the joint is achieved. Assume that all forces are parallel to the corresponding bars.

SOLUTION We start by drawing the force polygon from point A (Fig. 3-28*b*). Draw *AB* parallel and equal to F_1. Then add *BC* parallel and equal to F_2. The dotted line *AC* represents the resultant R_{12} of F_1 and F_2.

We now equilibrate this resultant by two components parallel to F_3 and F_4. Through endpoint A draw a parallel to F_3, and through point C draw a parallel to F_4. At their intersection we obtain point D, thus defining the magnitudes of F_3

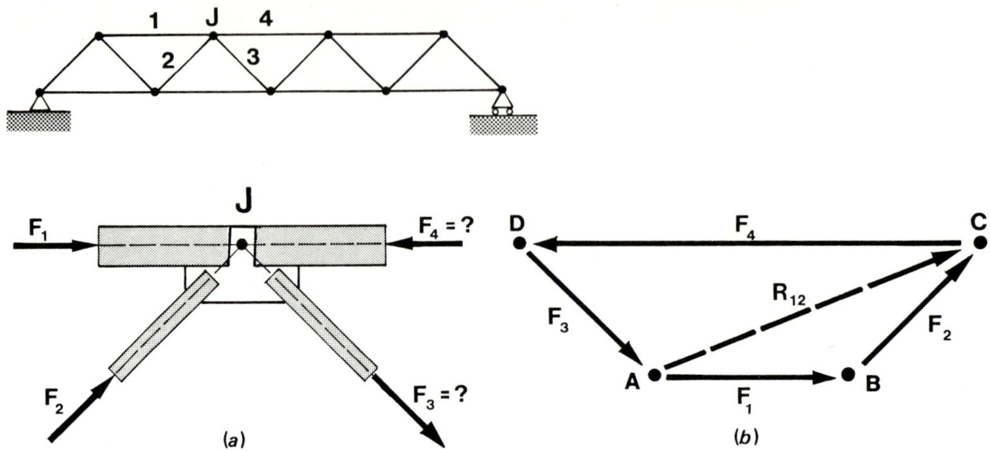

FIGURE 3-28 A roof truss in Illustrative Prob. 3-10. (*a*) Free body of joint *J*. (*b*) Force polygon. For equilibrium, the arrows of all forces acting on join *J* follow the same direction.

and F_4. The closed polygon *ABCD*, in which all the arrows run tip to tail, represents the equilibrium of the four forces. By transferring the arrows to the free body of joint *J* (Fig. 3-28*a*), we find that the force F_3 pulls away from, while F_4 pushes into, joint *J*. Consequently, F_3 is a tensile force while F_4 is a compressive force. From the scale of forces

$$F_3 = 1.5 \text{ kips} \quad (6.67 \text{ kN}) \quad \text{(tension)}$$

$$F_4 = 4.58 \text{ kips} \quad (20.4 \text{ kN}) \quad \text{(compression)}$$

New Terms Appearing in Chapter 3

Truss
Girder
Bridge crane
Crane runway
Column bracket
Collinear forces
Force, magnitude of a
Force, sense of a
Force, line of action of a
Force, point of application of a
Resultant
Vector
Scale of forces
Tip-to-tail sequence of force vectors
Concurrent forces
Parallelogram law
Force parallelogram
Force plan
Force triangle
Center of gravity
Wall
Pythagorean theorem
Equilibrant
Coplanar forces
Coplanar concurrent forces
Strut
Pier
Free body

Free-body diagram
Action and reaction
Principle of action and reaction
Fundamental laws of mechanics
Compressive axial force
Tensile axial force
Rope
Hanger
Rod
Pin
Components of a force
Inclined plane
Force polygon
Joint of a truss
Bar of a Truss

Important Structural Elements Appearing in Chapter 3

Truss. This is a structural system composed of long, relatively slender bars in a triangular pattern. The bars of a truss meet at joints, where they are connected to each other by pins, bolts, or welds.

Girder. This is a large main, or primary, beam usually supporting smaller, secondary beams or slabs.

Wall. This is a vertical structural element having length and height considerably larger than its thickness. Walls support roofs and floors or retain soil, water, and other materials or simply serve to enclose space.

Strut. This is a slender structural element stiff enough to transmit a compressive force. Similar to the column, it has cross-sectional dimensions much smaller than its length.

Pier. This is a massive structural element supporting loads from other structures resting on them.

Rope. This is a flexible structural element capable of transmitting tensile forces only. It is made by weaving or twisting individual fibers or groups of fibers. A rope of steel wires is called a *cable.*

Rod. This is a long, slender, solid structural element capable of transferring tensile forces only.

Quiz Number 2

A. TRUE OR FALSE

1. A force is completely determined by its magnitude and point of application.
2. A force may be considered as acting at any point on its line of action as long as its sense and magnitude are unchanged.
3. The force replacing the effect of two or more forces is called the *equilibrant.*
4. If a system of forces is in equilibrium, its resultant is equal to zero.
5. A vector is a quantity completely described by its magnitude only.
6. All quantities that possess magnitude and sense (direction) are called *polygons.*
7. If a force acts on a body, its effect will be the same whether it pushes or pulls on the body.

8. If two forces of the same direction or sense act on a point of the body, the magnitude of their resultant is equal to the sum of the magnitudes of the two forces; however, its sense will be opposite to the sense of the two forces.
9. If the scale of forces is 100 lb = 10 mm, a force of 250 lb is represented by a segment 25 mm long.
10. Forces are collinear if their magnitudes are equal.
11. In a concurrent force system, all forces intersect at one point.
12. The parallelogram law is valid because it was proved mathematically by Newton.
13. The sense of a force is indicated by the arrow on the force vector.
14. A boat under the forces of wind and current will move in the sense or direction of the equilibrant of these forces.
15. The center of gravity is the point in a body at which the weight of the body may be considered to act.
16. If the directions of two forces acting on a point make an angle of 45°, their resultant can be calculated by the pythagorean theorem.
17. In this chapter we determined the resultant of a nonconcurrent system of forces.
18. If a concurrent force system has zero resultant, the forces are in equilibrium.
19. Struts are elements capable of transferring tensile forces only.

B. MULTIPLE CHOICE

1. The principle of "action and reaction" was formulated by
   (*a*) Einstein (*b*) Newton (*c*) Pythagoras (*d*) none of these
2. What structural element resists forces opposite to those resisted by a rope?
   (*a*) hanger rod (*b*) cable (*c*) column (*d*) violin string
3. All forces acting on a free body at rest must be
   (*a*) in equilibrium (*b*) equal (*c*) opposite (*d*) nonconcurrent
4. A truss is a structural element consisting of
   (*a*) beams (*b*) concrete block (*c*) slabs (*d*) long, relatively slender bars
5. What is the correct term for a main beam supporting smaller, secondary beams?
   (*a*) strut (*b*) wall (*c*) girder (*d*) pier
6. What is the number of unknown forces we can determine in a coplanar concurrent force system in equilibrium?
   (*a*) one (*b*) two (*c*) three (*d*) four
7. Coplanar forces lie in one
   (*a*) line (*b*) space (*c*) plane (*d*) point
8. The forces due to an engine, a current, and the wind act on a boat. Under these forces the boat will
   (*a*) be in equilibrium
   (*b*) not move
   (*c*) move in the direction of the equilibrant
   (*d*) move in the direction of the resultant
9. Each joint separated from a truss as a free body and under the action of the bar axial forces meeting at the joint must
   (*a*) be in equilibrium (*b*) have a resultant (*c*) rotate (*d*) disintegrate

10. The arrows of all vectors in a force polygon representing a force system in equilibrium
    (*a*) must go in the same direction (tip to tail)
    (*b*) must go tip to tip
    (*c*) can be disregarded

C. SHORT ANSWERS

1. Forces which act along the same line are called ______________.
2. The force which replaces the effect of a given force system on a free body is called ______________.
3. The force which cancels the effect of a given force system on a free body is called ____________________.
4. A force can be resolved into ______________________________ in one plane.
5. A hanger is a structural element, which transfers ____________ axial forces.
6. If three concurrent forces are in equilibrium, all arrows of the force triangle must follow in _______ _______ _______ fashion.
7. When we isolate a portion of a structure to investigate its equilibrium under given forces, we call such a portion a ______________.
8. What forces must act on the free body shown in Fig. 3-29 for equilibrium to be achieved?

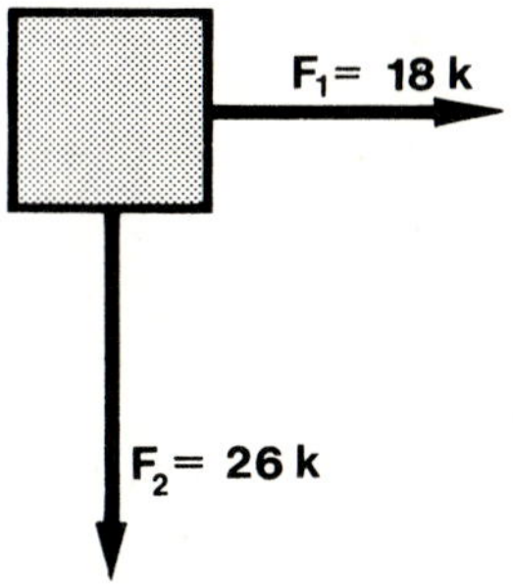

FIGURE 3-29

9. What single force will replace the effects of the forces acting on the free body shown in Fig. 3-30?

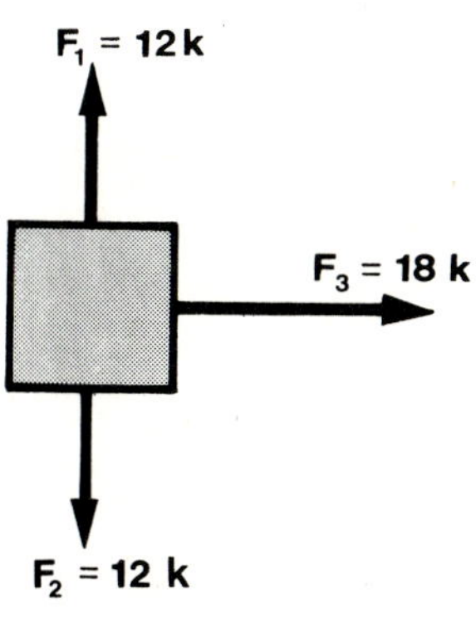

FIGURE 3-30

10. Close the force triangle in Fig. 3-31 by drawing the resultant and show its arrow.

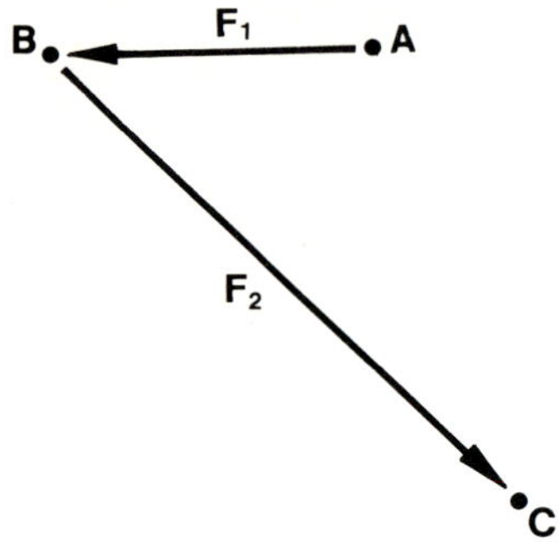

FIGURE 3-31

11. Close the force polygons in Fig. 3-32*a*, *b*, *c*, *d* by drawing the equilibrant and show its arrow.

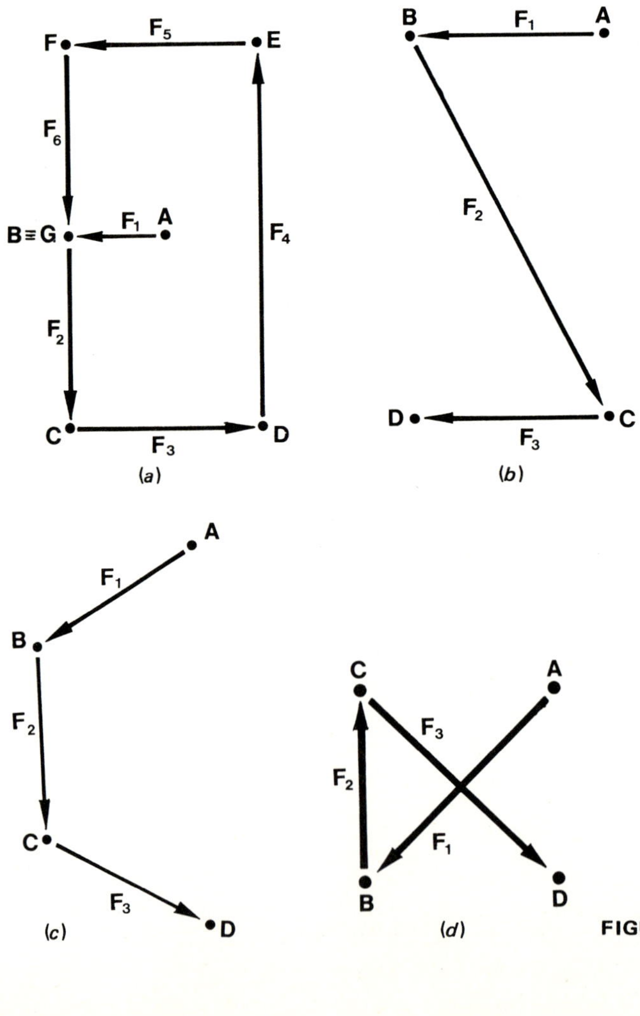

FIGURE 3-32

EXERCISES

3-1 A force F_1 = 100 kips acts at point M of the beam shown in Fig. 3-33. The lines of action of two other forces, F_2 and F_3, representing the reactions of the supports, are known. Find the magnitude and the sense of the two forces F_2 and F_3.

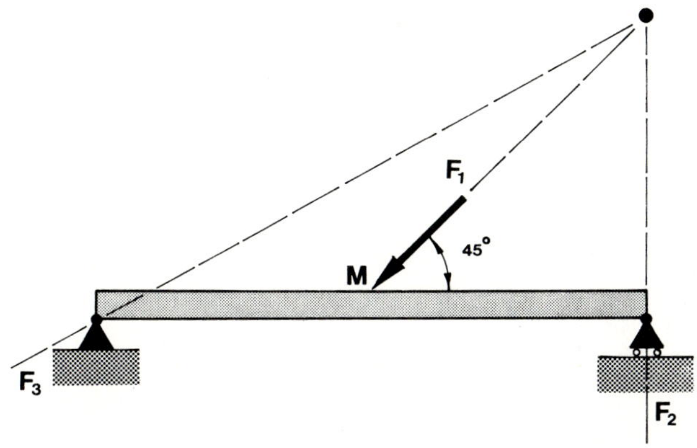

FIGURE 3-33

3-2 The column shown in Fig. 3-34 is loaded at the level of each floor by forces F_1 = 6 kips, F_2 = 10 kips, and F_3 = 15 kips. Find the total vertical force R acting on the footing.

Hint: Resolve F_1 and F_3 into the horizontal and vertical components. Then add up the vertical components with F_2 (since they all are collinear forces).

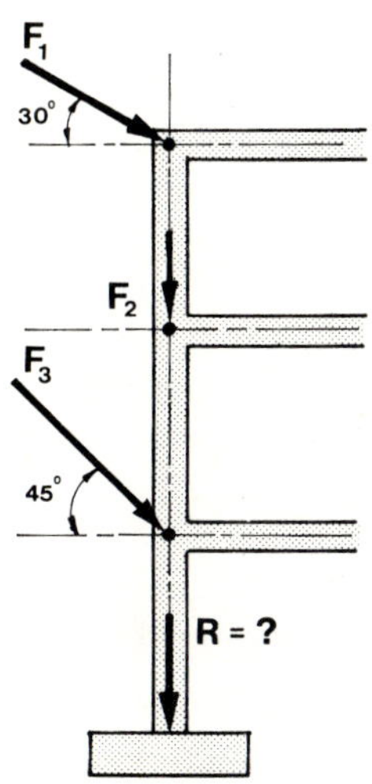

FIGURE 3-34

3-3 What are the horizontal and vertical components of the resultant of the two forces F_1 = 8 kips and F_2 = 16 kips, acting at the top of the pole shown in Fig. 3-35?

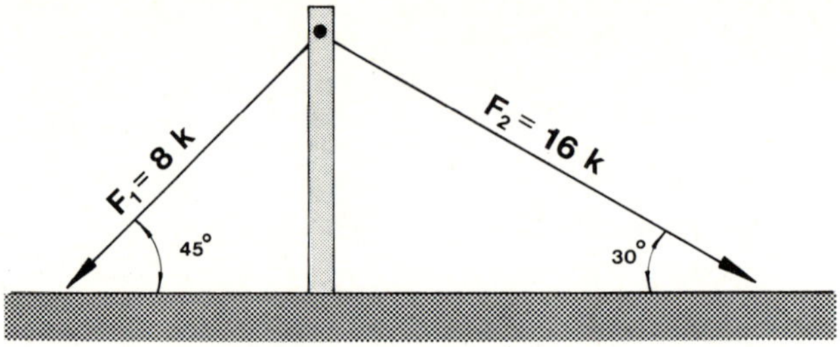

FIGURE 3-35

3-4 Find the axial forces in the members of the triangular truss shown in Fig. 3-36 that are due to the vertical force $F = 400$ lb at joint B.

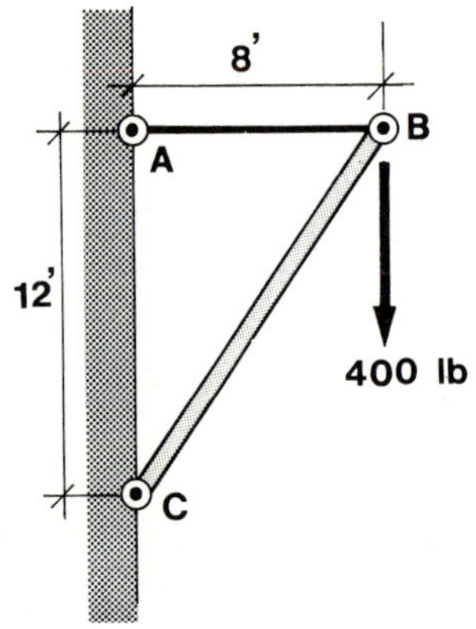

FIGURE 3-36

3-5 Find the magnitude and the direction of the resultant of the forces shown in Figs. 3-37 and 3-38.

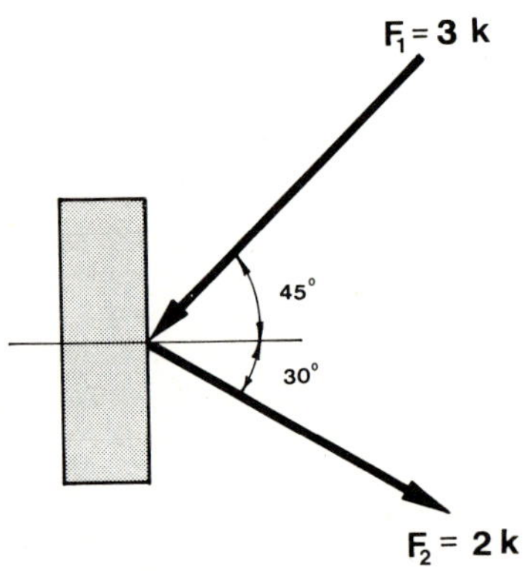

FIGURE 3-37

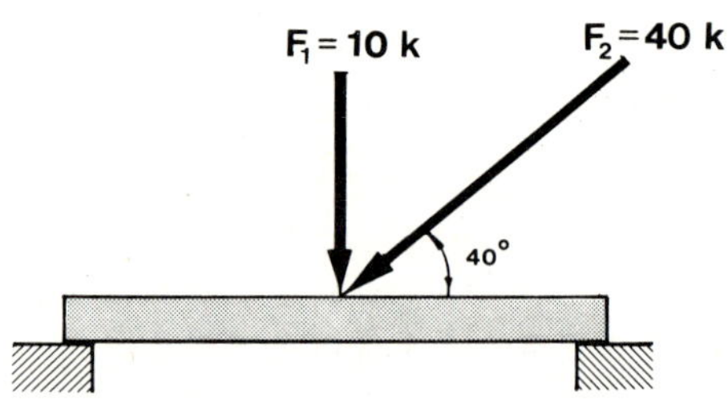

FIGURE 3-38

3-6 Two coplanar forces act on the flexible rod shown in Fig. 3-39. Find (*a*) the resultant of the two forces and the direction of the deflection of the rod tip and (*b*) a third force which would prevent the deflection of the rod.

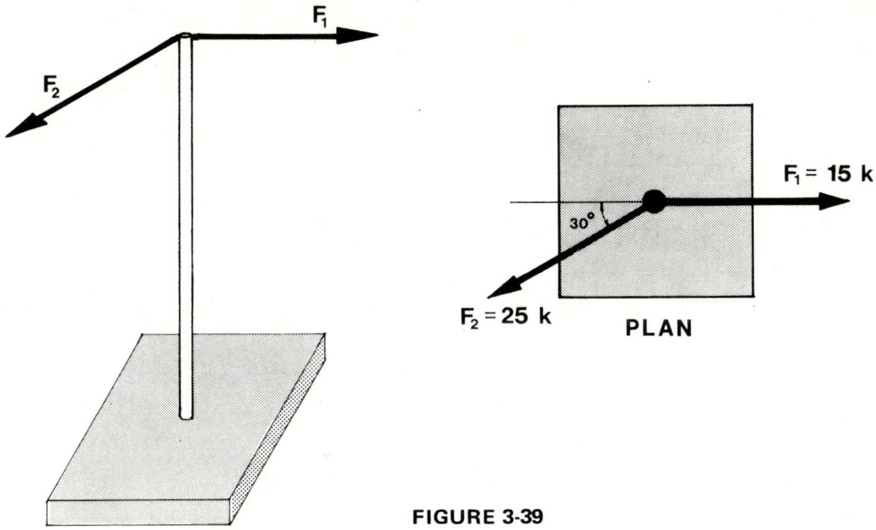

FIGURE 3-39

3-7 The column of a timber structure is loaded by three forces as shown in Fig. 3-40. Find the resulting force acting on the column.

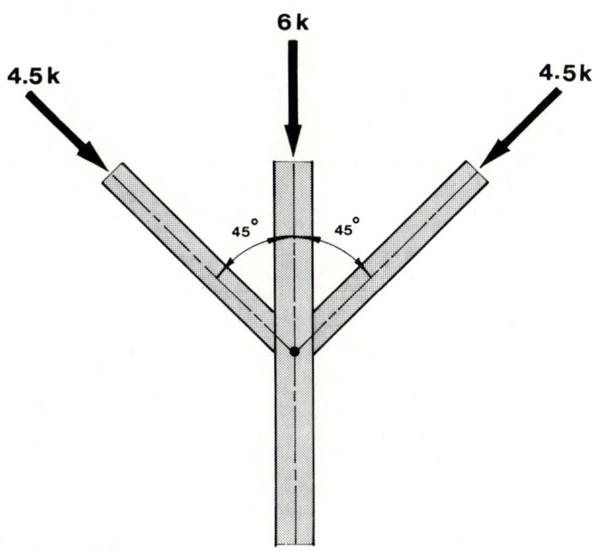

FIGURE 3-40

Steel frame of the Lutheran Brotherhood Building in Minneapolis. (*Photo by L. Cerny.*)

The Moment of a Force

4

If you try to tighten the nut of a bolt with your fingers, you may find that they are not strong enough to do this well. However, if you use a wrench, you will notice how little force you have to apply to its end to tighten the nut firmly. You will also find that the longer the wrench, the smaller the end force needed to tighten the nut; if the length of the wrench is increased, the *rotational effect* of the applied force increases. Finally you will realize that the wrench works best if the applied force is perpendicular to its longitudinal axis (Fig. 4-1).

The *perpendicular* distance r from the line of action of a force to the center of rotation (pivot) is called the *lever arm* of the force. The rotational effect of a force about a point is measured by the *moment of the force M*, which is the product of the force F and its lever arm r:

$$M = F \times r$$

Therefore, the units of moment are lb·in, lb·ft, kip·in, and kip·ft (or N·cm, N·m, and kN·m).

Since rotation can be in the clockwise or the counterclockwise sense, a moment also has sense; this can be represented graphically by a curved arrow or mathematically by a plus (+) or a minus (−) sign. Thus, according to their sense, moments can be positive or negative, depending on the accepted sign convention. Clockwise moments will be assumed as positive in this book.

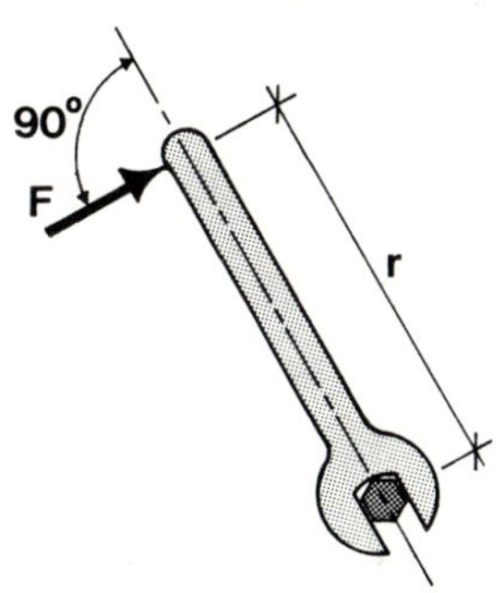

FIGURE 4-1 The moment of a force F. The lever arm r is perpendicular to the line of action of F.

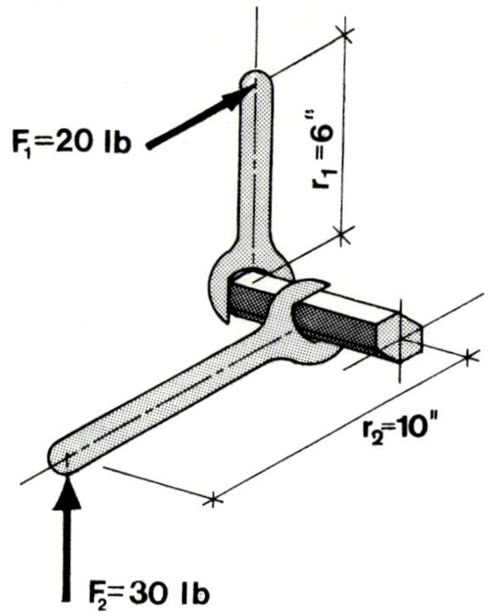

FIGURE 4-2

A long nut can be tightened by two wrenches (Fig. 4-2). Let the force on the first wrench be $F_1 = 20$ lb (89 N), that on the second $F_2 = 30$ lb (133.5 N), and their respective arms $r_1 = 6$ in (15.25 cm) and $r_2 = 10$ in (25.4 cm). Assume that the two wrenches are so close to each other that the forces can be considered coplanar. The moments of the two forces have the same clockwise sense, and their values are

$$M_1 = F_1 \times r_1 = 20 \times 6 = 120 \text{ lb}\cdot\text{in} \qquad (1\,357 \text{ N}\cdot\text{cm})$$

$$M_2 = F_2 \times r_2 = 30 \times 10 = 300 \text{ lb}\cdot\text{in} \qquad (3\,390 \text{ N}\cdot\text{cm})$$

The resulting tightening effect of the two forces—that is, their ***resultant moment***—is obtained by adding the two moments

$$M_R = M_1 + M_2 = 120 + 300 = 420 \text{ lb}\cdot\text{in} \qquad (4\,747 \text{ N}\cdot\text{cm})$$

The same resultant moment of 420 lb·in can also be developed by single forces with the different lever arms, provided the product of the force and the lever arm be equal to 420 lb·in.

For example

$$F = 10 \text{ lb} \qquad r = 42 \text{ in} \qquad M_R = 420 \text{ lb}\cdot\text{in}$$

$$F = 210 \text{ lb} \qquad r = 2 \text{ in} \qquad M_R = 420 \text{ lb}\cdot\text{in}$$

$$F = 840 \text{ lb} \qquad r = 0.5 \text{ in} \qquad M_R = 420 \text{ lb}\cdot\text{in}$$

If the available force is 60 lb (267 N), the lever arm needed to develop a moment of 420 lb·in is obtained from the equation

$$F \times r = M$$

which, when solved for r, gives

$$r = \frac{M}{F} = \frac{420 \text{ lb}\cdot\text{in}}{60 \text{ lb}} = 7 \text{ in} \qquad (17.8 \text{ cm})$$

Children playing on a seesaw know how to seek equilibrium, although they are totally ignorant about the rotational effects of forces, that is, about moments. They slide along the plank to increase or shorten the lever arm of their weights until the plank is horizontal. At that instant they attain *moment* or *rotational equilibrium* about the pivot point.

For moment equilibrium about a point, the sum of the moments of all forces about that point must be zero; in other words, the resultant moment about that point is zero.

To check moment equilibrium, we first choose one sense of rotation as positive, say, the clockwise sense.

The condition of moment equilibrium about a pivot point O of the two forces F_1 and F_2 in Fig. 4-3 is

$$F_1 \times r_1 - F_2 \times r_2 = 0 \qquad \text{or} \qquad F_1 \times r_1 = F_2 \times r_2$$

The principle of moment equilibrium is used, for example, to lift heavy weights by means of levers. The lever arm on the side of the weight to be lifted is smaller than the lever arm on the side of the applied force, so that the lifting force applied to the other end of the lever can be small. For example, given a fixed lever arm r_1 of 320 cm (126 in) and lifting force of 180 N, what should be the length of the lever arm r_2 on the weight side to lift 3 000 N?

From the moment equilibrium condition about the center of rotation we obtain

$$F_2 \times r_2 = F_1 \times r_1$$

$$r_2 = \frac{F_1 \times r_1}{F_2} = \frac{180 \text{ N} \times 320 \text{ cm}}{3\,000 \text{ N}} = \frac{37\,760 \text{ N}\cdot\text{cm}}{3\,000 \text{ N}} = 19.2 \text{ cm} \qquad (7.6 \text{ in})$$

The principle of moment equilibrium was known thousands of years ago, and the lever was used in the construction of the pyramids as well as in that of ingenious devices and mechanisms.

To pull a pail of water out of a well, a hemp rope is wound around a shaft. The shaft is rotated by the attached spokes, whose length r_1 is larger than the shaft radius r_2 (Fig. 4-4). Thus a relatively small force applied to the spokes can overcome the moment of the pail weight on the shaft. If the weight of the pail filled with water is W

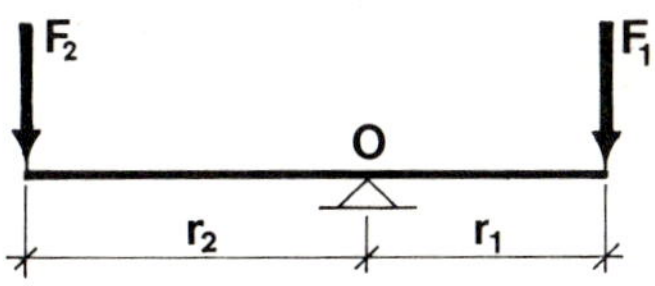

FIGURE 4-3

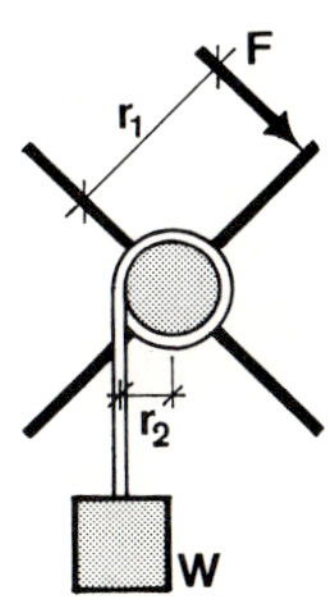

FIGURE 4-4

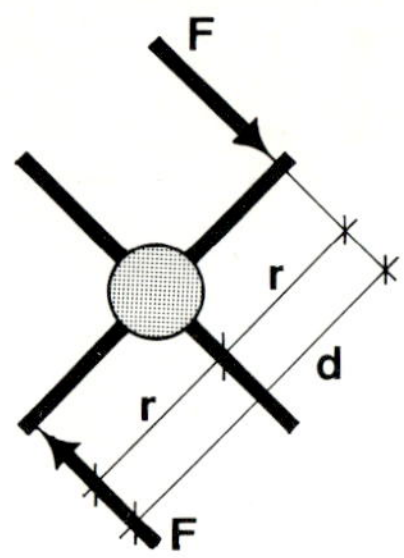

FIGURE 4-5 A force couple and its moment.

and the force applied to a spoke is F (both are assumed to be coplanar), the moment equilibrium condition about the center of the shaft is

$$F \times r_1 = W \times r_2$$

For example, with $W = 40$ lb (178 N), $r_1 = 24$ in (61 cm), and $r_2 = 6$ in (15.2 cm), the force required to lift the pail is

$$F = \frac{W \times r_2}{r_1} = \frac{40 \text{ lb} \times 6 \text{ in}}{24 \text{ in}} = 10 \text{ lb} \qquad (44.5 \text{ N})$$

Assume now that another force F of equal magnitude and opposite sense acts on the opposite spoke (Fig. 4-5). The resulting moment of both forces is

$$M_R = F \times r + F \times r = F \times 2r = F \times d$$

where $d = 2r$ is a perpendicular distance between the two forces.

Two parallel forces lying in a plane, equal in magnitude but of opposite sense, a perpendicular distance d apart, are called a *force couple* or a *couple*. The moment of a couple is equal to the product of one of its forces F and the perpendicular distance d between the forces

$$M = F \times d$$

We can easily see that the moment of a couple about *any* point in its plane is always the same and equal to $F \times d$. To this purpose, find the moment of the couple in Fig. 4-6 about point O:

$$M = +F(d + r) - F \times r = F \times d + F \times r - F \times r = F \times d$$

[The moment of the right force is negative ($-Fr$) because it is counterclockwise.]

The two parallel forces of a couple do not intersect at any point: the forces are *not* concurrent.

In a general *coplanar*, *nonconcurrent* force system the lines of action of all forces

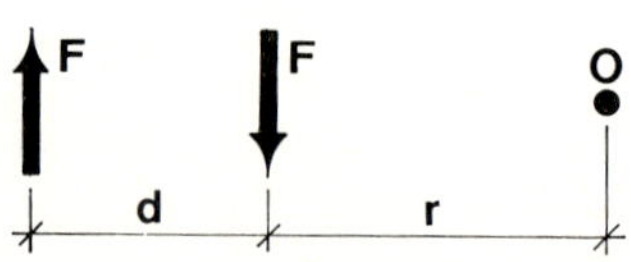

FIGURE 4-6 The moment of a couple about any point in the plane is always the same.

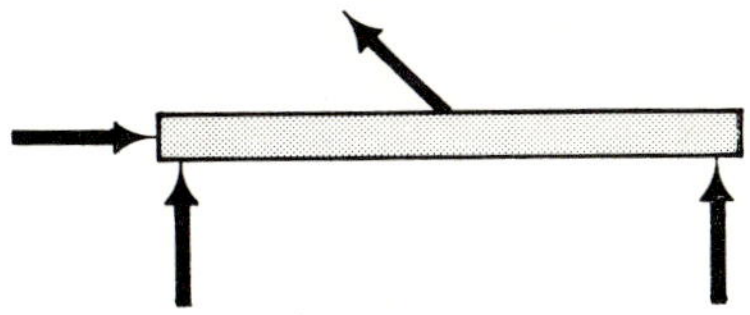

FIGURE 4-7 A coplanar nonconcurrent force system.

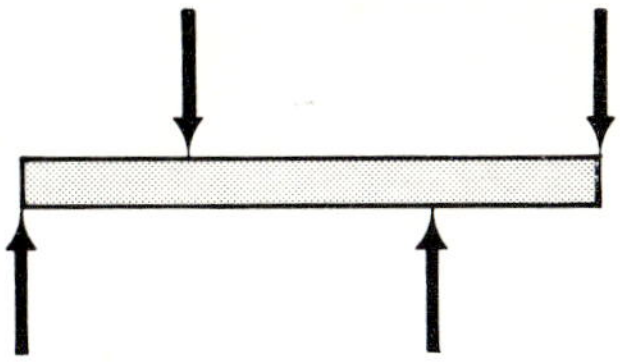

FIGURE 4-8 A coplanar parallel force system.

lie in the same plane but they do not *all* pass through a common point. Two examples of coplanar nonconcurrent forces are shown in Figs. 4-7 and 4-8, the latter being a special case of coplanar nonconcurrent force system—the *parallel force system.* Later in this chapter we will learn how to find the resultant and to check the equilibrium of coplanar nonconcurrent force systems. *For moment equilibrium of a coplanar nonconcurrent force system about any point in the plane, the sum of the moments of all forces about that point must be zero; that is, the resultant moment about that point is zero.*

Illustrative Problem 4-1

Find the magnitude of the force F_1 acting on the lever shown in Fig. 4-9 needed to achieve moment equilibrium about point O, if the force $F_2 = 160$ N (36 lb), the lever arm $r_1 = 50$ cm (20 in), and the lever arm $r_2 = 15$ cm (6 in). The force system is coplanar.

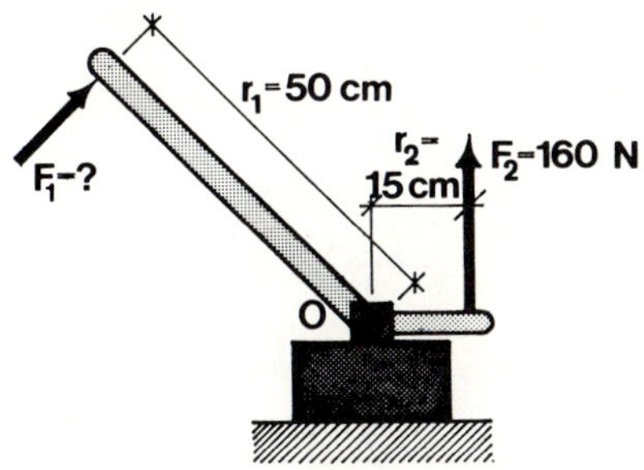

FIGURE 4-9 Moment equilibrium of a coplanar force system about point *O*.

SOLUTION The moment equilibrium condition about point O is

$$F_1 r_1 - F_2 r_2 = 0$$

This condition expresses the fact that the moments of the forces F_1 and F_2 have a zero resultant moment; that is, the rotational effect on the point O of the two forces is equal to zero. Substituting the numerical values shown in Fig. 4-9, we obtain

$$F_1 \times 50 \text{ cm} - 160 \text{ N} \times 15 \text{ cm} = 0$$

$$F_1 = \frac{160 \text{ N} \times 15 \text{ cm}}{50 \text{ cm}} = 48 \text{ N} \qquad (10.8 \text{ lb})$$

Illustrative Problem 4-2

Find the magnitude and the sense of the force F_3 needed to achieve moment equilibrium about point O of the coplanar nonconcurrent system of forces shown in Fig. 4-10. The known forces are $F_1 = 30$ kN (6.75 kips), $F_2 = 90$ kN (20.25 kips); the lever arms of the forces are $r_1 = 40$ cm (16 in), $r_2 = 20$ cm (8 in), and $r_3 = 60$ cm (24 in).

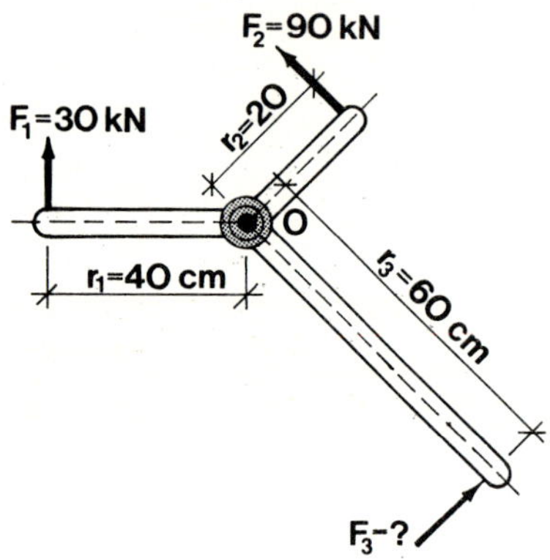

FIGURE 4-10

SOLUTION Assume an arbitrary sense for the unknown force F_3, for example, that is shown in Fig. 4-10. We can then write the moment equilibrium condition about the center of rotation O:

$$-F_3 \times r_3 - F_2 \times r_2 + F_1 \times r_1 = 0$$

from which

$$F_3 \times r_3 = F_1 \times r_1 - F_2 \times r_2$$

and

$$F_3 = \frac{F_1 \times r_1 - F_2 \times r_2}{r_3}$$

For the given values of forces and lever arms, we obtain

$$F_3 = \frac{30 \text{ kN} \times 40 \text{ cm} - 90 \text{ kN} \times 20 \text{ cm}}{60} = \frac{1\,200 \text{ kN}\cdot\text{cm} - 1\,800 \text{ kN}\cdot\text{cm}}{60}$$

$$= \frac{-600 \text{ kN}\cdot\text{cm}}{60 \text{ cm}} = -10 \text{ kN} \quad (2.25 \text{ kips})$$

The negative sign of the result indicates that our original assumption for the sense of the force F_3 was wrong. The correct sense of the force F_3 is shown in Fig. 4-11. To check results, we determine whether the sum of all the moments about O is zero:

$$F_1 \times r_1 - F_2 \times r_2 + F_3 \times r_3 = 30 \times 40 - 90 \times 20 + 10 \times 60$$

$$= 1\,200 - 1\,800 + 600 = 0$$

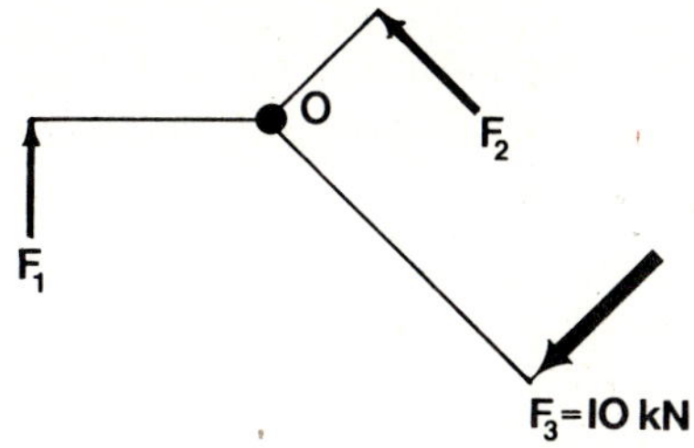

FIGURE 4-11 Moment equilibrium of a coplanar force system about point *O*.

Illustrative Problem 4-3

A beam with one end free and the other *clamped* or *fixed* in a rigid support is called a *cantilevered beam* or simply a *cantilever*. Assume that a concentrated load of 1 200 lb (5.34 kN) acts on the free end of the 6-ft (1.83-m)-long cantilever in Fig. 4-12. What is the moment of this force about the *root* of the cantilever, that is, point *O* at the fixed end?

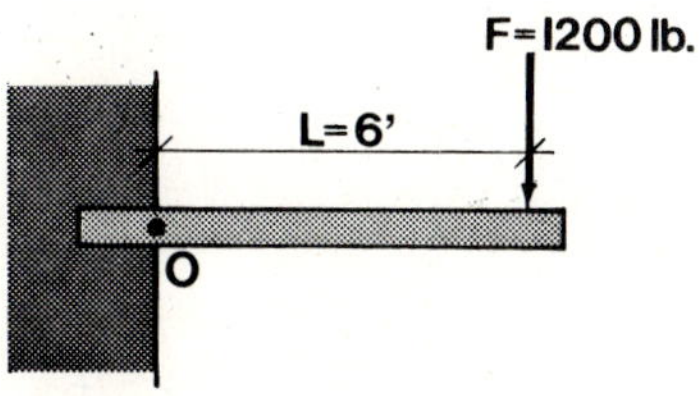

FIGURE 4-12 The cantilever in Illustrative Prob. 4-3.

SOLUTION

$$M = F \times L = 1\ 200 \text{ lb} \times 6 \text{ ft} = 7\ 200 \text{ lb·ft} \qquad (9.77 \text{ kN·m})$$

Illustrative Problem 4-4

The beam shown in Fig. 4-13 supports two concentrated vertical forces, F_1 = 3 000 N (675 lb) and F_2 = 6 000 N (1 350 lb), at the left and right third points of its length. The beam is 12 m (39.4 ft) long. The end supports exert the unknown vertical reactions R_A and R_B on the beam. Find the magnitude and sense of these reactions.

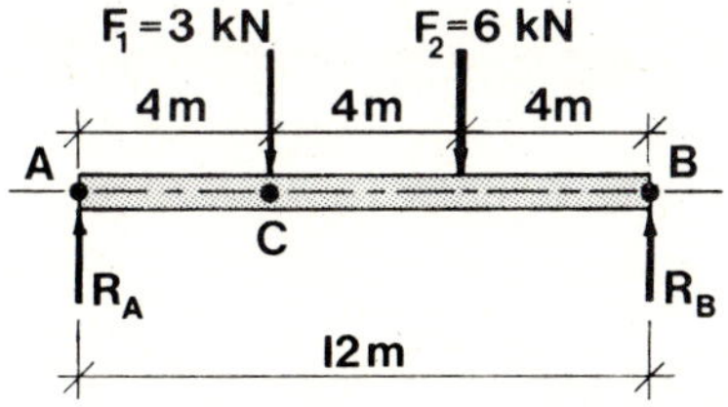

FIGURE 4-13 The beam in Illustrative Prob. 4-4.

SOLUTION The whole beam as a free body must be in equilibrium under the action of the forces F_1, F_2, R_A, and R_B. The rotational equilibrium of the beam means that the sum of the moments of all the forces of this system about *any* point in the plane must be equal to zero. If we take a moment equilibrium condition about point B of the beam, the lever arm of the unknown force R_B becomes zero and the moment of this force vanishes from the equilibrium equation. Assuming the sense of both unknown forces to be upward, we obtain

$$R_A \times 12\text{ m} - F_1 \times 8\text{ m} - F_2 \times 4\text{ m} = 0$$

or

$$R_A \times 12\text{ m} = F_1 \times 8\text{ m} + F_2 \times 4\text{ m}$$

$$R_A \times 12\text{ m} = 3\text{ kN} \times 8\text{ m} + 6\text{ kN} \times 4\text{ m}$$

$$R_A \times 12\text{ m} = 48\text{ kN·m}$$

and

$$R_A = \frac{48\text{ kN·m}}{12\text{ m}} = 4\text{ kN} \qquad (900\text{ lb})$$

Similarly, the force R_B can be computed by means of the moment equilibrium condition about point A:

$$-R_B \times 12\text{ m} + F_2 \times 8\text{ m} + F_1 \times 4\text{ m} = 0$$

or

$$R_B \times 12\text{ m} = F_2 \times 8\text{ m} + F_1 \times 4\text{ m}$$

$$R_B \times 12\text{ m} = 6\text{ kN} \times 8\text{ m} + 3\text{ kN} \times 4\text{ m}$$

$$R_B \times 12\text{ m} = 60\text{ kN·m}$$

and

$$R_B = \frac{60\text{ kN·m}}{12\text{ m}} = 5\text{ kN} \qquad (1\ 125\text{ lb})$$

The positive results of our computations indicate that our initial assumptions about the sense of the forces R_A and R_B were correct. The equilibrium system obtained has zero resultant moment, thus assuring that the beam will not rotate about *any* point in the plane. It is easy to see that this is also true for point C on the line of action of the force F_1. In fact, the resultant moment of the force system about C is

$$R_A \times 4\text{ m} + F_2 \times 4\text{ m} - R_B \times 8\text{ m}$$

or

$$4\text{ kN} \times 4\text{ m} + 6\text{ kN} \times 4\text{ m} - 5\text{ kN} \times 8\text{ m} = 16 + 24 - 40\text{ kN·m} = 0$$

4.1 THE RESULTANT OF A COPLANAR NONCONCURRENT FORCE SYSTEM

The concept of moment as a measure of the rotational effect of a force about a point can be used to find the resultant of a coplanar *nonconcurrent* force system.

The moment of the resultant of a coplanar nonconcurrent force system about any point in the plane is equal to the sum of the moments of the individual forces of the system about the same point.

First we will use this principle for a set of parallel forces, a special case of a nonconcurrent system. We wish to find the resultant of the parallel force system shown in Fig. 4-14. The force $F_1 = 10$ kN acts upward; the forces $F_2 = 40$ kN and $F_3 = 20$ kN both act downward. The perpendicular distances between F_1 and F_2 and between F_2 and F_3 are 4 m.

Remember that the effect of the resultant of a force system must replace the effects of all the forces in the system. The magnitude of the resultant in our example must, therefore, be equal to the sum of all the vertical forces and, hence, must also be vertical. Assuming downward forces as positive, we obtain

$$R = -F_1 + F_2 + F_3 = -10 + 40 + 20 = -10 + 60 = +50 \text{ kN}$$

The positive sign of R shows that it acts downward.

We must now find the line of action of the resultant $R = 50$ kN. For this purpose we choose an arbitrary point C, for example, on the line of action of the force $F_3 =$ 20 kN. By so doing we eliminate the moment of F_3 from our calculations, since the lever arm of F_3 about point C is zero. Then the distance x of the resultant R from point C is found from the condition that the moment of the resultant R about C must be equal to the sum of the moments of all the forces of the system about C. Assuming that the location of R is to the left of C, we obtain:

$$-R(x) = +F_1 \times 8 \text{ m} - F_2 \times 4\text{m}$$

$$-R(x) = +10 \text{ kN} \times 8 \text{ m} - 40 \text{ kN} \times 4 \text{ m}$$

$$-R(x) = +80 \text{ kN·m} - 160 \text{ kN·m}$$

$$-R(x) = -80 \text{ kN·m}$$

$$x = \frac{80 \text{ kN·m}}{R} = \frac{80 \text{ kN·m}}{50 \text{ kN}} = 1.60 \text{ m}$$

Hence the downward resultant $R = 50$ kN is 1.60 m to the left of point C.

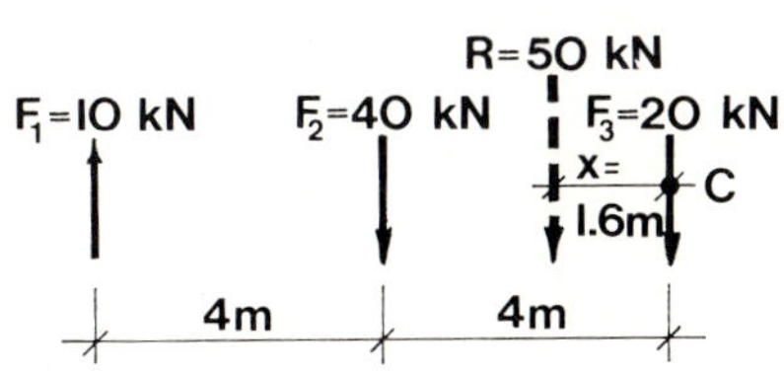

FIGURE 4-14 The resultant of a coplanar nonconcurrent force system.

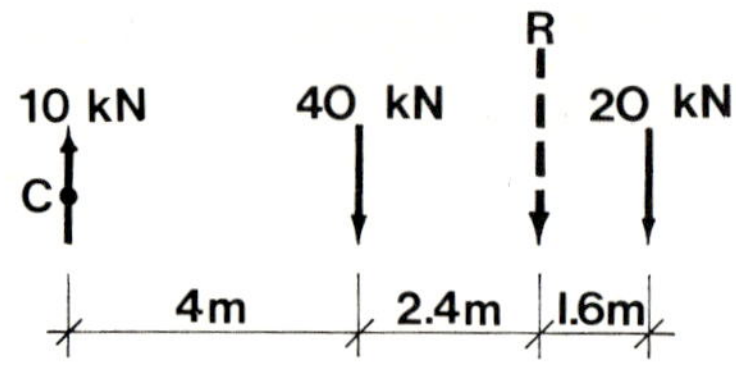

FIGURE 4-15 The resultant of a coplanar nonconcurrent force system.

If we now choose point C on the line of action of the force F_1 (Fig. 4-15), the moment of the resultant R about C is

$$M = R\,(4 + 2.4\text{ m}) = 50\text{ kN} \times 6.4\text{ m}$$
$$= 320\text{ kN·m}$$

This moment must be equal to the sum of the moments of all forces of the system about C:

$$M = F_2 \times 4\text{ m} + F_3 \times 8\text{ m}$$
$$= 40\text{ kN} \times 4\text{ m} + 20\text{ kN} \times 8\text{ m}$$
$$= 160 + 160 = 320\text{ kN·m}$$

Illustrative Problem 4-5

The concrete pier shown in Fig. 4-16 is loaded by two vertical forces, F_1 = 100 kips (445 kN) and F_2 = 60 kips (267 kN) and by two horizontal forces, F_3 = 20 kips (89 kN) and F_4 = 40 kips (178 kN). Find the magnitude and locate the line of action of the resultant of this force system.

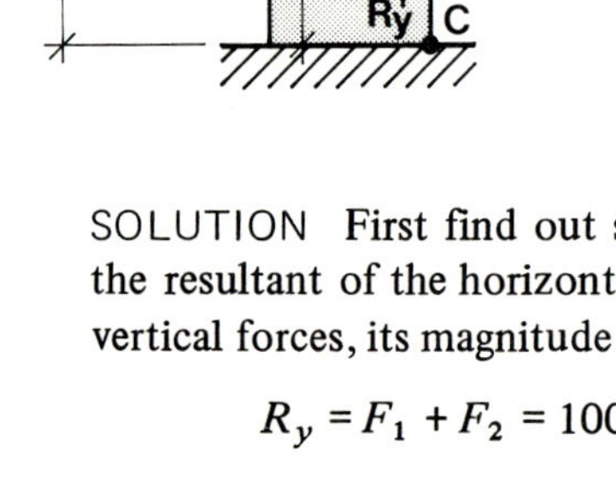

FIGURE 4-16 The concrete pier in Illustrative Prob. 4-5.

SOLUTION First find out separately the resultant of the vertical forces R_y and the resultant of the horizontal forces R_x. Since R_y has to replace the effects of all vertical forces, its magnitude is

$$R_y = F_1 + F_2 = 100 + 60 = 160\text{ kips} \qquad (712\text{ kN}) \qquad \text{(downward)}$$

The moment of R_y about C is equal to the sum of moments of F_1 and F_2 about C. Assuming that the location of R_y is to the left of C, we obtain

$$-R_y(x) = -F_1 \times 2\text{ ft} + F_2 \times 1.0\text{ ft} = -100\text{ kips} \times 2\text{ ft} + 60\text{ kips} \times 1\text{ ft}$$

$$-160\text{ kips }(x) = -200\text{ kip·ft} + 60\text{ kip·ft}$$

$$-160\text{ kips }(x) = -140\text{ kip·ft}$$

and

$$x = \frac{140\text{ kip·ft}}{160\text{ kips}} = \tfrac{7}{8}\text{ ft} = 10.5\text{ in} \qquad (26.7\text{ cm})$$

The line of action of R_y is located 10.5 in to the left of point C.

Similarly, R_x has to replace the effects of all horizontal forces, so that

$$R_x = F_3 + F_4 = 20 + 40 = 60\text{ kips} \qquad (267\text{ kN}) \qquad (\text{to the right})$$

The moment of R_x about C is equal to the sum of moments of F_3 and F_4 about C:

$$R_x(y) = F_3 \times 12\text{ ft} + F_4 \times 6\text{ ft}$$

$$R_x(y) = 20\text{ kips} \times 12\text{ ft} + 40\text{ kips} \times 6\text{ ft}$$

$$R_x(y) = 240\text{ kip·ft} + 240\text{ kip·ft}$$

$$R_x(y) = 480\text{ kip·ft}$$

$$y = \frac{480\text{ kip·ft}}{R_x} = \frac{480\text{ kip·ft}}{60\text{ kips}} = 8\text{ ft}$$

The line of action of R_x is located 8 ft above point C.

The resultant R or R_x and R_y goes through their intersection (Fig. 4-17) defined by the coordinates

$$x = \tfrac{7}{8}\text{ ft} \qquad \text{and} \qquad y = 8\text{ ft}$$

Its magnitude can be computed by using the force-triangle method or the pythagorean theorem:

$$R = \sqrt{R_x^2 + R_y^2} = \sqrt{60^2 + 160^2} = \sqrt{3\,600 + 25\,600}$$
$$= \sqrt{29\,200} = 170.9\text{ kips} \qquad (760\text{ kN})$$

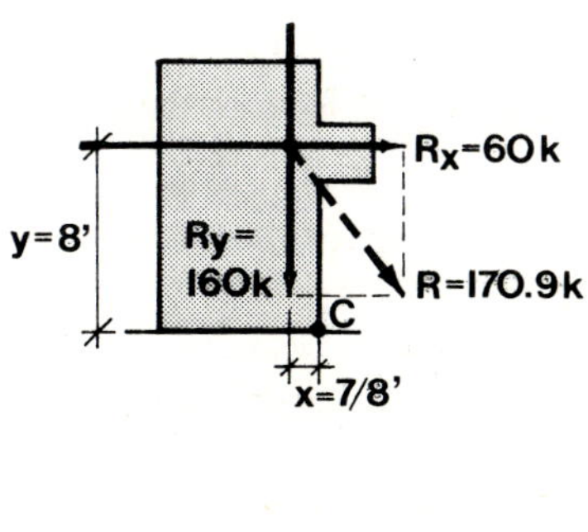

FIGURE 4-17

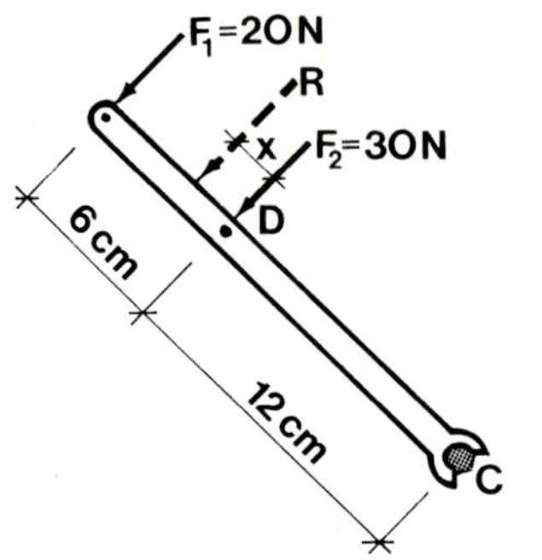

FIGURE 4-18 The resultant of two parallel forces.

Illustrative Problem 4-6

Two forces, $F_1 = 20$ N and $F_2 = 30$ N, act on the wrench shown in Fig. 4-18. Find their resultant and their resultant moment about point C.

SOLUTION The resultant R of F_1 and F_2 replaces their effect and

$$R = F_1 + F_2 = 20 + 30 = 50 \text{ N}$$

The line of action of R is somewhere between F_1 and F_2. To locate it, take moments about point D (this choice of point simplifies the calculations by eliminating the moment of F_2 from the equation). Since the moment of the resultant R of the two forces, F_1 and F_2, about any point in the plane is equal to the sum of the moments of the individual forces about that point, we have

$$-R(x) = -F_1 \times 6 \text{ cm} + F_2 \times 0$$

$$-R(x) = -20 \text{ N} \times 6 \text{ cm} = -120 \text{ N}\cdot\text{cm}$$

and

$$x = \frac{120 \text{ N}\cdot\text{cm}}{R} = \frac{120 \text{ N}\cdot\text{cm}}{50 \text{ N}} = 2.4 \text{ cm}$$

The moment of the forces F_1 and F_2 about C is then

$$M_c = -F_1 \times 18 \text{ cm} - F_2 \times 12 \text{ cm}$$

$$= -20 \text{ N} \times 18 \text{ cm} - 30 \text{ N} \times 12 \text{ cm} = -720 \text{ N}\cdot\text{cm} \qquad \text{(counterclockwise)}$$

The same counterclockwise moment is produced by the resultant R with a lever arm of $12 + 2.4 = 14.4$ cm:

$$M_c = -R \times 14.4 \text{ cm} = -50 \text{ N} \times 14.4 \text{ cm} = -720 \text{ N}\cdot\text{cm}$$

Illustrative Problem 4-7

A building 500 ft (152 m) tall has a square plan 120 ft (36.6 m) by 120 ft (Fig. 4-19). The weight of the building is 80 000 kips (356 000 kN). The maximum wind pressure on the building is considered as a uniformly distributed load

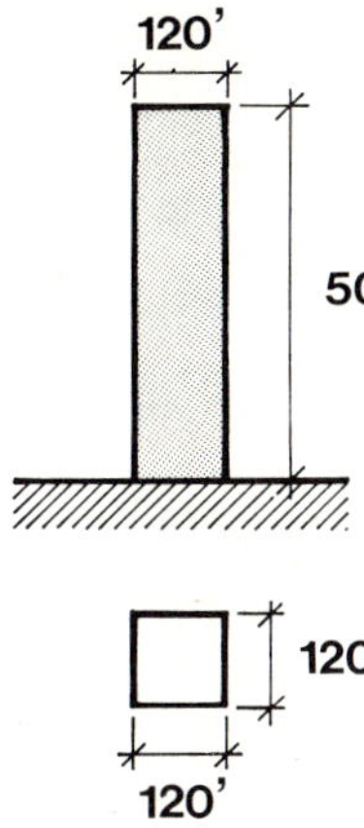

FIGURE 4-19

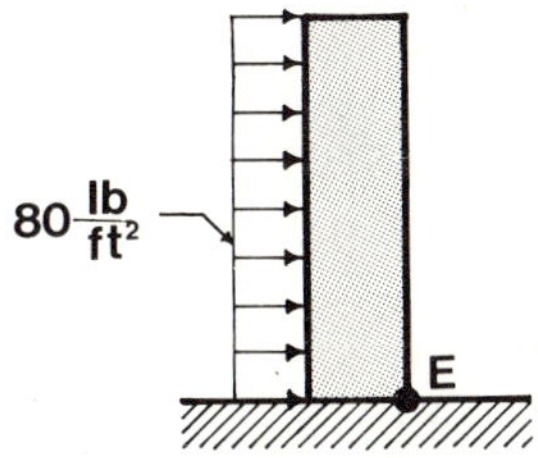

FIGURE 4-20

of 80 lb/ft² (3.83 kN/m²). Will the wind topple the building about the axis E at its base (Fig. 4-20)?

SOLUTION The action of wind can be represented by a single force H acting at midheight of the building and equal to the product of the unit wind pressure times the area of the surface exposed to wind:

$$H = 80 \frac{\text{lb}}{\text{ft}^2} \times 120 \text{ ft} \times 500 \text{ ft} = 4\,800\,000 \text{ lb} = 4\,800 \text{ kips} \qquad (21\,350 \text{ kN})$$

Its lever arm about the axis E is

$$r_1 = \frac{500}{2} = 250 \text{ ft} \qquad (76 \text{ m})$$

The vertical force representing the weight of the building is considered to act along the vertical axis of the building, and its lever arm about E is

$$r_2 = \frac{120}{2} = 60 \text{ ft} \qquad (18.3 \text{ m})$$

The two forces and their lever arms are shown in Fig. 4-21. The *overturning moment* of the wind force H is equal to

$$M_o = H \times r_1 = 4\,800 \times 250 = 1\,200\,000 \text{ kip·ft} \qquad (1\,627\,000 \text{ kN·m})$$

The *stabilizing moment* of the weight W of the building is equal to

$$M_s = W \times r_2 = 80\,000 \times 60 = 4\,800\,000 \text{ kip·ft} \qquad (6\,509\,000 \text{ kN·m})$$

Since the stabilizing moment M_s is much larger than the overturning moment M_o, there is no danger that the building will overturn as a result of wind action.

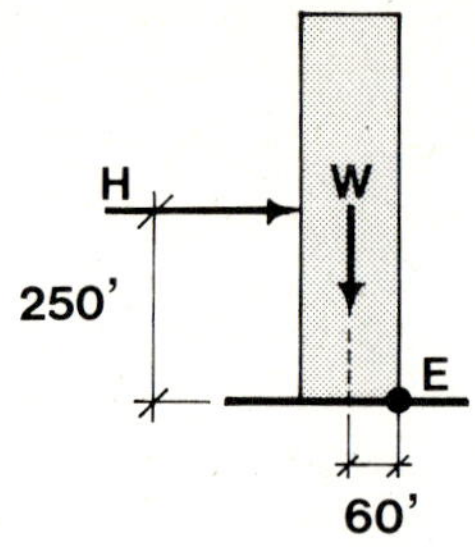

FIGURE 4-21

It is customary to express the margin of safety of a structure against overturning by a safety factor. The safety factor (SF) against overturning is *defined as the ratio of the stabilizing moment to the overturning moment.* In our case we have

$$SF = \frac{M_s}{M_o} = \frac{4\ 800\ 000}{1\ 200\ 000} = 4.0$$

In order to topple the building about the E axis, the overturning moment would have to be increased four times; that is, the wind pressure would have to quadruple. The investigation of the safety factor against overturning is also required for such structures as dams and retaining walls (Fig. 4-22).

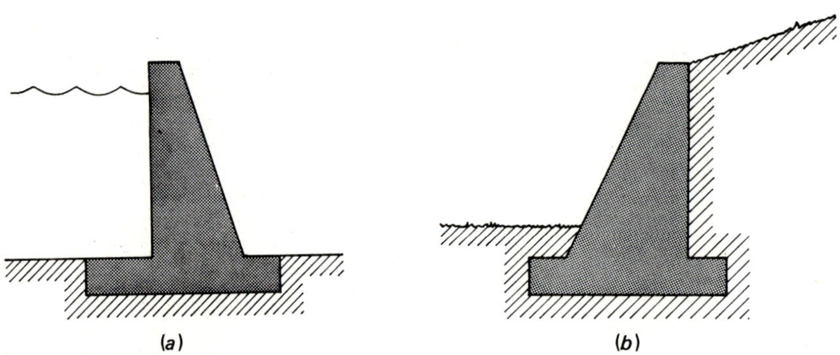

FIGURE 4-22 (*a*) Dam; (*b*) retaining wall.

4.2 THE RESULTANT OF THE GENERAL COPLANAR FORCE SYSTEM—GRAPHIC SOLUTION

In Illustrative Prob. 3-3 we found the resultant of two inclined coplanar forces by extending their lines of action to find their intersection, and by using a force triangle. We shall now find the resultant of several nonconcurrent coplanar forces acting at different points of a body by a similar graphic procedure, which is illustrated in the following example.

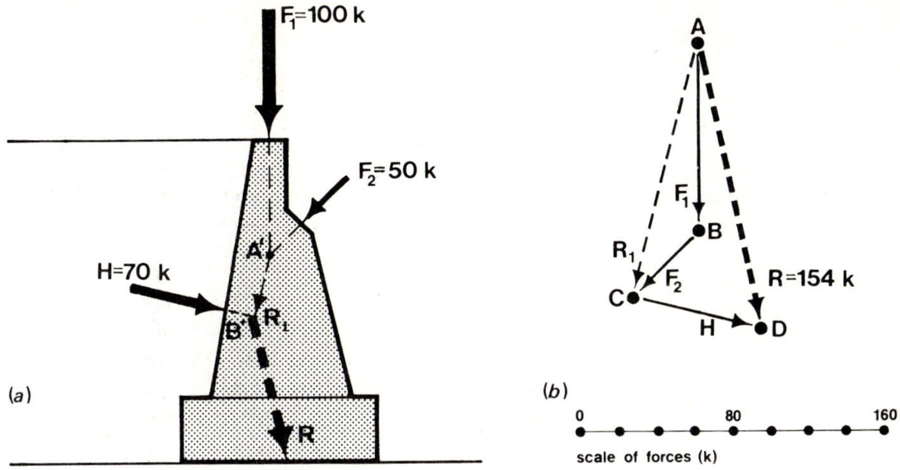

FIGURE 4-23 (*a*) Force plan of the forces acting on the pier. (*b*) Force polygon. The arrows of the resultants R_1 and R run against the arrows of the forces F_1, F_2, and H.

Illustrative Problem 4-8

Three forces, F_1, F_2, and H, act on the pier of a bridge (Fig. 4-23a). The force $F_1 = 100$ kips (445 kN) is vertical; it represents part of the weight of the bridge and of its live load. The force $F_2 = 50$ kips (222 kN) represents the compressive action of the bridge strut on the pier; the force $H = 70$ kips (311 kN) represents the pressure of the soil against the pier. Find the resultant of these three forces, that is, the magnitude and the direction of the resultant.

SOLUTION Draw the force polygon *ABCD* (Fig. 4-23*b*), where the vectors *AB*, *BC*, and *CD* represent in a certain scale the forces F_1, F_2, and H, respectively. The vector of the resultant R_1 of the forces F_1 and F_2 is obtained by connecting the tail of the force F_1 to the tip of the force F_2. The sense of this partial resultant is given by the arrow going from the tail of F_1 to the tip of F_2. This resultant is located on the pier (Fig. 4-23*a*) by noticing that it must pass through the point of intersection A' of the forces F_1 and F_2, and its line of action must be parallel to the vector *AC* in the force polygon (Fig. 4-23*b*).

The resultant R of R_1 and H is represented by the vector *AD* and its arrow runs from the tail of R_1 to the tip of H; R is the resultant of the three forces F_1, F_2, and H. To locate the resultant R on the pier, extend the line of action of R_1 until it intersects the line of action of H at point B' and draw a parallel to *AD* through point B'. This is the line of action of the resultant R on the pier. Its magnitude is obtained by measuring the length of the vector *AD* on the force scale. In our problem $R = 154$ kips (685 kN).

Thus, by the successive determination of the resultant of two and only two forces (where one of the forces is the resultant of two previously considered forces), we can find the overall resultant of a system of many forces.

4.3 THE FORCE POLYGON AND THE STRING POLYGON

When the point of intersection of any two forces of a system falls outside the paper, the resultant of a force system cannot be found by the method described in Sec. 4.2 but can be determined by the use of a *force polygon* and a so-called *string polygon*. The following example will illustrate this procedure and explain the meaning of these new terms.

Illustrative Problem 4-9

Find the resultant of the two coplanar oblique forces acting on the beam shown in Fig. 4-24*a* (force plan). Assume that the angle made by the lines of action of the two forces is small, so that we cannot plot their point of intersection on the paper. Let F_1 = 10 kN (2.25 kips) and F_2 = 8 kN (1.80 kips).

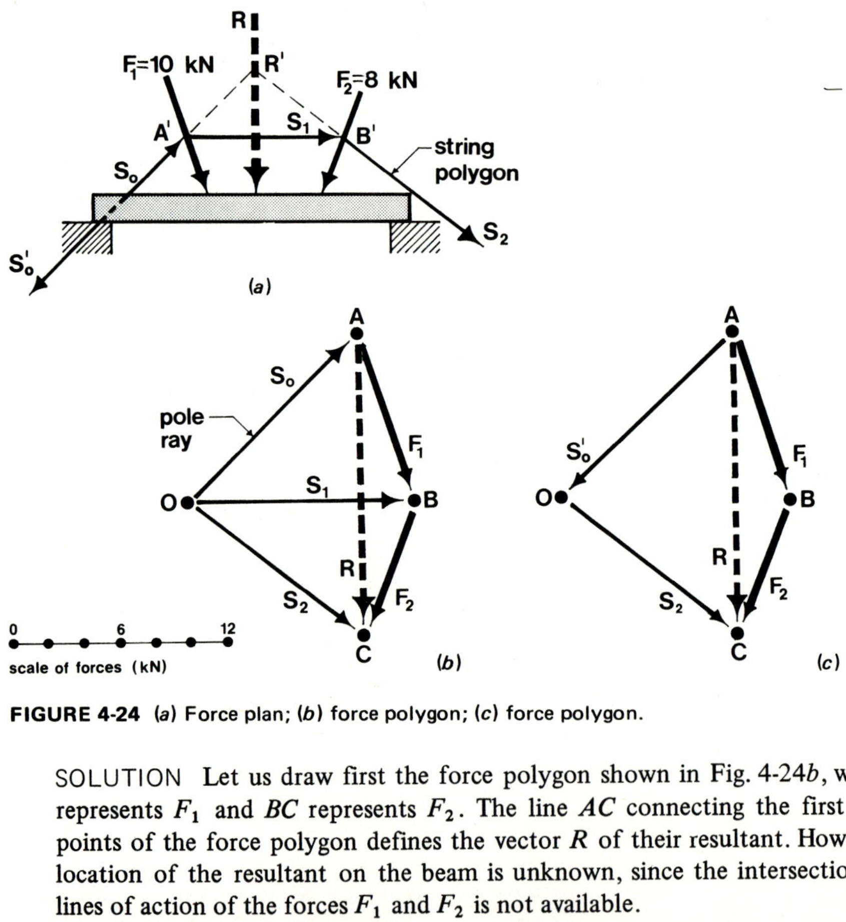

FIGURE 4-24 (*a*) Force plan; (*b*) force polygon; (*c*) force polygon.

SOLUTION Let us draw first the force polygon shown in Fig. 4-24*b*, where AB represents F_1 and BC represents F_2. The line AC connecting the first and last points of the force polygon defines the vector R of their resultant. However, the location of the resultant on the beam is unknown, since the intersection of the lines of action of the forces F_1 and F_2 is not available.

To overcome this difficulty, we use the basic principle that the resultant of two forces must pass through the point of intersection of the lines of action of its two force components and split the resultant R of F_1 and F_2 into two *other* components whose lines of action intersect within the paper. Since R can be split into two components in any number of ways, components intersecting within the paper can always be found.

To this purpose, we choose an arbitrary point O in the plane of the force polygon and draw the force vector OA. Let's call this vector S_o and assume that its arrow points from O to A. Then the vector OB, with its arrow pointing from O to B, represents the resultant S_1 of the forces S_o and F_1. Finally, the vector OC, pointing from O to C, is the resultant S_2 of the forces S_1 and F_2.

Let us now reverse the sense of the force S_o and call this force S'_o (Fig. 4-24*c*). The forces S'_o and S_2 are two *components* of the resultant R, as are the given forces F_1 and F_2 acting on the beam.

Let us now move to the force plan (Fig. 4-24*a*). Draw a parallel to S'_o from an arbitrary point A' on the line of action of F_1. Then draw a parallel to S_1 from point A' to meet the line of action of F_2 at B'. Finally, draw a parallel to S_2 from B'. Notice that three forces intersect at A': S'_o, F_1, and S_1. From the force polygon we know that S_1 is the resultant of S_o and F_1. Similarly, at point B', S_2 is the resultant of S_1 and F_2. Extend the lines of S'_o and S_2 in the force plan until they intersect at R'. This is the point through which the resultant R of the forces S'_o and S_2 must pass. Since R is also the resultant of F_1 and F_2, we have solved our problem. All that we have done by the choice of the point O is to substitute for the components F_1 and F_2 of R, two other components S'_o and S_2 which intersect within the paper. The greater the distance of O from R, the lesser the slope of S'_o and S_2 and the lower their intersection.

The vectors S'_o (S_o), S_1, and S_2 in the force polygon are only auxiliary partial resultants and are called *pole rays*, while their common point O is called the *pole* of the polygon. The broken line S_o, S_1, S_2 through points A' and B' in the force plan is called a *string polygon*. Thus, if we extend the first and last strings of the string polygon, we obtain R', the point of their intersection, through which the line of action of the resultant R of the given force system must pass.

The string polygon takes its name from the fact that it is the shape of a string suspended from two fixed points from which the forces of the system hang—you may see this by turning Fig. 4-24*a* upside down and imagining the forces F_1 and F_2 to have opposite sense.

Note: The pole and the pole rays are drawn in the force polygon diagram while the string polygon is drawn in the force plan.

Illustrative Problem 4-10

Find the magnitude and the direction of the resultant of the forces acting on the bridge pier of Illustrative Prob. 4-8 by means of a force polygon and a string polygon.

SOLUTION First draw the force polygon (Fig. 4-25*b*), choosing arbitrarily the pole O and drawing the pole rays OA, OB, OC, and OD. Now transfer to the force plan—the pier with its forces—and draw the strings of the string polygon parallel

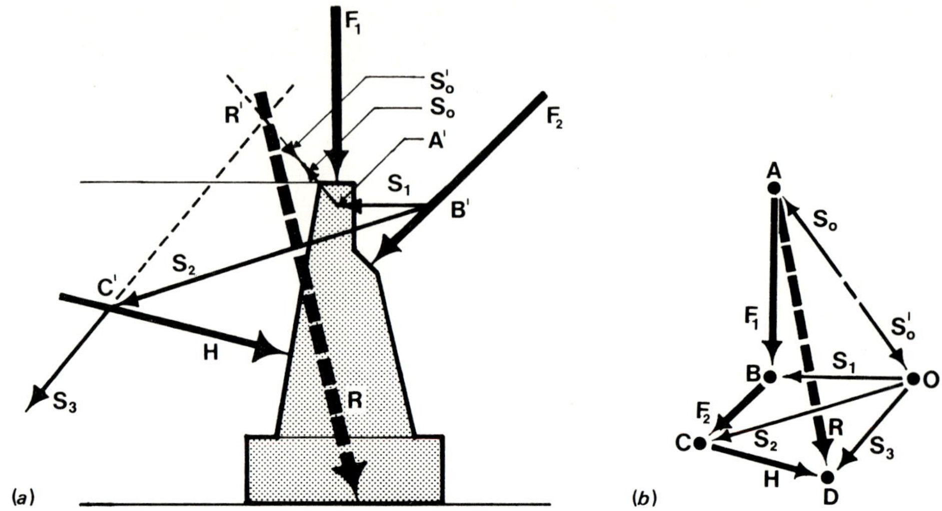

FIGURE 4-25 (*a*) Force plan; (*b*) force polygon.

to the pole rays S_o, S_1, S_2, S_3 through A', B', C' (Fig. 4-25*a*). Extend the first and last strings of the string polygon until they intersect at point R'. The resultant R must pass through this point. The vector AD of the resultant R in the force polygon defines its magnitude and direction. Draw the parallel to the vector AD through the point R', show the sense of R by means of an arrow, and write next to it its magnitude measured on the force scale.

New Terms Appearing in Chapter 4

Rotational effect (of a force)
Lever arm of a force
Moment of a force
Moment equilibrium
Lever
Force couple
Nonconcurrent force system
Parallel force system
Cantilever
Resultant moment
Dam
Retaining wall
Overturning moment
Stabilizing moment
Safety against overturning
Safety factor (SF)
Pole
Pole rays
String polygon

Important Structural Elements Appearing in Chapter 4

Cantilever. This is a beam with one end free (unsupported) and the other clamped or fixed firmly into a rigid support.

Lever. This is a mechanism based on the moment principle and used to magnify the effect of an applied force, such as for lifting heavy weights.

Dam. This is a wall used to retain liquid materials.

Retaining wall. This is a wall used to retain soil or other solid materials.

Quiz Number 3

A. TRUE OR FALSE

1. The moment of a force is the time of application of a force.
2. The unit of moments is the foot.
3. The moment of a force about a point measures the rotational effect of the force about that point.
4. Moments can be clockwise or counterclockwise.
5. The resultant of two moments is a force.
6. If a system of coplanar forces is in equilibrium its rotational effect about any point in the plane is equal to zero.
7. If a coplanar force system acting on a body is in equilibrium, does the body start to rotate as a result of the resultant moment?
8. Clockwise moments must always have positive signs.
9. The longer the arm of a constant force with respect to a point O, the smaller the moment of this force about O.
10. The lever arm of a force with respect to a given point is the perpendicular distance from the line of action of the force to the given point.
11. Two equal perpendicular forces are called a *force couple.*
12. The moment of a force couple about a point in its plane is the same regardless of the point.
13. The moment of a force couple is the product of the magnitude of one of the forces and the perpendicular distance between the forces.
14. A parallel force system is a special case of a nonconcurrent force system.
15. All the forces in a nonconcurrent force system intersect at a common point.
16. In a coplanar force system, all the forces lie in the same plane.
17. The moment of the resultant of a coplanar force system about any point in its plane is equal to the sum of the moments of the individual forces of the system about the point.
18. The resultant of two parallel downward forces acts downward and is equal to the difference between the two forces.

B. MULTIPLE CHOICE

1. The safety factor against overturning of a body under the action of a force system is the ratio of
   (*a*) the largest force to the smallest force of the system
   (*b*) the stabilizing moment to the overturning moment

(*c*) the resultant of the system to the arm of the resultant about the point of overturning
(*d*) the weight of the body to the applied overturning forces

2. The pole and pole rays are always drawn
(*a*) on the force plan
(*b*) on the force polygon
(*c*) on neither of the above
3. The pole is chosen
(*a*) arbitrarily
(*b*) on the line of action of the resultant
(*c*) 4 ft from the largest force in the force polygon
(*d*) always to the right of the force polygon
4. The resultant of two parallel forces $F_1 = 40$ lb (downward) and $F_2 = 20$ lb (upward) is
(*a*) $R = 60$ lb downward
(*b*) $R = 20$ lb downward
(*c*) $R = 20$ lb upward
(*d*) $R = 40$ lb downward
5. The string polygon is drawn with strings parallel to the pole rays:
(*a*) on the force plan
(*b*) on the force polygon
(*c*) on neither of the above
6. The resultant of three equal downward parallel forces is
(*a*) zero
(*b*) a horizontal force equal to the sum of the given forces
(*c*) a sloping force equal to the sum of the given forces
(*d*) none of the above
7. The resultant moment of two moments $M_1 = 200$ lb·ft clockwise and $M_2 =$ 200 lb·ft counterclockwise is
(*a*) zero
(*b*) 200 lb·ft, clockwise
(*c*) 200 lb·ft, counterclockwise
(*d*) 400 lb·ft, clockwise
8. A force couple consists of two equal and opposite forces $F = 10$ lb at a perpendicular distance of 12 in. What is the moment of the couple?
(*a*) zero (*b*) 1 200 lb·in (*c*) 120 lb·ft (*d*) 10 lb·ft
9. The cantilever beam is
(*a*) supported on two ends
(*b*) supported only in the middle
(*c*) free on one end and fixed into the support on the other end
(*d*) none of the above

C. SHORT ANSWERS

1. The product of a force and its lever arm is called ________.
2. Coplanar forces which do not have any rotational effect about any point in their plane are in rotational or moment ______________.

3. The moment of the resultant of a force system about a point is equal to ________
____
____.
4. If the stabilizing moment of the forces on a body is 100 kip·ft and the overturning moment is 200 kip·ft, what is the safety factor against overturning?

   SF = ______
5. Is the SF in question 4 adequate? ______
6. Will the body in question 4 really overturn? ______
7. To lift a heavy weight (Fig. 4-26), the lever arm r_1 close to the weight should be __________ than the lever arm r_2 of the lifting force.

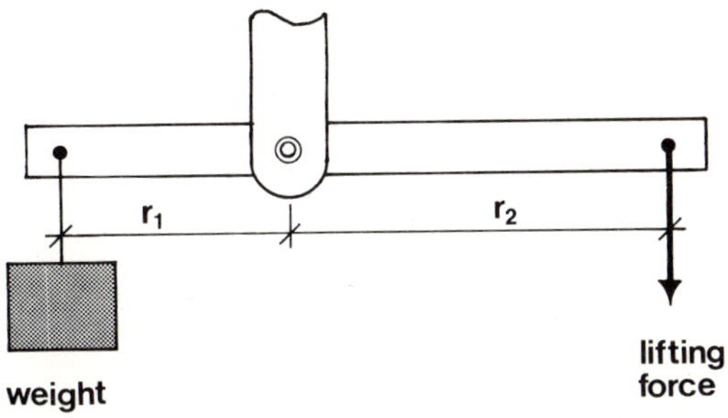

FIGURE 4-26

8. Parallels to the pole rays drawn in the force plan constitute a ____________
____________.
9. The point of intersection of all the pole rays is called a ________.
10. Does it matter whether the pole is chosen to the right or to the left of the force polygon? ______

EXERCISES

4-1 Compute the moments about point C and about point B of the forces acting on the beam shown in Fig. 4-27. Is this force system in equilibrium?

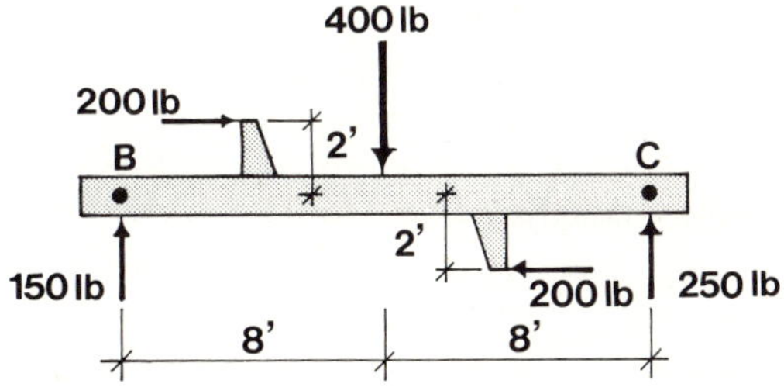

FIGURE 4-27

4-2 Calculate the moments about points B and C of the forces acting on the cantilever shown in Fig. 4-28.

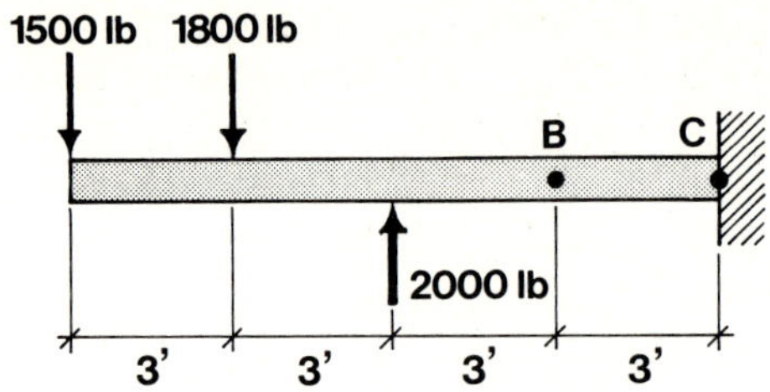

FIGURE 4-28

4-3 The boom of a crane is 40 ft long. Its balancing arm is 10 ft long and carries a 40-kip counterweight at its end (Fig. 4-29). What load should be placed on the boom end for moment equilibrium about point O to be satisfied?

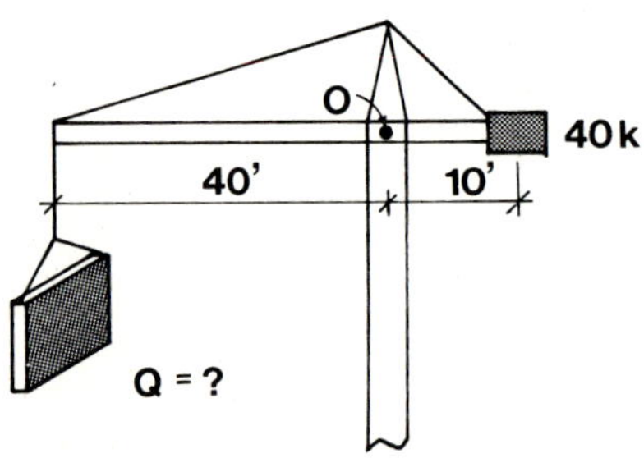

FIGURE 4-29

4-4 What must be the magnitude of the force T to hold a plank weighing 20 lb in the position shown in Fig. 4-30? Note that the lever arms of the forces T and F are 6 ft and 10 ft, respectively.

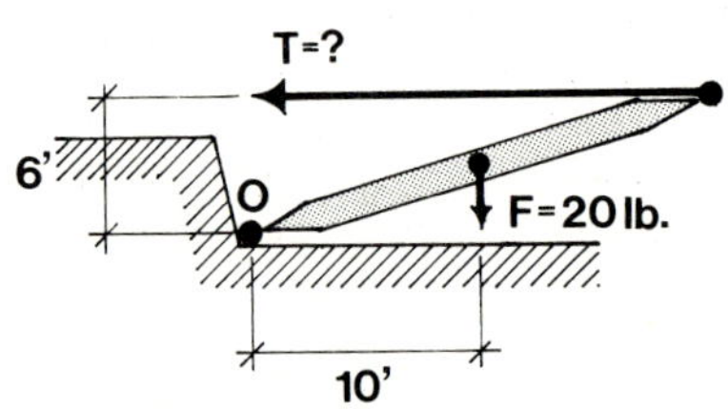

FIGURE 4-30

4-5 What pressure can be developed on the 3-in-diameter piston in Fig. 4-31 by a horizontal force T = 10 000 lb?

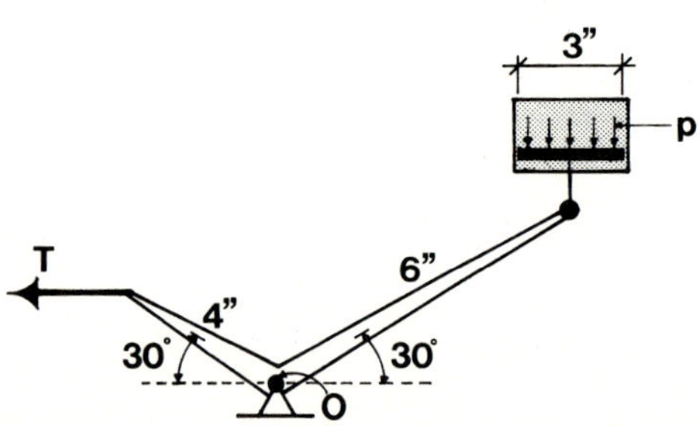

FIGURE 4-31

Hint: From the moment equilibrium about point O find, first, the force P on the piston rod. Then distribute this force uniformly across the circular area of the piston to obtain the pressure on the piston.

4-6 A cylindrical tank (Fig. 4-32) is 8 ft in diameter and weighs 500 lb. It must be raised over a 2-ft step by a horizontal force T. Find the magnitude of T.

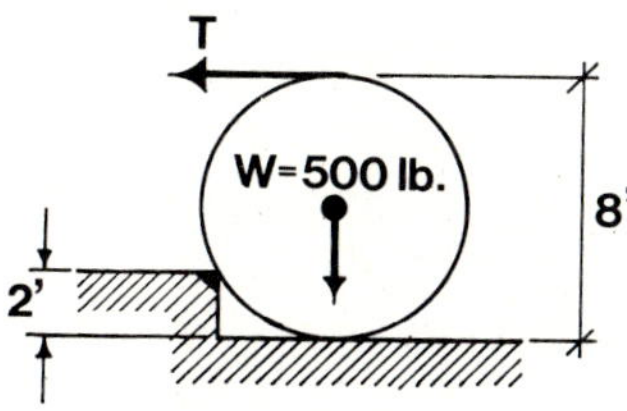

FIGURE 4-32

4-7 Find the resultant (magnitude and direction) of the force system acting on the arch structure shown in Fig. 4-33 by means of a force polygon and a string polygon.

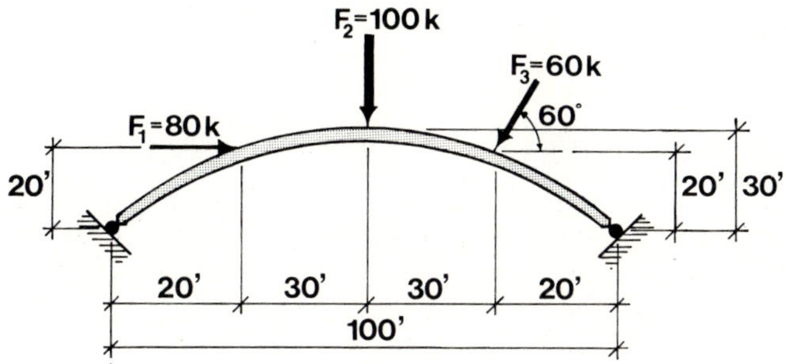

FIGURE 4-33

4-8 Find the resultant of the parallel force system acting on the beam shown in Fig. 4-34 by means of a force polygon and a string polygon.

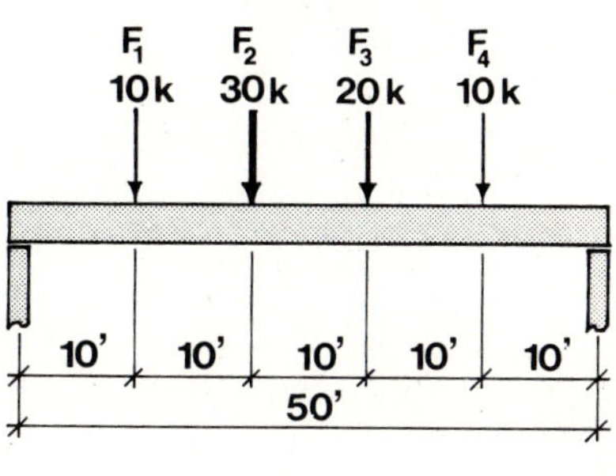

FIGURE 4-34

Concrete wall panel on the Northwestern Life building in Minneapolis. (*Photo by L. Cerny.*)

The Center of Gravity: The Centroid

5

5.1 THE CENTER OF GRAVITY OF A BODY

Consider a square plate of constant thickness and of weight W (Fig. 5-1), made of a homogeneous material. Small hooks are attached at various points of the plate boundary. First suspend the plate from the hook located at corner A (Fig. 5-1*a*). The plate assumes a position in which the prolongation of the string axis coincides with one of the plate diagonals and divides the plate into two parts of equal weight $W/2$. The line of action of the resultant W of these two forces $W/2$ must coincide with the axis of the string, that is, with the diagonal. If the plate is now suspended from point B, the line of action of the resultant W will coincide with the other diagonal of the plate (Fig. 5-1*b*). The lines of these two resultants intersect at a point inside the body called the *center of gravity* or *centroid* of the plate. If the plate is suspended from the hook at any boundary point C, the prolonged string axis passes again through the center of gravity (Fig. 5-1*c*).

If a body has an irregular shape such as in Fig. 5-1*d*, the forces W_1 and W_2 are not necessarily equal. However, for moment (rotational) equilibrium of the plate about any point on the resultant W,

$$W_1 \times r_1 = W_2 \times r_2$$

The axes of the prolonged string are called *centroidal axes*. To locate the center of gravity or the centroid of a body, we must locate the point of intersection of two centroidal axes.

The center of gravity or centroid of the body (made of homogeneous material) is the point in which we can consider the entire weight of the body to be concentrated in solving problems of statics.

The force due to gravity acts on each elementary particle of the body, and the resultant of all these elementary forces goes through its center of gravity or centroid. The centroid of a body is thus located by finding the resultant of the forces acting on its individual parts. By analyzing bodies that have one axis of symmetry, we find that

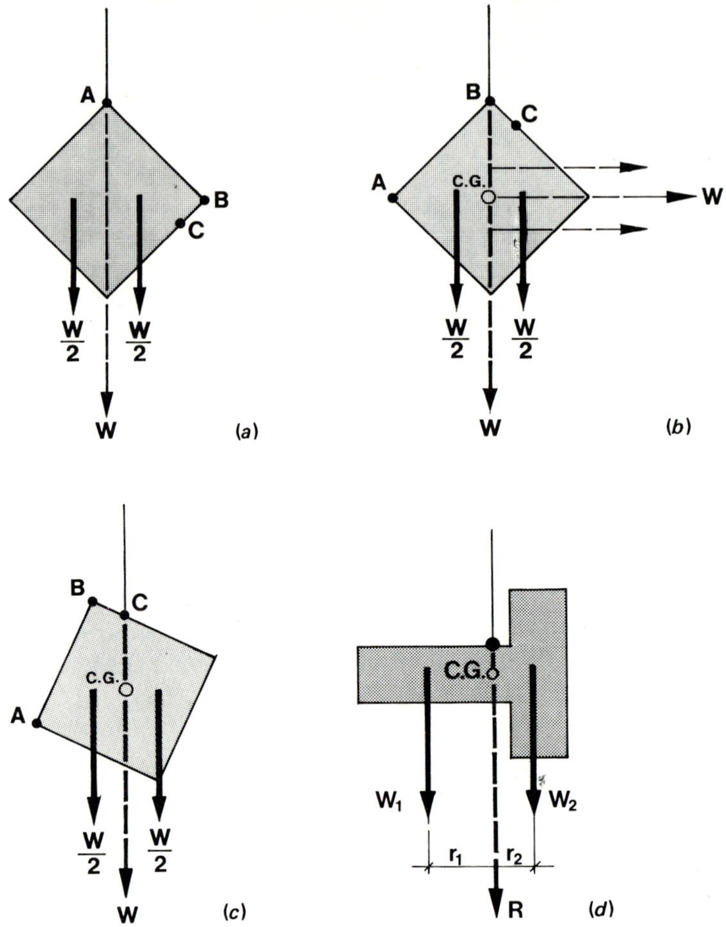

FIGURE 5-1 The center of gravity (CG) of (*a*, *b*, *c*) a thin square plate and (*d*) a thin T-shape plate.

the centroid is located on that axis. Hence, if a body has more than one axis of symmetry, the center of gravity must lie at their intersection (Fig. 5-2).

If we decrease the thickness of the plate, its weight also decreases. A plate with zero thickness has zero weight and, instead of a three-dimensional body, is a plane figure, an area. Squares, rectangles, triangles, trapezoids, and circles are examples of plane figures or areas. From these basic figures we can compose more complicated figures (Fig. 5-3).

With zero thickness, the center of gravity or centroid of a body (plate) changes to the *centroid of an area*.

The location of the centroid of some simple figures is given in Table A-10. The horizontal and vertical distances from a chosen set of x, y axes to the centroid are indicated by $\bar{x}$ (read as "x bar") and $\bar{y}$ (read as "y bar"), respectively.

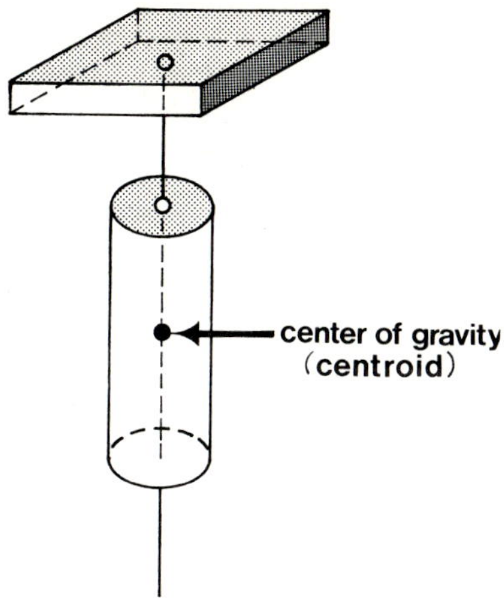

FIGURE 5-2 The center of gravity of a cylinder.

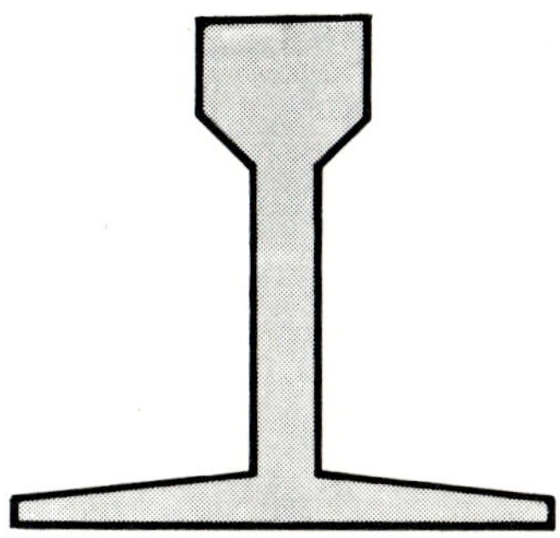

FIGURE 5-3

5.2 THE MOMENT OF AN AREA

We will introduce the concept of the moment of an area in order to simplify the calculation of the centroid of a plane figure. Imagine a horizontal force F_x of as many pounds as there are square inches in the area A of a rectangle and applied at the centroid of the rectangle (Fig. 5-4). The moment of the area about the x axis is

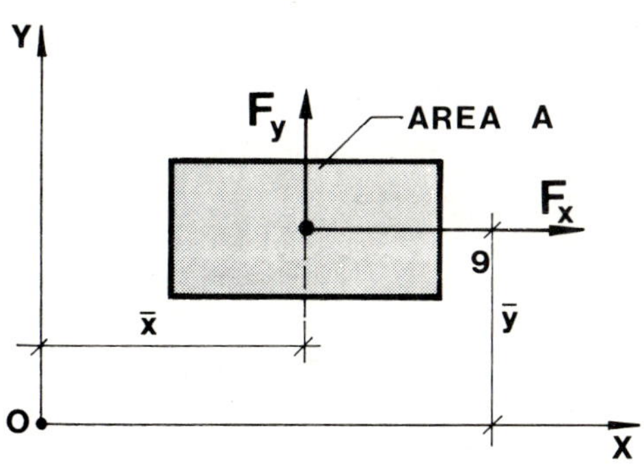

FIGURE 5-4 The moment of an area.

defined as the product of the area and the perpendicular distance from its centroid to the x axis,

$$M_x = F_x\bar{y} = A\bar{y}$$

Similarly, using a vertical force $F_y = A$, we define the moment of the area about the y axis as

$$M_y = F_y\bar{x} = A\bar{x}$$

The units of the moment of an area are thus in^2 × in = in^3 (or ft^3, cm^3, and so on).

Illustrative Problem 5-1

Determine the moment of the triangular area shown in Fig. 5-5 about the x and y axes.

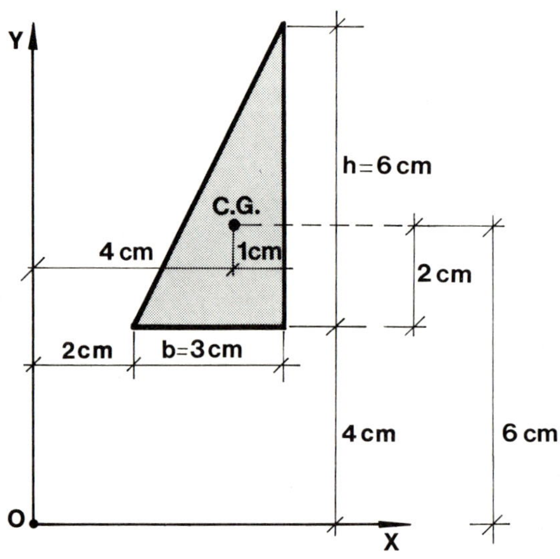

FIGURE 5-5 The centroid of a triangular area.

SOLUTION The area of the triangle is

$$A = \frac{b \times h}{2} = \frac{3 \times 6 \text{ cm}}{2} = 9 \text{ cm}^2$$

The distance from the centroid (CG) to the horizontal side of the triangle is (see Table A-10)

$$\frac{h}{3} = \frac{6}{3} = 2 \text{ cm}$$

and the distance from the centroid to the x axis is

$$\bar{y} = 4 + 2 = 6 \text{ cm}$$

thus the moment of the area about the x axis is

$$M_x = A\bar{y} = 9 \text{ cm}^2 \times 6 \text{ cm} = 54 \text{ cm}^3$$

The distance from the centroid to the vertical side of the triangle is

$$\frac{b}{3} = \frac{3}{3} = 1 \text{ cm}$$

and the distance from the centroid to the y axis is

$$\bar{x} = 2 + 2 = 4 \text{ cm}$$

The moment of the area about the y axis is then

$$M_y = A\bar{x} = 9 \text{ cm}^2 \times 4 \text{ cm} = 36 \text{ cm}^3$$

5.3 CENTROIDS OF COMPOSITE AREAS

In finding the centroid of a composite area, we may follow either a graphical or a numerical approach. Let us demonstrate both approaches by means of the following example.

To find the centroid of the plane figure representing the cross section of the concrete T beam shown in Fig. 5-6, divide the figure into simple areas, in this case, two

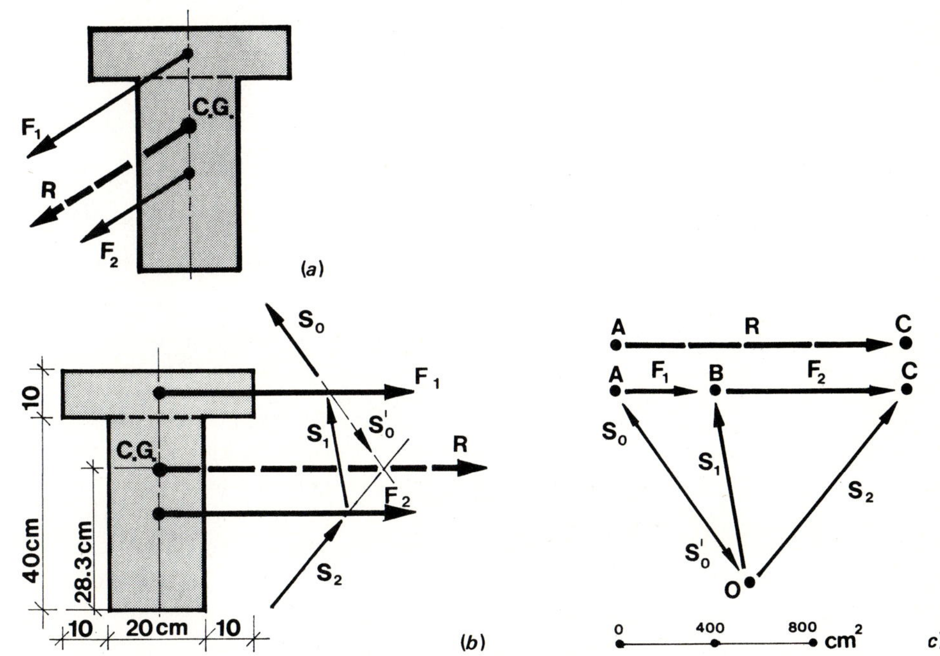

FIGURE 5-6 The centroid of an area representing the cross section of a concrete T beam.

rectangles. Apply two parallel forces, F_1 and F_2, in an arbitrary direction at the centroids of the two rectangles (Fig. 5-6*a*). The line of action of the resultant of these two forces will act along a centroidal axis of the composite figure. Since the cross section is symmetrical about the vertical axis, we know that the latter is also a centroidal axis. The centroid of the entire section lies at the intersection of these two centroidal axes.

It is convenient to determine first the horizontal centroidal axis of this area. The two horizontal forces acting at the centroids of the two rectangles are chosen proportional to their areas (Fig. 5-6*b*),

$$F_1 = A_1 = 40 \times 10 = 400 \text{ cm}^2 \qquad \text{(or units of force)}$$

$$F_2 = A_2 = 40 \times 20 = 800 \text{ cm}^2 \qquad \text{(or units of force)}$$

Choose the force scale and draw the force polygon (Fig. 5-6*c*); then choose arbitrarily the pole O and draw the pole rays. Draw the string polygon to the forces F_1 and F_2 in Fig. 5-6*b* with sides parallel to the pole rays. The intersection of the prolonged first and last sides or strings of the string polygon defines a point on the resultant R of the two forces. Since both F_1 and F_2 are horizontal, their resultant must also be horizontal. The line of action of R is a centroidal axis, the vertical axis of symmetry is another. The intersection of the two centroidal axes defines the centroid (CG) of the cross section.

The numerical approach to the determination of the centroids of plane figures is based on the moment principle described in Chap. 4 (Sec. 4.1). The moment of the resultant of a coplanar force system about any point or axis is equal to the sum of the moments of the individual forces of the system about the same point or axis.

In our problem the magnitude of the resultant is

$$R = F_1 + F_2 = 400 + 800 = 1\,200 \text{ cm}^2$$

and its moment about the arbitrarily selected x axis (Fig. 5-7) is

$$M_x = R\bar{y} = 1\,200\bar{y}$$

The sum of the moments of the two forces, F_1 and F_2, about the x axis is

$$M_x = F_1 \times y_1 + F_2 \times y_2 = 400 \times 45 + 800 \times 20 = 34\,000 \text{ cm}^3$$

where

$$y_1 = 40 + \tfrac{10}{2} = 45 \text{ cm}$$

$$y_2 = \tfrac{40}{2} = 20 \text{ cm}$$

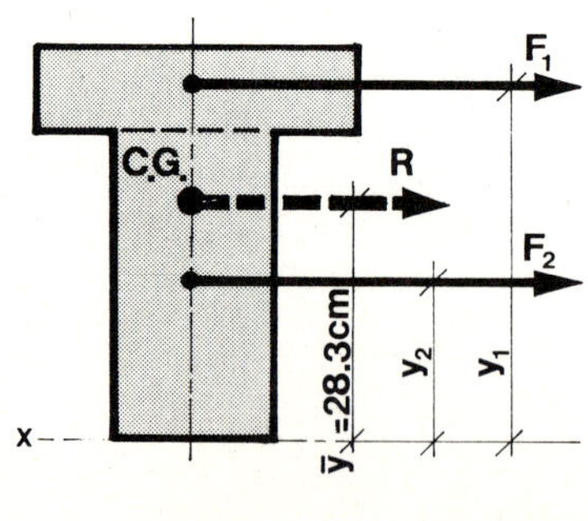

FIGURE 5-7

Then

$$R\bar{y} = 34\ 000\ \text{cm}^3$$

or

$$1\ 200\bar{y} = 34\ 000\ \text{cm}^3$$

and

$$\bar{y} = \frac{34\ 000}{1\ 200} = 28.3\ \text{cm}$$

The centroid of the cross section is located on the axis of symmetry 28.3 cm above the x axis.

Illustrative Problem 5-2

Find the centroid of the trapezoidal cross section of the concrete retaining wall shown in Fig. 5-8. Use both the graphical and the numerical approaches.

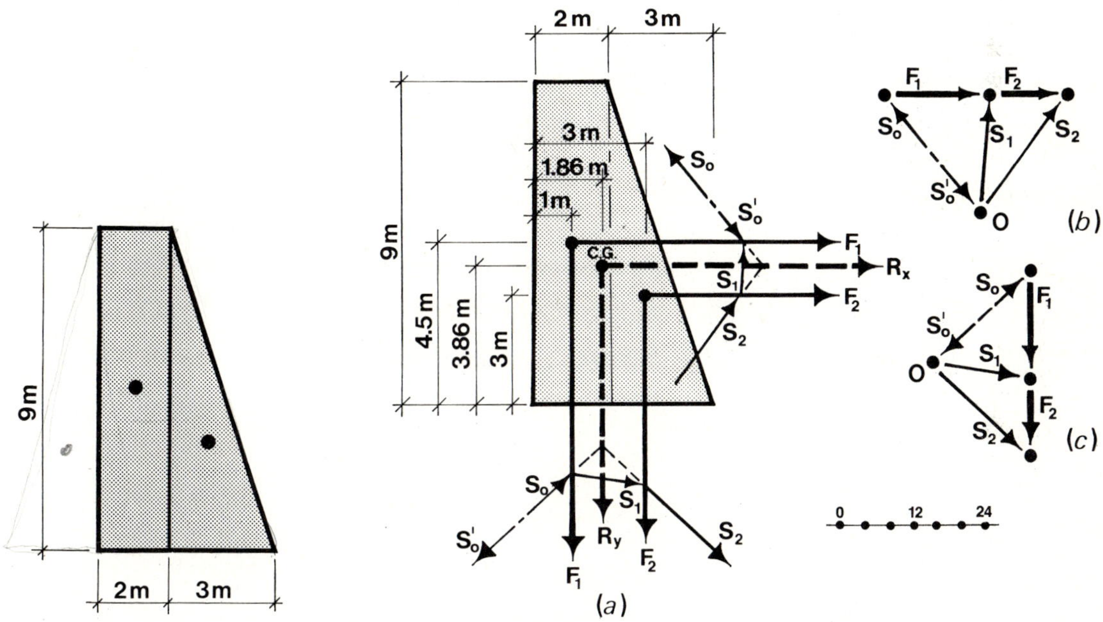

FIGURE 5-8 The centroid of a trapezoidal area, graphical solution.

SOLUTION The cross section does not have any axis of symmetry. Therefore, we must find two centroidal axes, for example, the horizontal and vertical centroidal axes.

Divide the cross section into a rectangle and a triangle as shown in Fig. 5-8. Their areas are

$$\text{Rectangle:}\quad A_1 = 2 \times 9 = 18\ \text{m}^2$$

$$\text{Triangle:}\quad A_2 = \tfrac{1}{2} \times 3 \times 9 = 13.5\ \text{m}^2$$

The corresponding forces proportional to the areas are

$$F_1 = 18 \text{ m}^2$$

$$F_2 = 13.5 \text{ m}^2$$

(a) Graphical Solution The forces F_1 and F_2 are applied at the centroids of the component areas in a horizontal direction (Fig. 5-8*a*). Their resultant is

$$R_x = F_1 + F_2 = 18 + 13.5 = 31.5 \text{ m}^2$$

and its line of action is found by means of a force and string polygon (Fig. 5-8*b*). Thus the horizontal centroidal axis is determined.

The same forces F_1 and F_2 are then applied again at the centroids of the individual areas, but in the vertical direction, and the location of their resultant R_y is similarly found (Fig. 5-8*c*). Thus the vertical centroidal axis is found. The point of intersection of the two centroidal axes defines the centroid of the whole section.

(b) Numerical Solution It is convenient to choose the origin of the coordinate system at the left bottom corner of the section, with the x axis horizontal and the y axis vertical (Fig. 5-9). The lever arms of the horizontal forces are thus

$$y_1 = \tfrac{9}{2} = 4.5 \text{ m}$$

$$y_2 = \tfrac{9}{3} = 3.0 \text{ m}$$

The sum of the moments of the forces F_1 and F_2 about the x axis is then

$$M_x = F_1 \times y_1 + F_2 \times y_2 = 18 \times 4.5 + 13.5 \times 3 = 121.5 \text{ m}^3$$

The moment of their resultant R_x about the x axis is

$$M_x = R_x \bar{y} = 31.5\bar{y}$$

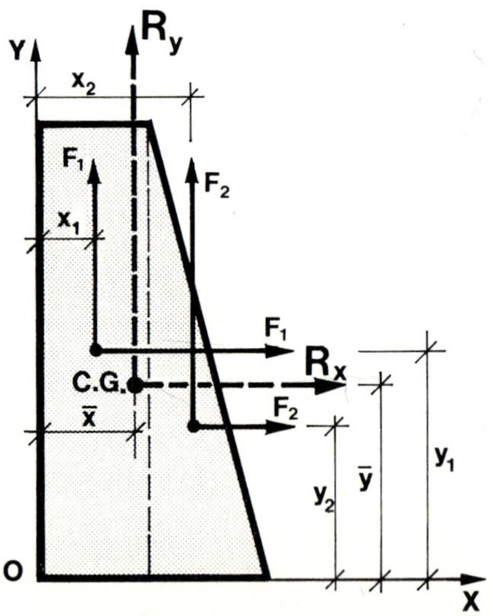

FIGURE 5-9 The centroid of a trapezoidal area, numerical solution.

or

$$31.5\bar{y} = 121.5$$

and

$$\bar{y} = \frac{121.5}{31.5} = 3.86 \text{ m}$$

This procedure is then applied to the forces F_1 and F_2 acting vertically at the centroids of the component areas. The lever arms of the forces with respect to the y axis are

$$x_1 = \tfrac{2}{2} = 1.0 \text{ m}$$

$$x_2 = 2 + \tfrac{1}{3}\,3 = 2 + 1 = 3.0 \text{ m}$$

Then

$$M_y = F_1 \times x_1 + F_2 \times x_2 = 18 \times 1.0 + 13.5 \times 3 = 58.5 \text{ m}^3$$

The moment of the resultant about the y axis is

$$M_y = R_y\bar{x} = 31.5\bar{x}$$

The moment of the resultant is equal to the sum of the moments of the two forces, F_1 and F_2:

$$31.5\bar{x} = 58.5$$

from which

$$\bar{x} = \frac{58.5}{31.5} = 1.86 \text{ m}$$

Illustrative Problem 5-3

Find the centroid of the steel beam section consisting of a wide-flange section **W** 12 × 120 and a channel section **C** 12 × 30 (Fig. 5-10). Perform all calculations in metric units.

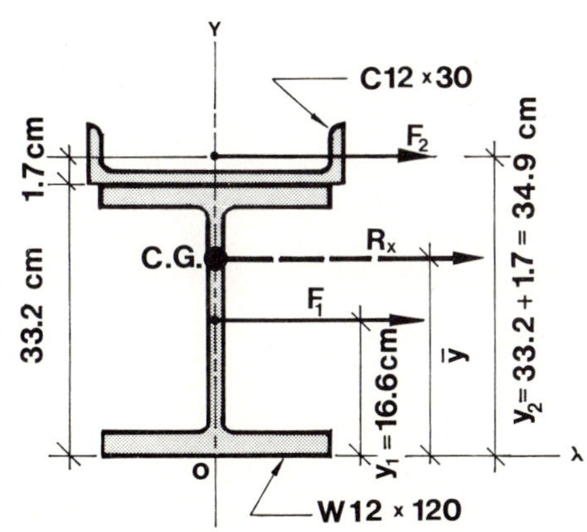

FIGURE 5-10 The centroid of a steel beam cross section.

SOLUTION The metric equivalent to a **W** 12 X 120 is a **W** 330 X 1 750 (see Table A-5), and its area is $A_1 = 228\ \text{cm}^2$. The metric equivalent to a **C** 12 X 30 is a **C** 300 X 440 (see Table A-8), and its area is $A_2 = 57\ \text{cm}^2$. One of the two centroidal axes of the compound section is its vertical axis of symmetry. The horizontal centroidal axis is found as the resultant of the two horizontal forces

$$F_1 = 228\ \text{cm}^2 \qquad \text{and} \qquad F_2 = 57\ \text{cm}^2$$

corresponding to the areas of the individual sections and acting horizontally at their centroids. The dimensions necessary for the location of these centroids are also found in Tables A-5 and A-8. The centroid of the wide-flange section is on its vertical axis of symmetry at the midpoint of its depth:

$$y_1 = \frac{332\ \text{mm}}{2} = 166\ \text{mm} = 16.6\ \text{cm}$$

The centroid of the channel is on its axis of symmetry, at a distance of 17 mm = 1.7 cm from its longer side. Figure 5-10 shows the arbitrarily chosen x axis and the lever arms of the forces about this axis. The magnitude of the resultant R_x is

$$R_x = F_1 + F_2 = 228 + 57 = 285\ \text{cm}^2$$

The moment of the resultant about the x axis is equal to the sum of the moments of the forces F_1 and F_2 about this axis, or

$$R_x\bar{y} = F_1 \times y_1 + F_2 \times y_2$$

$$285\bar{y} = 228 \times 16.6 + 57 \times 34.9$$

$$285\bar{y} = 5\ 774\ \text{cm}^3$$

$$\bar{y} = \frac{5\ 774}{285} = 20.26\ \text{cm} \qquad (8\ \text{in})$$

The centroid of the compound section is located on the vertical axis of symmetry, 20.26 cm from the x axis.

Illustrative Problem 5-4

Find the centroid of the section shown in Fig. 5-11, consisting of a rectangle with a circular hole 10 cm in diameter. The center of the hole is located on the vertical axis of symmetry, 10 cm from the bottom.

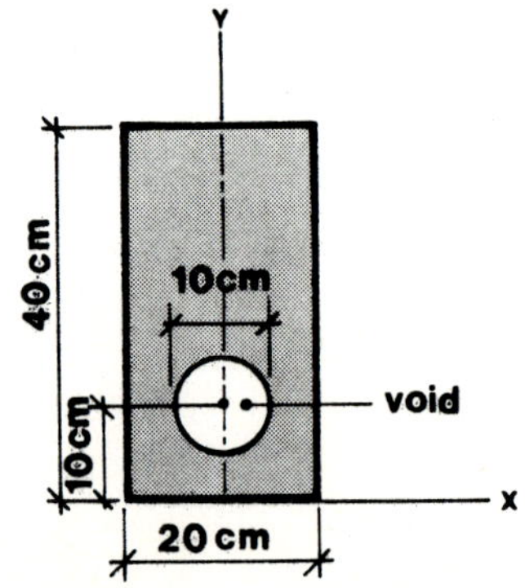

FIGURE 5-11 The centroid of a rectangular area with a circular hole.

SOLUTION Select the x axis arbitrarily at the base of the rectangle. If there were no hole, the force representing a solid rectangle would be

$$F_1 = 20 \times 40 = 800 \text{ cm}^2$$

and it would act at the middle of the depth of the rectangle, that is, with $y_1 = 40$ cm/2 = 20 cm. The effect of the opening can be accounted for by considering a negative force proportional to the area of the circle

$$F_2 = \frac{\pi \times 10^2}{4} = 78.5 \text{ cm}^2$$

and acting in the opposite sense to that of the force F_1. The resultant R_x of the two opposite forces F_1 and F_2 is

$$R_x = F_1 - F_2 = 800 - 78.5 = 721.5 \text{ cm}^2$$

Its location is found from the moment equation about the x axis located at the base of the section (Fig. 5-12),

$$R_x\bar{y} = F_1 \times y_1 - F_2 \times y_2$$

where

$$y_1 = \frac{40}{2} = 20 \text{ cm} \qquad \text{and} \qquad y_2 = 10 \text{ cm}$$

Then

$$721.5\bar{y} = 800 \times 20 - 78.5 \times 10$$

$$721.5\bar{y} = 15\,215$$

$$\bar{y} = \frac{15\,215}{721.5} = 21.0 \text{ cm} \qquad (8.3 \text{ in})$$

The presence of the opening caused the shift of the centroid of the section 1.0 cm above the centroid of the solid rectangle without the hole.

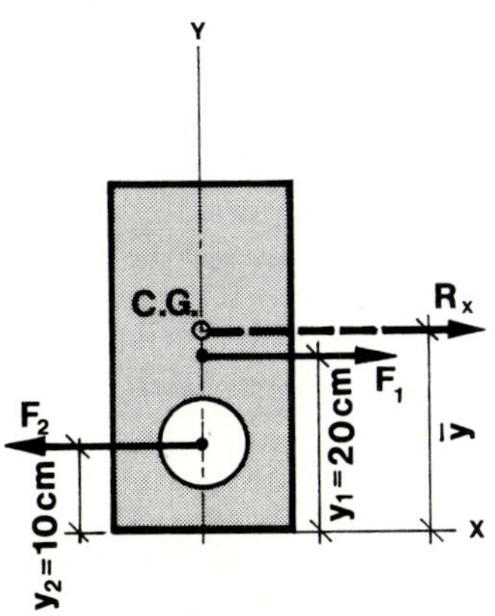

FIGURE 5-12

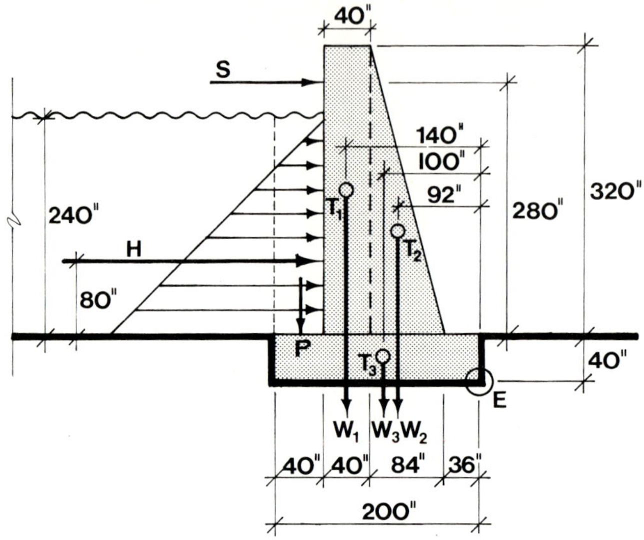

FIGURE 5-13 The centroid of a concrete dam section.

Illustrative Problem 5-5

The concrete dam shown in Fig. 5-13 resists the horizontal pressure of the water, and the horizontal force $S = 4$ kips/ft of the dam (58.4 kN/m). The unit weight of concrete $\gamma = 150$ lb/ft^3 (23.5 kN/m^3), and the unit weight of water $\gamma_w = 62.4$ lb/ft^3 (9.8 kN/m^3). Compute the resultant moment of all the forces about point E at the right bottom edge of the footing, considering 1 ft of the dam, and check the stability of the dam against overturning.

SOLUTION Determine first the distribution of the lateral water pressure (the so-called hydrostatic pressure) acting on the dam. The pressure of water increases linearly with its depth: for each 1-ft increase in water depth, the lateral water pressure increases by the weight of 1 ft^3 of water, that is, 62.4 lb/ft^2 (9.8 kN/m^2 for each meter of depth increase).

If the depth of water is 240 in = 20 ft (6.1 m), the maximum water pressure at the bottom of the dam is

$$62.4 \times 20 = 1\,248 \ \frac{\text{lb}}{\text{ft}^2} \quad \left(9.8 \times 6.1 = 59.75 \ \frac{\text{kN}}{\text{m}^2}\right)$$

Since the variation of the horizontal pressure of the water against the dam is linear, it can be represented by an inclined straight line. The resultant lateral force H due to water pressure is equal to the volume of the loading prism ABC in Fig. 5-14. The height of the prism is 20 ft (6.1 m), the base represents the maximum water pressure of 1 248 lb/ft^2 (59.75 kN/m^2), and the length of the prism is 1 ft (1 m in metric calculations):

$$H = \tfrac{1}{2} \times 1.0 \text{ ft} \times 20 \text{ ft} \times 1\,248 \ \frac{\text{lb}}{\text{ft}^2} = 12\,480 \text{ lb} \qquad (55.5 \text{ kN})$$

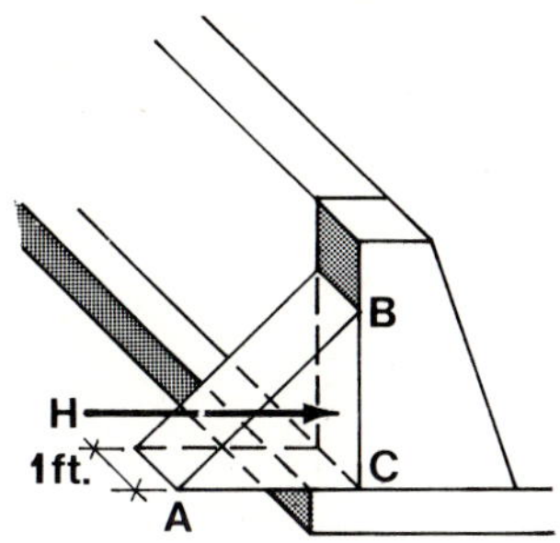

FIGURE 5-14 Distribution of water pressure on a vertical wall.

or in the metric system

$$H = \tfrac{1}{2} \times 1.0 \text{ m} \times 6.1 \text{ m} \times 59.75 \frac{\text{kN}}{\text{m}^2} = 182.2 \text{ kN}$$

a force distributed over a length of dam equal to 1 m (3.28 ft). The force per foot of the length of the wall is therefore

$$H = \frac{182.2}{3.28} = 55.5 \text{ kN} = 12\,480 \text{ lb}$$

This is the result for H we obtained by using U.S. units. This force acts at the center of gravity of the prism, which is located at one third of the prism height from its bottom, that is, $6\frac{2}{3}$ ft (2.03 m) above the bottom.

To compute the weight of the wall, divide it into three parts (Fig. 5-15): two parallelograms, *DEFG* and *KLMN*, and one prism, *LNO*. The weight of the vertical parallelogram *KLMN* is

$$W_1 = V\gamma = \frac{40 \times 320}{144} \times 1.0 \times 150 = 13\,333 \text{ lb} \qquad (59.3 \text{ kN})$$

where V is the volume of the parallelogram and γ is the unit weight of the material of the wall. The weight W_2 of the parallelogram *DEFG* (the footing) is

$$W_2 = \frac{200 \times 40}{144} \times 1.0 \times 150 = 8\,333 \text{ lb} \qquad (37.0 \text{ kN})$$

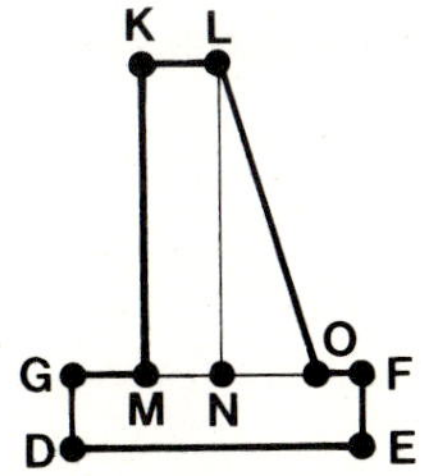

FIGURE 5-15

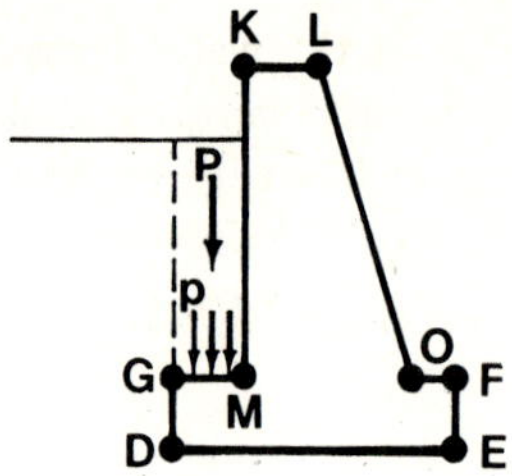

FIGURE 5-16

The weight of the prism *LNO* is

$$W_3 = \frac{1}{2} \times \frac{84 \times 320}{144} \times 1.0 \times 150 = 14\,000 \text{ lb} \qquad (62.3 \text{ kN})$$

A water pressure p acts also on the portion MG of the footing (Fig. 5-16), as a result of the water column 20 ft deep

$$p = 62.4 \times 20 = 1\,248 \ \frac{\text{lb}}{\text{ft}^2}$$

The total force P due to the vertical pressure p is the product of this pressure and the area $MG \times 1.0$ ft

$$P = 1\,248 \times \frac{40}{12} \times 1.0 = 4\,160 \text{ lb} \qquad (18.5 \text{ kN})$$

All the forces act at the centroids of the corresponding bodies, or loading blocks. The forces, their lever arms to point E, and the moments of the forces about E are listed in the following table.

FORCE, lb	LEVER ARM, in	MOMENT ABOUT E, lb·in
W_1 = 13 333	140	−1 866 620
W_2 = 8 333	100	− 833 300
W_3 = 14 000	$36 + \frac{2}{3} 84 = 92$	−1 288 000
H = 12 480	$\frac{1}{3} 240 = 80$	+ 998 400
P = 4 160	180	− 748 800
S = 4 000	320	+1 280 000

The sum of the negative moments is $(-)M_E = -4\,736\,720$ lb·in; the sum of the positive moments is $(+)M_E = 2\,278\,400$ lb · in.

The resultant moment about E is $M_E = -2\,458\,320$ lb·in (277.8 kN·m). The negative sign shows that the resultant moment of the force system about point E

is counterclockwise. It pushes the dam against the water pressure and supporting soil; that is, it helps to stabilize the dam against overturning about point E. The safety of the retaining wall is expressed by the factor of safety (SF) against overturning, which is defined as the ratio of the stabilizing moment to the overturning moment of the forces acting on the dam. The stabilizing moment in our problem is negative and equals

$$(-)M_S = 4\ 736\ 720 \text{ lb}\cdot\text{in}$$

The overturning moment is positive and equals

$$(+)M_o = 2\ 278\ 400 \text{ lb}\cdot\text{in}$$

Hence, the coefficient of safety or the safety factor is

$$\text{SF} = \frac{4\ 736\ 720}{2\ 278\ 400} = 2.08$$

Our dam is sufficiently safe against overturning, since codes usually require safety factors against overturning equal to or greater than 2. In our case with SF = 2.08, the dam would overturn if we increased the overturning moment 2.08 times.

New Terms Appearing in Chapter 5

Center of gravity	Composite area
Centroid	Water pressure
Centroid of a body	Water pressure on a wall
Centroid of an area	Triangular pressure diagram
Moment of an area	Safety factor (SF)
Stability	Stabilizing moment
Safety against overturning	Overturning moment

Quiz Number 4

A. TRUE OR FALSE

1. The moment of an area about the x axis is equal to the product of the area times the perpendicular distance from the centroid of the area to this axis.
2. The centroid of a rectangle is located at the bottom third point of its depth.
3. The centroid of a right triangle is located at the bottom third point of its height.
4. Centroidal axes do not necessarily pass through the centroid.
5. To find the centroid of an area of general shape, we need to find one and only one centroidal axis.
6. If an area has two axes of symmetry, the centroid of the area is at their intersection.

7. If we suspend a body from its centroid, it will start to rotate until it assumes the equilibrium position.
8. If we suspend a triangular thin plate by means of a string from one of its corners, the prolongation of the string line will pass through the centroid of the plate.
9. If areas are measured in square meters and lever arms in meters, the dimension of the moment of an area is in meters to the fourth power.
10. The moment of the resultant of a force system about any point is equal to the sum of the moments of the individual forces of the system about that point.
11. The centroid of a cross section can be found by a graphical as well as a numerical approach.
12. A body composed of two different materials is called a *homogeneous body*.
13. A channel steel shape has two axes of symmetry.
14. The pressure of water against a vertical wall varies linearly with the depth of water. Therefore, the pressure diagram can be represented by a right triangle.

B. MULTIPLE CHOICE

1. The centroid of a rectangle is located
   (*a*) at its base
   (*b*) at a corner
   (*c*) at the intersection of its diagonals
   (*d*) at the lower third point of its depth
2. If an area has one axis of symmetry, how many additional centroidal axes have to to be found to locate its centroid?
   (*a*) none (*b*) one (*c*) two (*d*) three
3. If the unit weight of a liquid is 100 lb/ft^3, the pressure in lb/ft^2 of this liquid against a vertical wall, 5 ft below the surface is
   (*a*) 500 (*b*) 100 (*c*) zero (*d*) 5 000
4. If the stabilizing moment on a structure is equal to 3 000 kN·m and the overturning moment to 1 000 kN·m, the factor of safety against overturning is equal to
   (*a*) zero (*b*) 3.0 (*c*) 1.0
5. If the depth of water retained by a dam is 10 ft and the unit weight of water is 62.4 lb/ft^3, the maximum lateral pressure on the dam is
   (*a*) 124.8 lb/ft^2 (*b*) 0.624 kips/ft^2 (*c*) 6 240 lb/ft^2
6. The point at which the total weight of a body may be considered concentrated is called its
   (*a*) rotational axis
   (*b*) moment center
   (*c*) center of gravity

C. SHORT ANSWERS

1. If the stabilizing moment on a body is M_s and the overturning moment is M_0, what is the safety factor against overturning? SF = ________ .
2. Any straight line passing through the centroid is called the ______________ ____________ .

3. The diagram of the lateral water pressure on a vertical wall has the shape of a ____________________.
4. The greater the depth below the surface of water, the ____________ the water pressure.
5. A vertical wall retains a depth of water of 20 ft. What is the total horizontal force H on the wall due to the water pressure? $H =$ ____________________.
6. What are the coordinates $\bar{x} =$ _______ and $\bar{y} =$ _______ of the centroid of the triangle shown in Fig. 5-17?

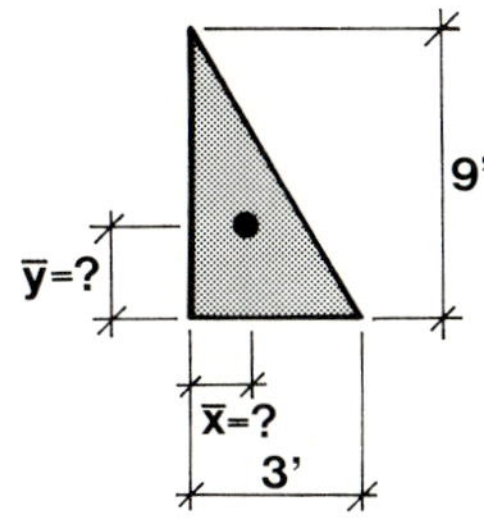

FIGURE 5-17

7. Is the centroid of the tubular section shown in Fig. 5-18 located at the point marked CG? ________ .

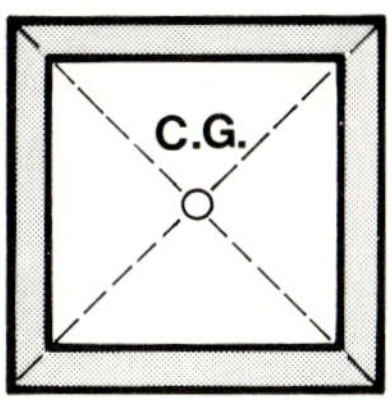

FIGURE 5-18

EXERCISES

5-1 Find the centroid of the hollow section shown in Fig. 5-19.

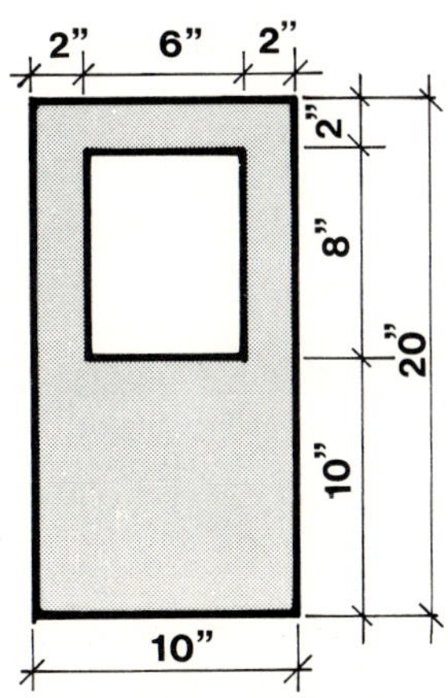

FIGURE 5-19

5-2 Find the centroid of the sections shown in Figs. 5-20*a* through 5-20*c*.

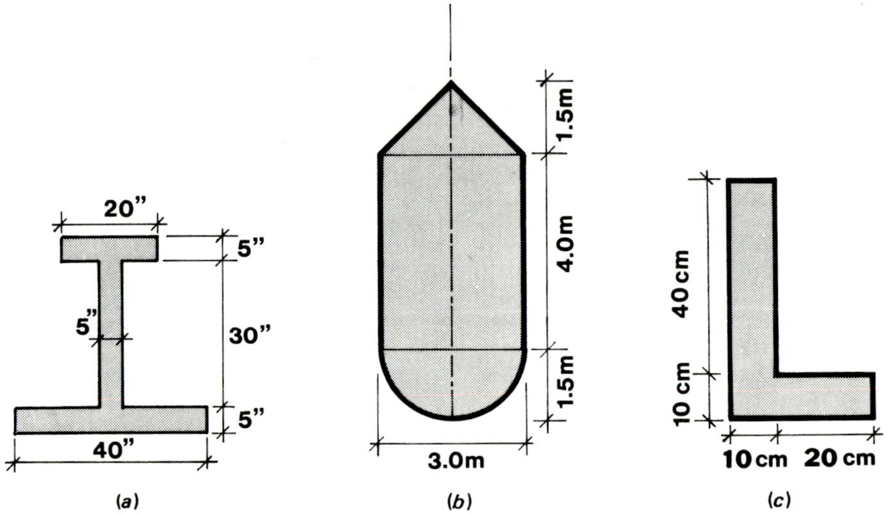

FIGURE 5-20

5-3 Determine the factor of safety against overturning of the retaining wall or dam shown in Fig. 5-21. The unit weight of concrete is 150 lb/ft^3. The horizontal resultant of the soil pressure is H = 15 260 lb, and the vertical resultant is V = 10 190 lb.

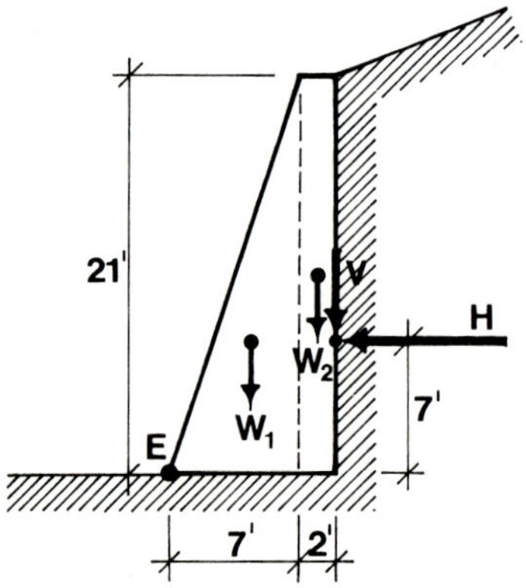

FIGURE 5-21

Concrete beams supported by a girder (*Courtesy of the Prestressed Concrete Manufacturers Association of California, Inc.*).

Beams and Their Support Reactions

6

6.1 EQUILIBRIUM OF A BODY IN A PLANE

If the resultant of a force system acting on a body is zero, the body does not have cause to translate, but it still could rotate about a point. To prevent rotation of the body (about any point), the resultant moment of the force system about any point must be zero. Thus we arrive at the following *equilibrium conditions for a force system* acting on a body:

1. The resultant of a force system acting on a body must be equal to zero; in other words, the sum[†] of the forces is zero

$$\Sigma F = 0$$

2. The sum of the moments of all the forces of the system about any point must be zero

$$\Sigma M = 0$$

When we solved the problem of the motion of a boat under the action of the stream current and the propulsive force of its motor (p. 41), we found that the resulting displacement of the boat could be resolved into two component displacements, one caused by the stream current and the other by the push of the motor. The translational motion of a body in a plane can similarly be resolved into any two component displacements in the plane; for example, into horizontal and vertical components. The resultant displacement of the body shown in Fig. 6-1 from A to C can be achieved by pushing it first horizontally from A to B and then vertically from B to C. Similarly, any force in a plane can be resolved into two components acting in the same plane (Fig. 6-2), say, its horizontal and vertical components. Thus the first equilibrium condition can be resolved into two conditions:

[†]The Greek capital letter "sigma" (Σ) stands for "the summation of."

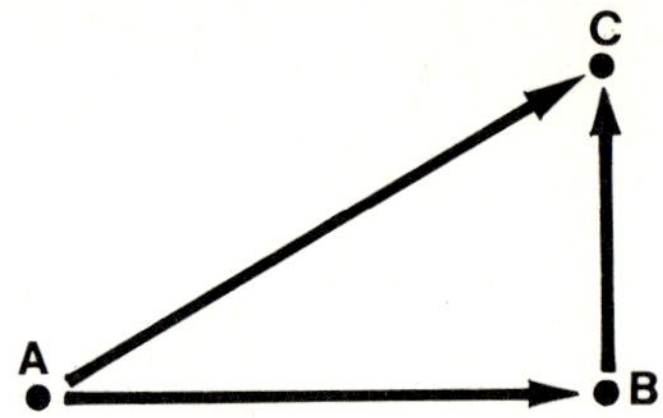

FIGURE 6-1 Displacement of a body from point A to point C.

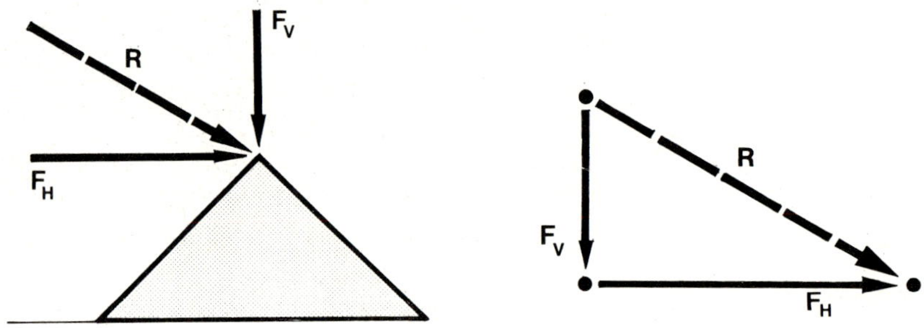

FIGURE 6-2 A force in a plane can be resolved into two components acting in the same plane.

1*a*. The sum of the horizontal components F_H of all the forces acting on a body in a plane must be zero, or

$$\Sigma F_H = 0$$

1*b*. The sum of the vertical components F_V of all the forces acting on a body in a plane must be zero, or

$$\Sigma F_V = 0$$

Conditions 1*a*, 1*b*, and 2 are the three equilibrium conditions for a force system acting in a plane. The first two are *force equilibrium conditions* and guarantee that the body will not move horizontally or vertically, and hence will not translate. The third is the *moment equilibrium condition*, which guarantees that the body will not rotate.

6.2 BEAM SUPPORTS

Structural elements must be supported so as not to translate or rotate under the action of the loads they support. The supports, in turn, must be able to transmit the loads to the foundations and to the subgrade soil. Thus, at the points of support, support reactions act on the structural element and must be such as to assure its equilibrium under the given loads. We have learned that three equilibrium conditions exist for a coplanar force system consisting of the loads and the reactions acting on a plane structural element.

The *simply supported beam* is one of the most fundamental structural elements. It is a body supported at two points, whose supports guarantee that the beam will

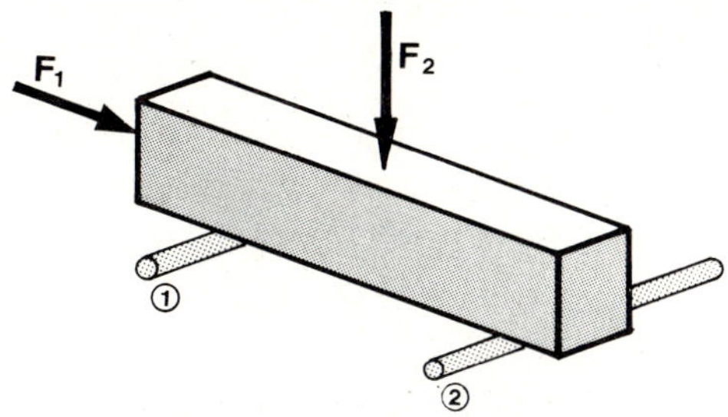

FIGURE 6-3 A beam supported on two rollers is not stable.

remain at rest under the effect of any loading system. If we place a beam on two rollers and apply a horizontal force to it (Fig. 6-3), it will start to slide horizontally as a result of this force. Supports which allow movement of a structure in a given direction are called *sliding* or *roller supports.* Obviously, no reaction can be developed by a sliding support in the direction in which it allows the body to slide. To prevent sliding of the beam in Fig. 6-3, a third support must be furnished, in addition to the two already provided (Fig. 6-4). This third support must be able to develop a horizontal reaction of such magnitude that equilibrium of all the horizontal forces acting on the beam is achieved. This reaction must be equal and opposite to the horizontal force F_1, so that

$$F_1 - R_c = 0$$

and

$$R_c = F_1$$

The third support (in our case the wall) must be strong enough to develop the reaction $R_c = F_1$.

Instead of furnishing a third sliding support, one of the two sliding supports can be modified to prevent sliding. This can be achieved by a *pin* or a *hinge* (Fig. 6-5).

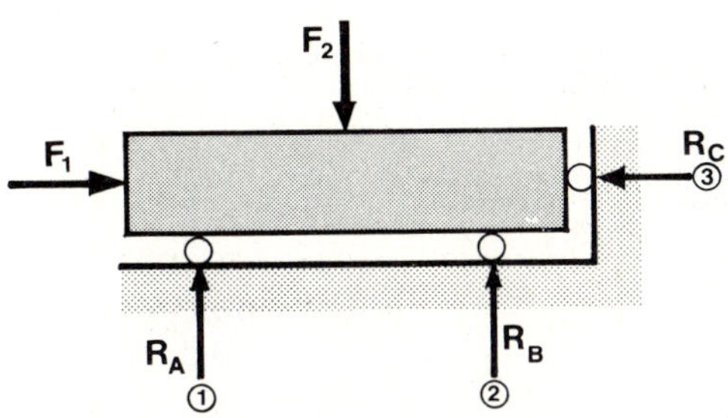

FIGURE 6-4 A simply supported beam in equilibrium under the action of a force system.

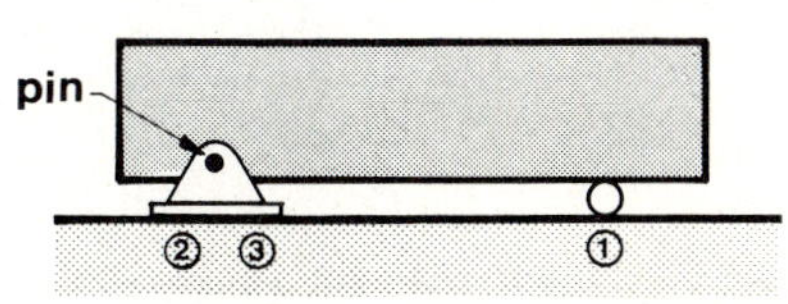

FIGURE 6-5 A simply supported beam with one hinge (pin) and one sliding support.

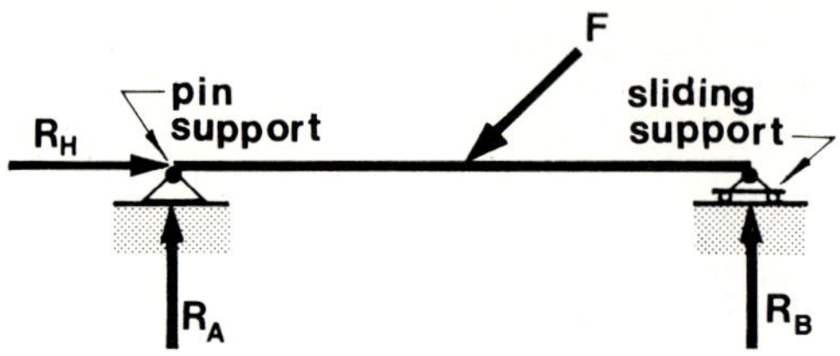

FIGURE 6-6 Schematic drawing of a simply supported beam.

This type of support, called a *pin* or *hinge support*, must develop the horizontal reaction needed to achieve horizontal equilibrium, but also a vertical reaction which, together with the vertical reaction of the sliding support, will equilibrate the vertical loads. A structural element with one hinge support and one roller support is called a *simply supported beam.* A simply supported beam is schematically shown in Fig. 6-6. It has one pin support, where both a horizontal *and* a vertical component of the reaction can develop, and one sliding or roller support capable of developing only a vertical reaction (perpendicular to the sliding plane).

The sliding support also has the important function of allowing the beam to change its length freely when its temperature changes. (The beam contracts when its temperature drops and expands when its temperature rises.) Moreover, the beam bends and deflects under the action of transverse loads; if its curved length is to remain equal to its straight length (as it does), the distance between its supports must decrease. This decrease is also made possible by the presence of the sliding support (Fig. 6-7). The sliding support allows the beam to "breathe" freely.

Sliding and pin supports may often be seen in steel bridges. Sliding supports may consist of steel rollers with bottom and top bearing plates. The top plate may be provided with a projection fitting into a notch in the beam *bearing plate* so that the beam end can rotate when the beam deflects (Fig. 6-8). A pin or hinge support prevents the beam from sliding but allows the beam end to rotate when the beam deflects (Fig. 6-9). In many engineering structures ideal sliding or pin supports cannot always

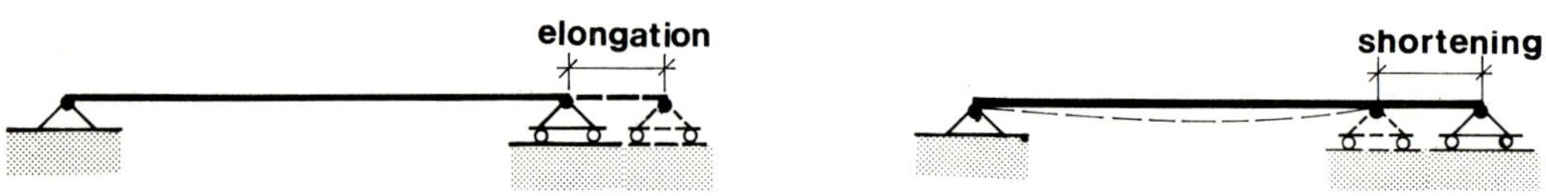

FIGURE 6-7 Elongation and shortening of a simply supported beam.

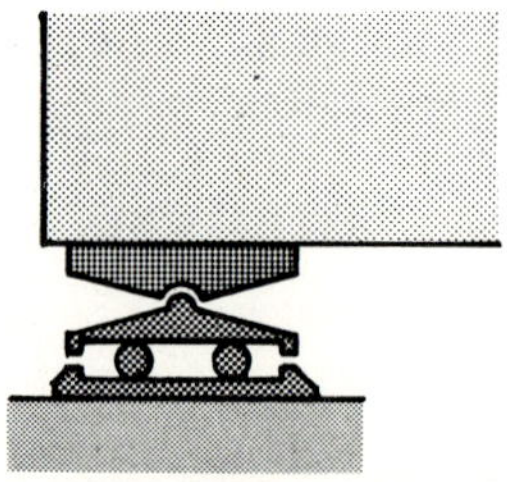

FIGURE 6-8 Sliding support of a beam.

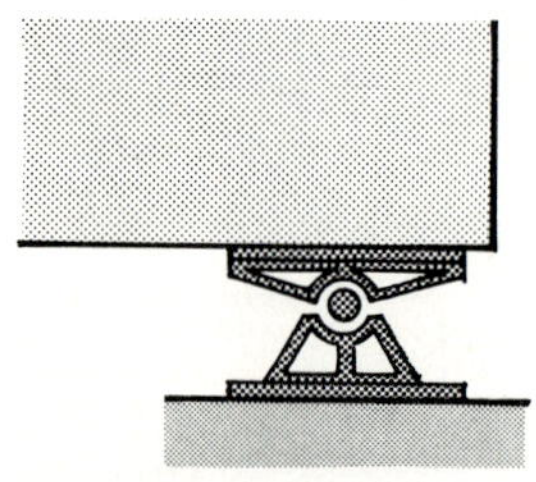

FIGURE 6-9 Hinge (pin) support of a beam.

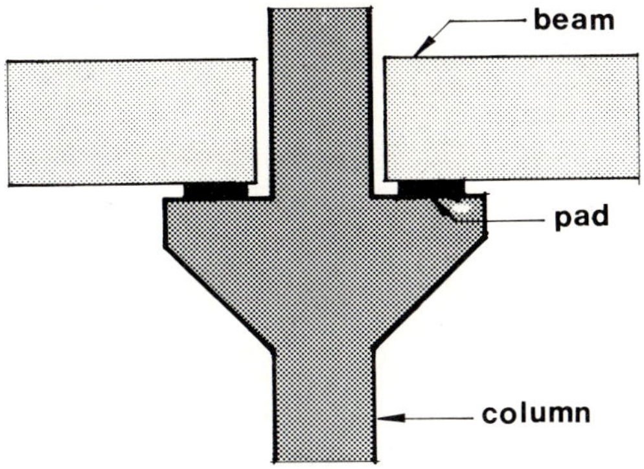

FIGURE 6-10 Precast concrete beams resting on the brackets of a column.

be assured. For example, a *precast concrete beam*† in a parking garage structure rests on neoprene pads set on brackets in the columns. The pad supports do not allow ideally frictionless sliding and rotational motion of the beam at the supports but develop a small amount of friction (Fig. 6-10). *Lintels*, which support masonry above the doors and windows, are often considered to act as simply supported beams, although their supports on masonry are far from "ideal" sliding or pin supports (Fig. 6-11). A similar situation occurs at the supports of floor beams set in the pockets of the *bearing walls* (Fig. 6-12).

The supports of simply supported beams may be placed at different points along the length of the beam. The basic type of simply supported beam has its supports at the ends (Fig. 6-13*a*). The beam shown in Fig. 6-13*b* is called a *simply supported beam* with one *overhanging end* or an *overhang*, while that in Fig. 6-13*c* has two overhangs.

A beam may have one end unsupported and the other built into a rigid support; this fixed or built-in support prevents the beam from sliding or rotating as a whole.

†Precast concrete beams are usually cast in forms at a precasting plant and then transported to the construction site and placed in the structure.

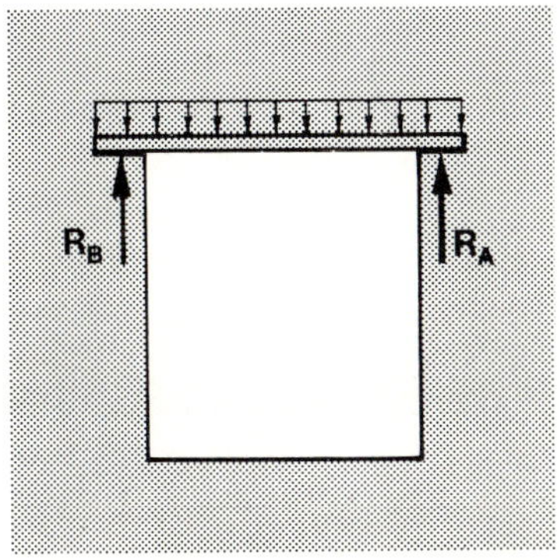

FIGURE 6-11 A lintel supporting masonry above the doors and windows.

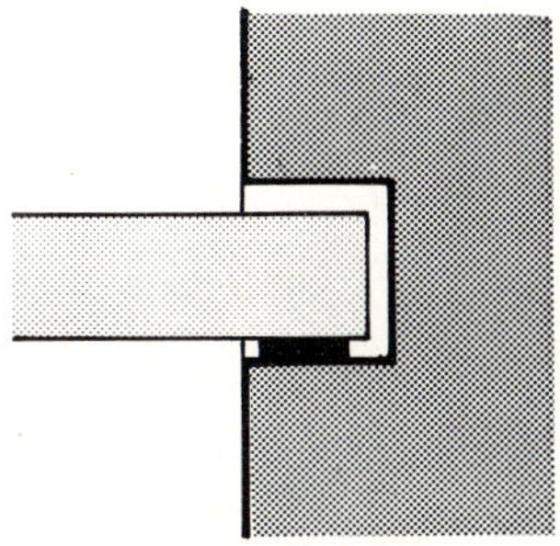

FIGURE 6-12 A beam set in the pocket of a bearing wall.

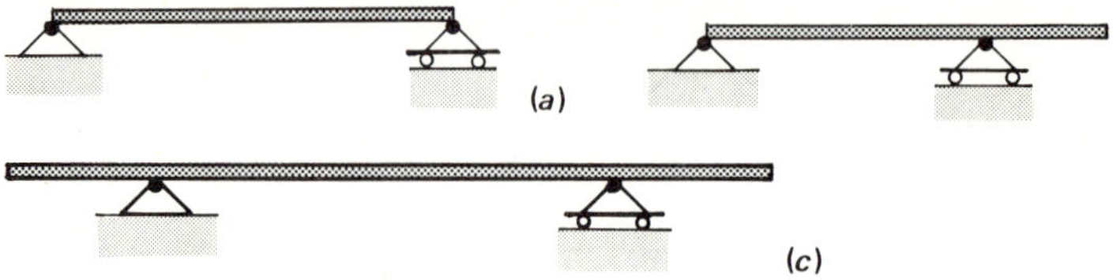

FIGURE 6-13 (*a*) The basic type of a simply supported beam; (*b*) a simply supported beam with one overhanging end; (*c*) a simply supported beam with two overhanging ends.

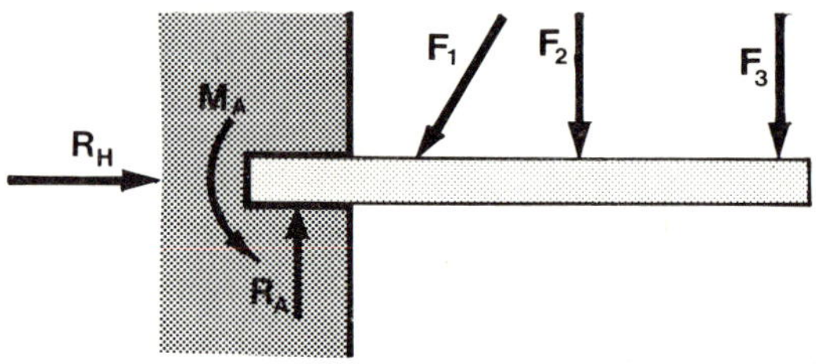

FIGURE 6-14 A cantilever is fixed at one end and free at the other end.

This type of beam is called a *cantilever.* If a cantilever beam were supported by a single pin rather than being fixed at one end, the resulting structure would not be stable; because of its own weight, the beam would rotate at the pin until its free end finds a "support" capable of developing the needed stabilizing reaction. If no support were found, the beam would hang vertically from the pin.

A fixed end resists horizontal and vertical forces as well as the moments of the lateral forces, which exert a rotational action at the fixed end (Fig. 6-14).

Beams may also be supported, for example, by two fixed supports, by two pin supports, or by one fixed and one pin support. In all these cases the beam end reactions cannot be evaluated by the three equations of equilibrium on p. 112. Such beams are briefly treated in Chap. 12.

6.3 BEAM REACTIONS

As stated in Sec. 6.1, three equilibrium conditions are available to evaluate the reactions of a beam acted on by a coplanar force system:

1. The sum of all horizontal forces acting on the beam (including the reactions) must be zero, or $\Sigma F_H = 0$. This condition guarantees that the beam as a whole will not move in a horizontal direction.
2. The sum of all vertical forces acting on the beam (including the reactions) must be zero, or $\Sigma F_V = 0$. This condition guarantees that the structure as a whole will not move in a vertical direction.
3. The sum of moments of all the forces acting on the beam (including the reactions) about any point in the plane must be zero, or $\Sigma M = 0$. This condition guarantees that the beam as a whole will not rotate about any point in the plane.

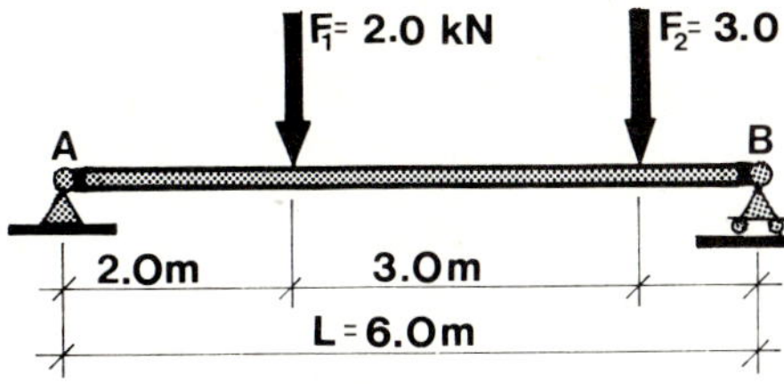

FIGURE 6-15

Note again that all the forces acting on the beam are assumed to act in the same (vertical) plane.

Let us use these three equilibrium conditions to evaluate the reactions of the simply supported beam shown in Fig. 6-15. The distance between its supports is called the *span* of the beam: our beam has a span $L = 6.0$ m (19.7 ft). It is loaded by two vertical forces, $F_1 = 2.0$ kN (0.45 kips) and $F_2 = 3.0$ kN (0.675 kips).

At the pin support A, two components of the total reaction can develop, the vertical component R_A and the horizontal component R_H, while only a vertical reaction, R_B, can develop at the sliding support B. The positive sense of these reactions (chosen arbitrarily) is shown in Fig. 6-16.

1. Let us use the first force equilibrium condition,

$$\Sigma F_H = 0$$

The only horizontal force acting on the beam is the reaction R_H. Since the sum of all the horizontal forces acting on the beam must be zero and the beam has no horizontal loads, the horizontal reaction is zero:

$$R_H = 0$$

2. Let us now use the second force condition of equilibrium $\Sigma F_V = 0$, considering upward forces as positive:

$$R_A - F_1 - F_2 + R_B = 0$$

$$R_A - 2 - 3 + R_B = 0$$

$$R_A + R_B - 5 = 0$$

or

$$R_A + R_B = 5 \text{ kN}$$

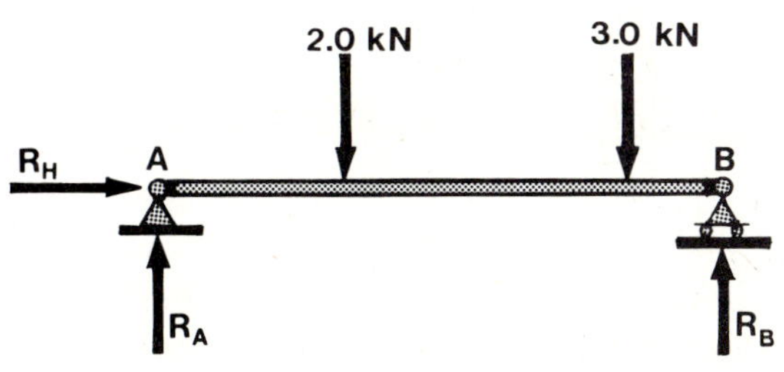

FIGURE 6-16 Reactions of a simply supported beam.

We are still unable to evaluate either R_A or R_B from this simple condition of equilibrium, since they are both unknown.

3. Therefore, we turn to the third, the moment condition of equilibrium, $\Sigma M = 0$. Write, for example, the moment equilibrium condition about the point B, thus eliminating R_B from the equation (since it has zero lever arm about the point B). Considering clockwise moments as positive, we obtain

$$R_A \times 6 - F_1 \times 4 - F_2 \times 1.0 = 0$$

$$6R_A - 2 \times 4 - 3 \times 1 = 0$$

$$6R_A - 8 - 3 = 0$$

$$6R_A - 11 = 0$$

from which

$$R_A = \tfrac{11}{6} \text{ kN}$$

The positive sign of the result suggests that our original assumption about the sense of R_A was correct. Finally, knowing that $R_A = \frac{11}{6}$ kN, we compute R_B from condition 2:

$$R_A + R_B = 5$$

$$\frac{11}{6} + R_B = 5$$

$$R_B = 5 - \frac{11}{6} = \frac{30 - 11}{6} = \frac{19}{6} \text{ kN}$$

Since moment equilibrium requires the resultant moment to be zero about any point in the plane, we can, alternatively, evaluate R_B directly by means of a moment equilibrium condition about A, eliminating R_A from the equation (since its lever arm about A is now zero):

$$-R_B \times 6 + F_1 \times 2 + F_2 \times 5 = 0$$

$$-6R_B + 2 \times 2 + 3 \times 5 = 0$$

$$-6R_B + 4 + 15 = 0$$

$$-6R_B + 19 = 0$$

$$-6R_B = -19$$

$$R_B = +\tfrac{19}{6} \text{ kN}$$

which checks the result obtained from condition 2.

Illustrative Problem 6-1

Find the reactions of a simply supported concrete beam (Fig. 6-17) with a span of 20 ft (6.10 m) due to its own weight. The beam has a rectangular cross

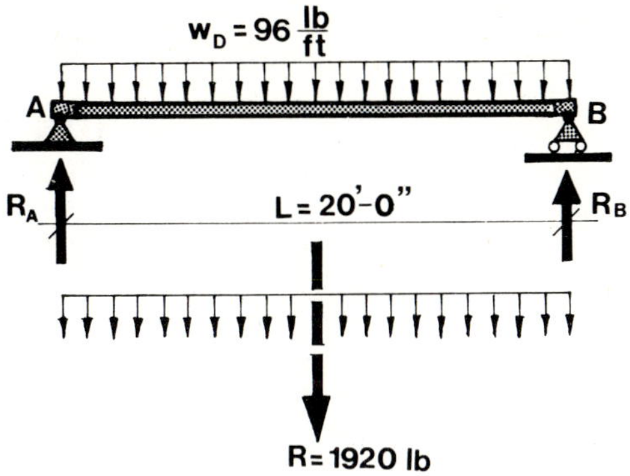

FIGURE 6-17 Reactions of the beam in Illustrative Prob. 6-1.

section 8 X 12 in (20 X 30 cm). Assume the unit weight of concrete to be $\gamma =$ 144 lb/ft³ (22.6 kN/m³).

SOLUTION The weight of 1 ft of beam is

$$w_D = \frac{8 \times 12}{144} \times 144 = 96 \frac{\text{lb}}{\text{ft}} \qquad \left(1.40 \frac{\text{kN}}{\text{m}}\right)$$

Since the dead load is uniformly distributed along the beam, the resultant R of this load is located at the centroid of the loading diagram, that is, at midspan of the beam. The resultant R is called a *statically* equivalent concentrated force, since it can replace the effect of the uniformly distributed load on the beam for the purposes of calculating the reactions. If the length of the loading diagram is L, then

$$R = w_D \times L = 96 \times 20 = 1\,920 \text{ lb} \qquad (8\,540 \text{ N})$$

Since the beam is loaded only by a vertical load, the horizontal reaction R_H in the pin support is zero. Considering both vertical reactions to act upward on the beam, the summation of all vertical forces gives

$$R_A + R_B - R = 0$$

$$R_A + R_B = R$$

or

$$R_A + R_B = 1\,920 \text{ lb}$$

Using a moment equilibrium condition about B, we have

$$R_A \times 20 - R \times 10 = 0$$

$$20 R_A = 10R$$

or

$$R_A = \frac{10R}{20} = \frac{R}{2} = \frac{1\,920}{2} = 960 \text{ lb} \qquad (4\,270 \text{ N})$$

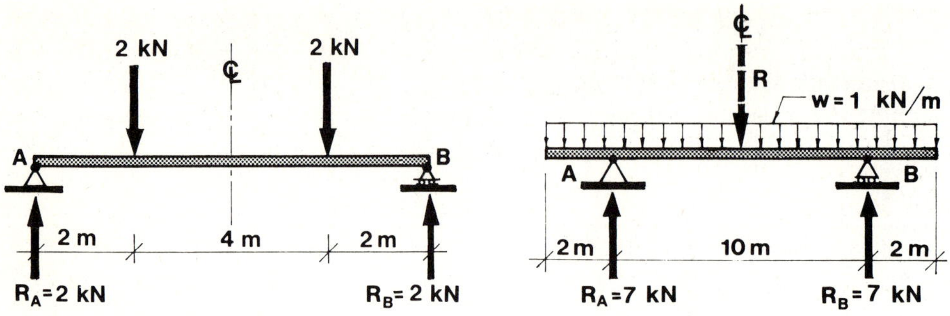

FIGURE 6-18 Reactions of symmetrical beams loaded by symmetrical loading.

Substituting $R_A = 960$ lb into the equilibrium equation for the vertical forces, we obtain

$$R_A + R_B = 1\,920$$

$$960 + R_B = 1\,920$$

$$R_B = 1\,920 - 960 = 960 \text{ lb}$$

This example shows how a geometrically symmetrical beam with symmetrical loading has symmetrical reactions, each equal to one-half the total load on the beam. This observation allows the reactions on the beams shown in Fig. 6-18 to be found directly without calculations.

Illustrative Problem 6-2

Find the reactions of a simply supported beam with one overhang, loaded by a uniformly distributed load $w = 200$ N/m over its entire length (Fig. 6-19).

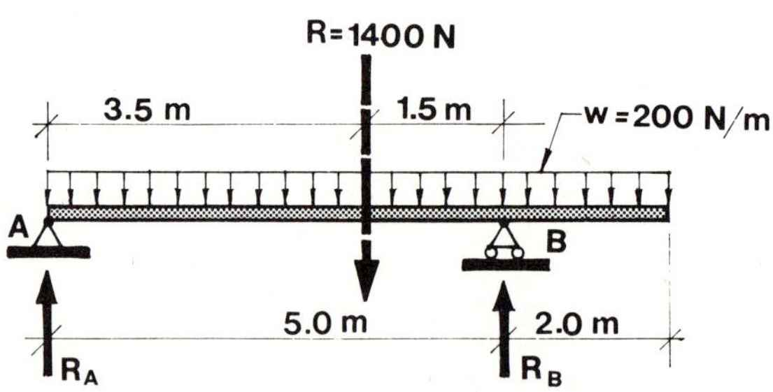

FIGURE 6-19 Reactions of the beam in Illustrative Prob. 6-2.

SOLUTION The equivalent concentrated force R is located at the midpoint of the loading diagram and has a magnitude

$$R = w \times 7 = 200 \times 7 = 1\,400 \text{ N}$$

Choose the positive sense of the reactions to be upward and write the equilibrium condition for the vertical forces:

A concrete beam with one overhanging end (*Photo by L. Cerny*).

$$R_A + R_B - R = 0$$

$$R_A + R_B - 1\,400 = 0$$

$$R_A + R_B = 1\,400 \text{ N}$$

The reaction R_A is obtained from a moment equilibrium condition about B:

$$R_A \times 5 - R \times 1.5 = 0$$

$$R_A \times 5 - 1\,400 \times 1.5 = 0$$

$$R_A \times 5 - 2\,100 = 0$$

$$R_A = \frac{2\,100}{5} = 420 \text{ N}$$

Similarly, R_B is found from a moment equilibrium condition about A:

$$-R_B \times 5 + R \times 3.5 = 0$$

$$-R_B \times 5 + 1\,400 \times 3.5 = 0$$

$$-R_B \times 5 + 4\,900 = 0$$

$$R_B = +\frac{4\,900}{5} = 980 \text{ N}$$

The equilibrium condition of the vertical forces may now be used as a check:

$$R_A + R_B = 1\,400 \text{ N}$$

$$420 + 980 = 1\,400$$

$$1\,400 = 1\,400$$

Illustrative Problem 6-3

Compute the reactions of a simply supported beam with one overhang (Fig. 6-20) loaded by a concentrated force $F = 1$ kN at its free end.

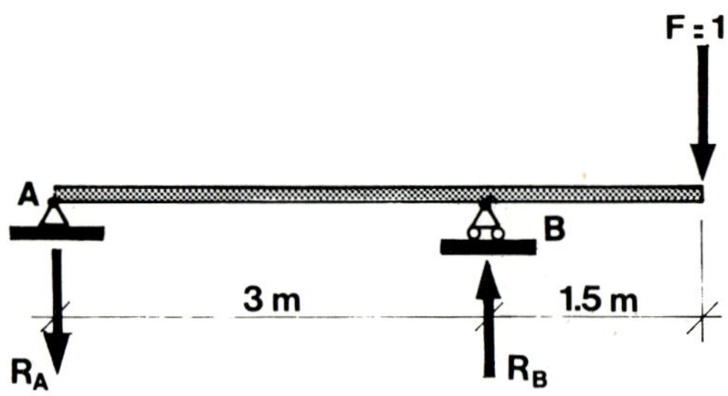

FIGURE 6-20 Reactions of the beam in Illustrative Prob. 6-3.

SOLUTION Choose the positive sense of the reaction R_A to be downward (since the reaction prevents the support A from lifting) and the positive sense of R_B to be upward. From the moment equilibrium condition about B we obtain

$$-R_A \times 3 + F \times 1.5 = 0$$

$$3R_A = 1.5$$

$$R_A = \frac{1.5}{3} = 0.5 \text{ kN}$$

Then, from the equilibrium condition for the vertical forces,

$$-R_A + R_B - F = 0$$

$$-0.5 + R_B - 1.0 = 0$$

$$R_B = +1.5 \text{ kN}$$

The senses of R_A and R_B were correctly chosen as shown by the positive values of the reactions.

The sense of the reaction R_A can be arbitrarily chosen upward. The moment equilibrium condition about B is now

$$R_A \times 3 + F \times 1.5 = 0$$

$$3R_A = -1.5$$

or

$$R_A = -\frac{1.5}{3} = -0.5 \text{ kN}$$

The negative sign of the result indicates that our assumed upward sense of R_A is not correct. Therefore, R_A is downward.

Illustrative Problem 6-4

Compute the reactions of the cantilever beam shown in Fig. 6-21 under the load shown.

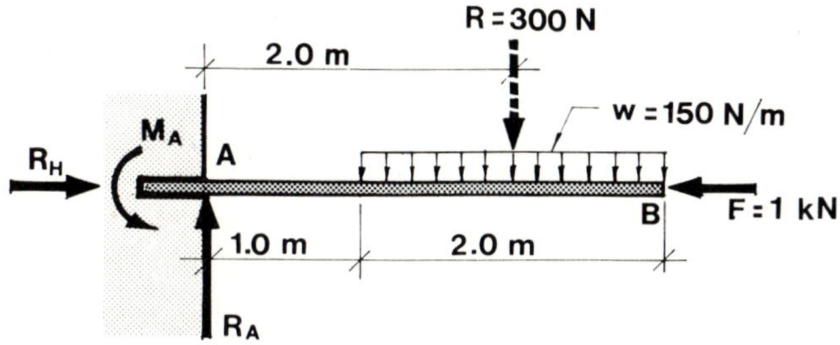

FIGURE 6-21 Reactions of the cantilever in Illustrative Prob. 6-4.

SOLUTION Choose the positive sense of the reactions as shown in Fig. 6-21 and notice that besides the horizontal reaction R_H and the vertical reaction R_A, a moment M_A must develop at the fixed support to obtain rotational equilibrium.

The equilibrium condition for the horizontal forces (assuming positive forces to act to the right) yields

$$R_H - 1.0 = 0$$

$$R_H = 1.0 \text{ kN}$$

The positive result confirms that the sense of R_H was correctly chosen.

The equilibrium condition for the vertical forces gives

$$R_A - R = 0$$

$$R_A - 150 \times 2 = 0$$

$$R_A = 300 \text{ N}$$

Finally, the moment M_A is found from a moment equilibrium condition about A

$$-M_A + R \times 2 = 0$$

$$-M_A + 300 \times 2 = 0$$

$$M_A = 600 \text{ Nm}$$

The positive sign of the result confirms the correctness of the sense chosen for the moment M_A.

In all the previous examples of simply supported and cantilever beams we had to evaluate a maximum of three unknown reaction components and had three static equilibrium conditions to do it with.

Structures in which all the reactions can be computed from static conditions of equilibrium are called *statically determinate structures*. More complicated systems can be built by means of simple statically determinate structural elements, which may be statically determinate. For example, the structure shown in Fig. 6-22*a* is composed of two cantilever beams and one simply supported beam resting on their free ends.

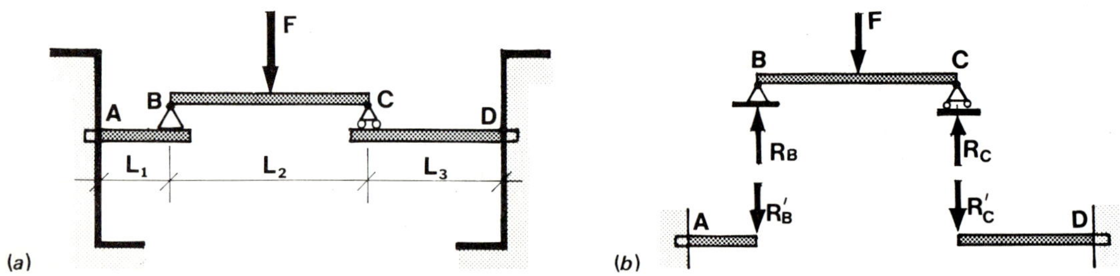

FIGURE 6-22 (*a*) Structural system consisting of simple statically determinate elements; (*b*) determination of reactions.

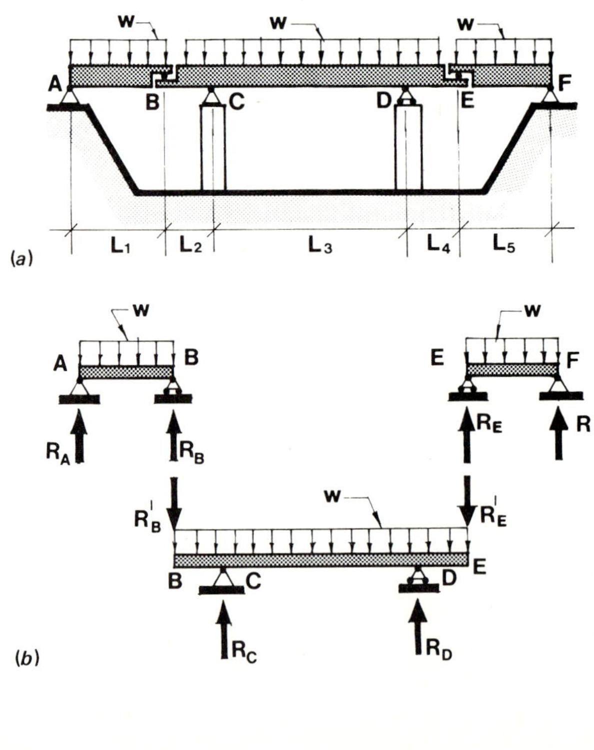

FIGURE 6-23 (*a*) Structural system consisting of simple statically determinate elements; (*b*) determination of reactions.

The solution of this problem is started by considering the simply supported beam as a separate structure or a free body as shown in Fig. 6-22*b*. Once the reactions R_B and R_C of the simply supported beam are found, they will be reversed to act as actions, R'_B and R'_C, on the cantilever end beams. Thus, instead of a complex statically determinate structure, three simple statically determinate elements must be analyzed.

Similarly, the structure shown in Fig. 6-23*a* can be resolved into one simply supported beam with two overhangs and two simply supported beams, each with one end supported on the abutment and the other on an overhang. To evaluate the reactions, we start with the side beams. After computing their reactions, R_A, R_B and R_E, R_F, we change the sense of R_B and R_E and let these forces, R'_B and R'_E, act at the ends of the beam overhangs (Fig. 6-23*b*). Only then can we compute the reactions R_C and R_D.

Some interesting structural systems can be developed by combining simply supported and cantilever beams; try to invent some of them. Such structural systems are important in *prefabricated construction*, in which a structure is assembled from components manufactured in a plant and transported to the site. Such prefabricated elements, usually made of concrete, are called *precast concrete elements.* Construction using such elements is called *precast concrete construction.*

Statically determinate structural elements do not have to be horizontal or straight. *Curved* simply supported or cantilever beams are often used to build aesthetically pleasing structures, as those shown in Fig. 6-24. Structures made out of straight members, like those shown in Fig. 6-25, are called *frames.* The points where two members of a frame meet are called *joints*. If the frames have *rigid joints*, where their members are "welded" together, they are called *rigid frames* (Fig. 6-26).

Such structures are *at times* statically determinate, and their reactions can be computed from the static conditions of equilibrium.

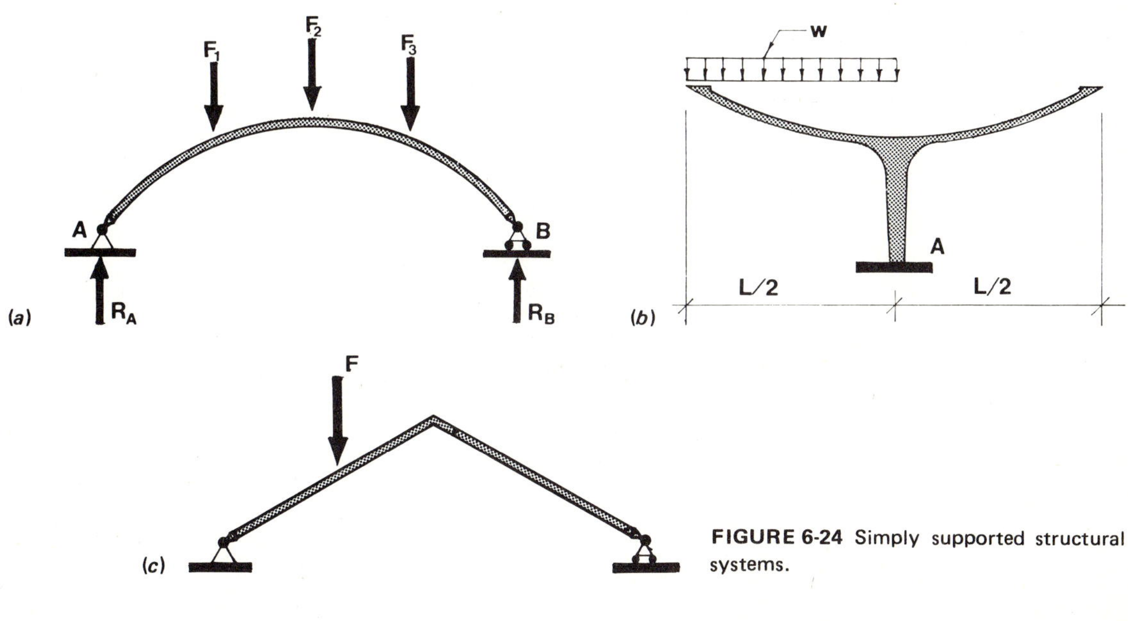

FIGURE 6-24 Simply supported structural systems.

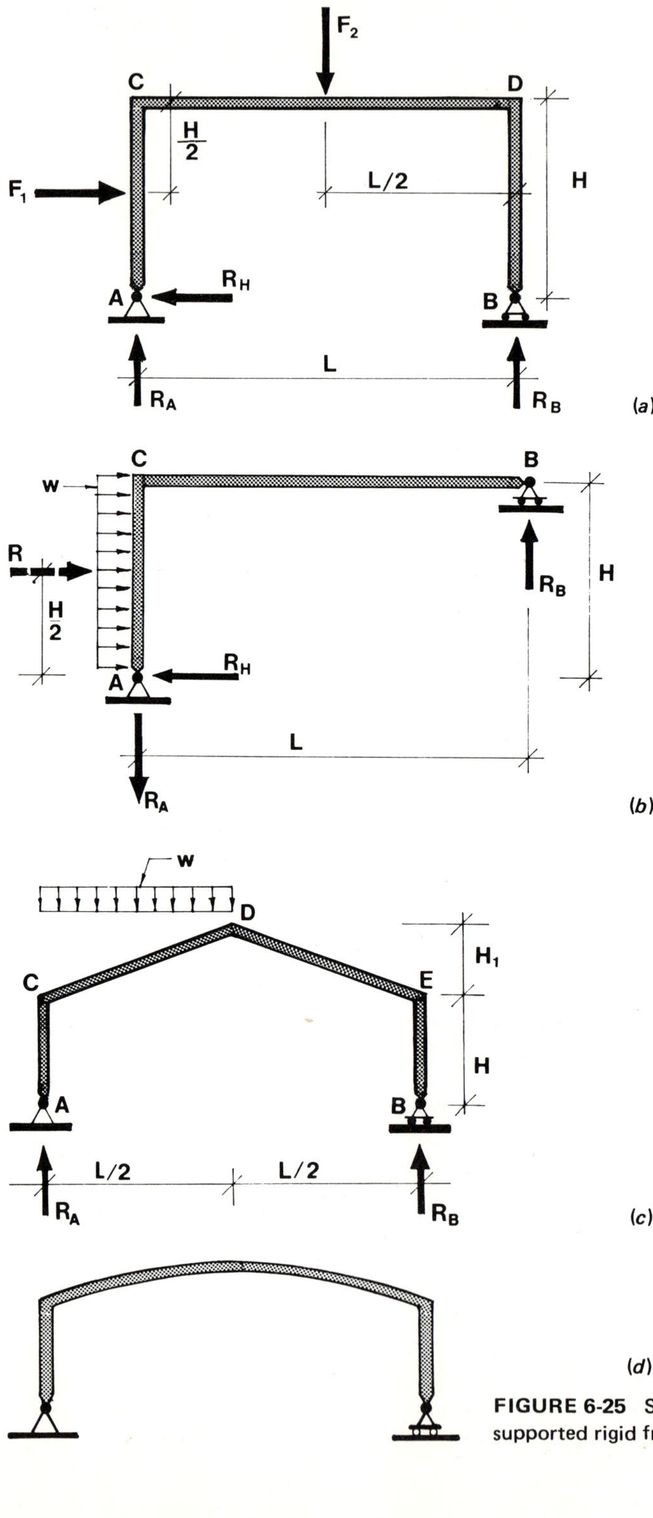

FIGURE 6-25 Statically determinate simply supported rigid frames.

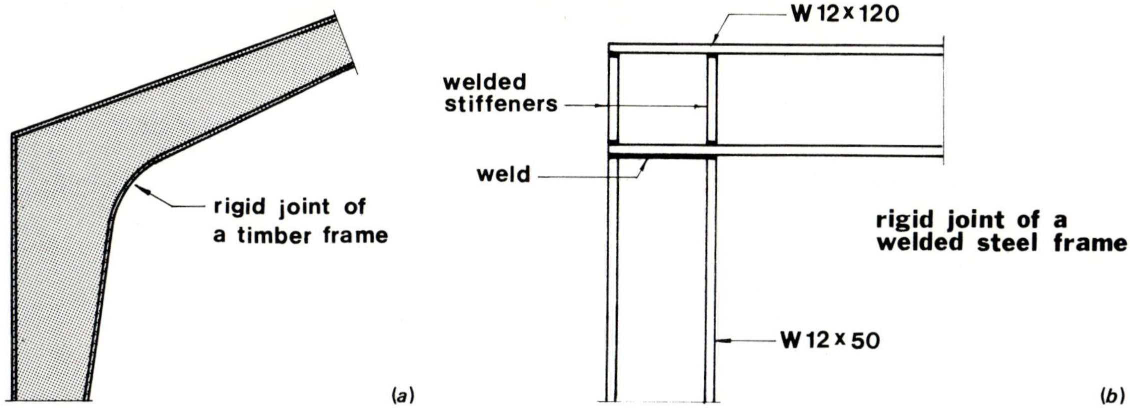

FIGURE 6-26 Rigid joints of rigid frames.

Illustrative Problem 6-5

Find the reactions of the structure shown in Fig. 6-23, which is loaded by a uniformly distributed load $w = 2$ kips/ft (29.2 kN/m). The structure is symmetrical and has spans $L_1 = L_5 = 40$ ft (12.2 m), $L_2 = L_4 = 20$ ft (6.1 m), and $L_3 = 100$ ft (30.5 m).

SOLUTION Solve first for the reactions of the simply supported side beams. These beams as well as their loads are symmetrical. Hence (Fig. 6-23*b*)

$$R_A = R_B = \frac{wL_1}{2} = \frac{2 \times 40}{2} = 40 \text{ kips} \qquad (178 \text{ kN})$$

Similarly

$$R_E = R_F = 40 \text{ kips}$$

Let us now proceed to the center beam with the overhangs, which are also symmetrical and symmetrically loaded. The two actions R'_B and R'_E must be applied at the overhangs. Thus the total load on the beam is

$$R'_B + R'_E + w(L_2 + L_3 + L_4) = 40 + 40 + 2\,(20 + 100 + 20)$$
$$= 80 + 280 = 360 \text{ kips}$$

This total load on the beam is reacted equally at the two supports; hence the reaction at each support is

$$R_C = R_D = \frac{360}{2} = 180 \text{ kips} \qquad (800 \text{ kN})$$

Illustrative Problem 6-6

Compute the reactions at the fixed end section A of the cantilever beam of the roof frame shown in Fig. 6-24*b*. The overhanging arms have a horizontal span

$L/2 = 12$ ft (3.66 m). The snow load, accumulated only on one side of the roof, has a magnitude $w = 1\,200$ lb/ft (17.5 kN/m) on the horizontal projection of the roof.

SOLUTION The three reaction components at the fixed end section A of the roof structure are R_A, R_H, and M_A (Fig. 6-27). From the equilibrium condition for the horizontal forces

$$R_H = 0$$

since there is no horizontal load on the structure.

The summation of the vertical forces gives

$$R_A - R = 0$$

where R is the equivalent concentrated force

$$R = w \times \frac{L}{2} = 1\,200 \times 12 = 14\,400 \text{ lb}$$

Thus

$$R_A = R = 14\,400 \text{ lb} \qquad (64 \text{ kN}) \qquad (\text{upward})$$

Finally, from the moment equilibrium condition about A, considering moments positive when clockwise, we obtain

$$+M_A - R \times 6 = 0$$

$$M_A = R \times 6 = 14\,400 \times 6 = 86\,400 \text{ lb·ft } (117 \text{ kN·m})$$

The assumed sense of M_A is correct since M_A is positive.

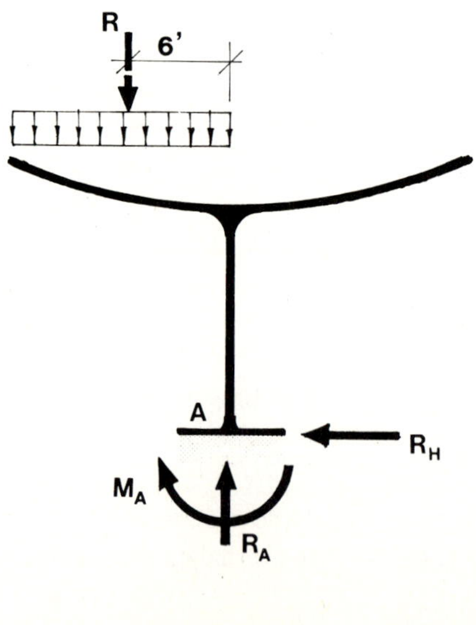

FIGURE 6-27

Illustrative Problem 6-7

Compute the reactions of the simply supported statically determinate frame shown in Fig. 6-25*b*, loaded by a horizontal wind load w = 800 lb/ft (11.7 kN/m). The span L = 80 ft (24.4 m) and the height H = 40 ft (12.2 m).

SOLUTION Assume the sense of the reactions as shown in Fig. 6-25*b*. The three static conditions of equilibrium give

1. $\Sigma F_H = 0$

$$w \times H - R_H = 0$$

$$R_H = w \times H = 800 \times 40 = 32\ 000 \text{ lb } (142.3 \text{ kN})$$

2. $\Sigma F_V = 0$

$$-R_A + R_B = 0$$

or

$$R_A = R_B$$

From this condition we still cannot determine either R_A or R_B.

3. $\Sigma M_A = 0$ (the sum of the moments of all the forces on the frame about A is zero)

$$-R_B \times L + R \times \frac{H}{2} = 0$$

where R is the concentrated force equivalent to the uniformly distributed wind load

$$R = w \times H$$

and the lever arm of R about A is $H/2$. Thus

$$-R_B \times L + w \times H \times \frac{H}{2} = 0$$

$$-R_B \times 80 + 800 \times 40 \times 20 = 0$$

$$R_B = \frac{800 \times 40 \times 20}{80} = 8\ 000 \text{ lb} \qquad (35.6 \text{ kN})$$

New Terms Appearing in Chapter 6

Translational effect of a force
Rotational effect of a force
Horizontal component of a force
Vertical component of a force
Three conditions of equilibrium
Force equilibrium condition
Moment equilibrium condition
Supports of beams
Simply supported beam
Roller support

Sliding support
Pin support
Hinge support
Precast concrete beam
Neoprene pads
Bearing wall
Beam with one overhang (simply supported)
Beam with two overhangs (simply supported)
Stable structure
Span of a beam
Statically equivalent concentrated force
Symmetrical beams
Symmetrical loading
Symmetrical reactions
Statically determinate structure
Prefabricated construction
Precast concrete construction
Curved beams (simply supported)
Bent beams (simply supported)
Frames (statically determinate)
Joints of frames
Rigid joints
Rigid frames

Important Structural Elements Appearing in Chapter 6

Simply supported beam. This is a structural element supported on one pin (hinge) support and one sliding (roller) support. It is a statically determinate structure, since its reactions can be computed from the conditions of equilibrium.

Curved simply supported beam. This is a beam with a curved shape.

Frame (statically determinate). This is a structural element whose length is broken into several straight or curved segments. The points where the segments meet or intersect are called *joints*. The frame is so supported that all the reactions can be computed from the conditions of equilibrium.

Lintel. This is a beam over a window or door opening.

Bearing wall. This carries loads from the roof and floors of a building. Walls which do not carry loads and serve only to separate spaces are called *partitions*.

Quiz Number 5

A. TRUE OR FALSE

1. Three equilibrium conditions can be written for a coplanar force system in equilibrium.
2. Three static equilibrium conditions are used to evaluate the reactions of simply supported and cantilever beams loaded by a coplanar force system.
3. A body in equilibrium does not slide; however, it may rotate as a whole about a point.
4. Only two roller supports must be provided to achieve stability for a beam under a general coplanar force system.
5. A pin (hinge) support can develop a reactive moment.
6. A pin support alone is sufficient to support a statically determinate beam.
7. With a drop in temperature, a simply supported beam contracts. Such a shortening can freely develop if one of its supports is a sliding support.

8. A simply supported beam deflects under load and its span tends to shorten.
9. The span of all simply supported beams is equal to the distance between their ends.
10. A pin support does not allow for any rotation of a beam at the point of support.
11. The fixed support of a cantilever assures its stability.
12. If a cantilever is loaded by a general force system acting in a vertical plane, only two reaction components at the fixed end have to be evaluated.
13. Bodies used as structures must be supported so as not to translate or rotate.
14. Lintel beams are supported by ideal pin and roller supports.
15. An almost ideal behavior of roller and pin supports is achieved in the bearing assemblies of steel bridge supports.
16. Neoprene pads used in the supports of precast concrete beams only approximate the ideal, frictionless behavior of sliding and hinge supports.
17. A symmetrical beam loaded by a symmetrical loading does not necessarily have symmetrical reactions.
18. Structural systems made up of statically determinate elements may be statically determinate.
19. The reactions of statically determinate structures can be evaluated by using only equilibrium equations.
20. A horizontal wind load on a statically determinate frame causes a horizontal reaction to develop at its pin support.
21. A statically determinate frame cannot resist a wind load, since it would slide in the direction of the wind.

B. MULTIPLE CHOICE

1. A simply supported beam loaded by a general coplanar force system has a maximum of ________ reaction components.
   (*a*) one (*b*) two (*c*) three (*d*) four
2. A pin support can develop ________ reaction components.
   (*a*) zero (*b*) one (*c*) two (*d*) three
3. A roller support can develop ________ reaction components.
   (*a*) zero (*b*) one (*c*) two (*d*) three
4. A fixed support can develop ________ reaction components.
   (*a*) zero (*b*) one (*c*) two (*d*) three
5. Which of these beams is not statically determinate?
   (*a*) a cantilever
   (*b*) a simply supported beam with one overhang
   (*c*) a beam whose two ends are fixed into two rigid supports
6. The support conditions in ________ are the closest to satisfying the ideal behaviors of a pin support and a roller support.
   (*a*) lintels
   (*b*) steel bridges
   (*c*) precast concrete beams on neoprene pads
   (*d*) timber beams in masonry bearing walls

7. Which of the following is not an equilibrium condition for a simply supported beam, loaded by a general coplanar force system consisting of forces F?
   (*a*) $\Sigma F = R$ (*b*) $\Sigma F_H = 0$ (*c*) $\Sigma F_V = 0$ (*d*) $\Sigma M = 0$
8. If a uniformly distributed load with an intensity of $w = 40$ lb/ft is applied over a length of 5 ft, its statically equivalent concentrated force is
   (*a*) 2 000 lb (*b*) 200 lb·in (*c*) 800 lb (*d*) 0.2 kips
9. If a wind load of 400 lb/ft acts on a frame 30 ft high, the statically equivalent concentrated load is
   (*a*) 12 kips (*b*) 12 kips/ft (*c*) 120 kips·in (*d*) 400/30 lb
10. A horizontal, simply supported beam is loaded by a vertical concentrated force $F = 100$ kips at midspan. Which of the following statements does not apply to the beam reactions?
    (*a*) both are equal
    (*b*) symmetrical
    (*c*) both are equal to 50 kips
    (*d*) they are not equal

C. SHORT ANSWERS

1. A force has a ____________________ and a ________________________ effect on a body.
2. A body, under the effect of load, _______________ and _______________.
3. There are ________ equilibrium conditions for the evaluation of the reactions in a simply supported beam loaded by a general coplanar force system.
4. If a simply supported beam is loaded by an inclined force, whose horizontal component is equal to 10 kips, the horizontal component of the reaction at the pin support is R_H = ________ kips.
5. A ________ support is capable of developing two reaction components. It allows for a rotation of the beam at the point of support.
6. A ____________________________ support can develop only one reaction component, which acts at right angles to the direction of sliding.
7. A __________ support is capable of developing three reaction components.
8. A concrete beam, fabricated in a plant and shipped to the construction site, is called a ____________ concrete beam.
9. Beams whose reactions can be evaluated by means of three equilibrium conditions are called ____________________________________ beams.
10. A lintel usually supports _________________________ over __________ and __________.

EXERCISES

6-1 Compute the reactions of the simply supported beam loaded as shown in Fig. 6-28, if $a = 12$ m, $b = 6$ m, $c = 6$ m, $w = 2$ kN/m, $F_1 = 15$ kN, and $F_2 = 30$ kN.

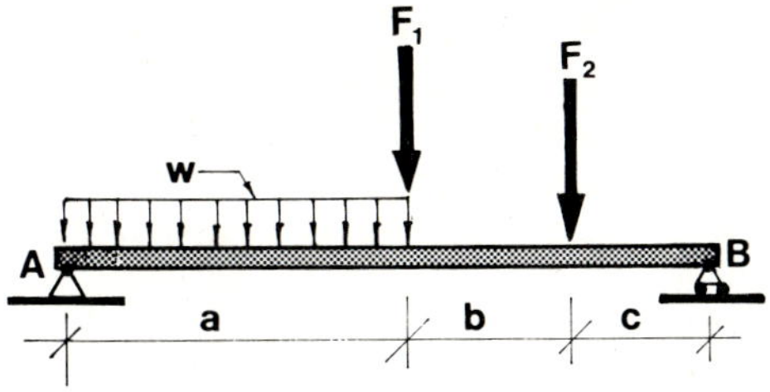

FIGURE 6-28

6-2 Compute the reactions R_A, R_H, and R_B of the beam system shown in Fig. 6-29, if $F_1 = 2$ kN, $F_2 = 3$ kN, $F_3 = 12$ kN, $a = 8$ m, $b = 4$ m, and $c = 2$ m.

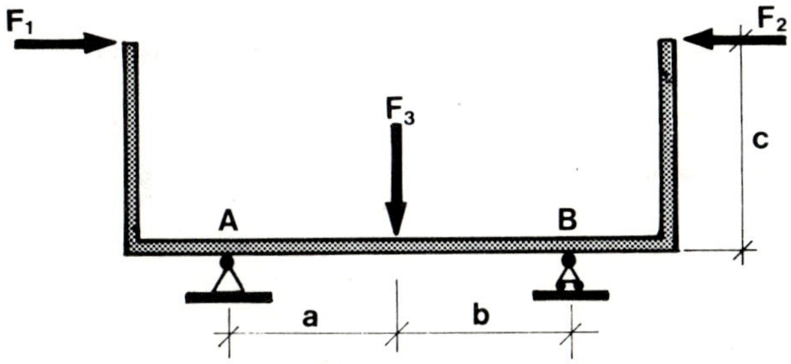

FIGURE 6-29

6-3 Compute the reactions R_A, R_H, and M_A of the cantilever shown in Fig. 6-30, if $F_1 = 100$ kN, $F_2 = 50$ kN, $F_3 = 60$ kN, $a = 4$ m, $b = 8$ m, and $c = 20$ m.

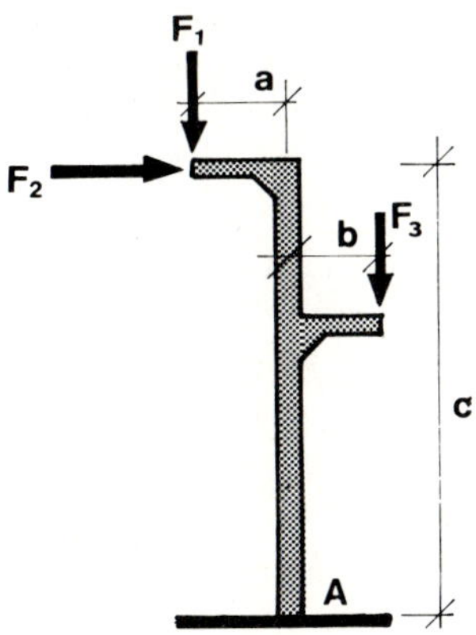

FIGURE 6-30

6-4 Compute the reactions at the fixed support of a horizontal cantilever with a span $L = 8$ m, loaded by a uniformly distributed load $w = 2$ kN/m over its entire length.

6-5 Compute the reactions R_A, R_H, and R_B of the frame in Fig. 6-25*a*, if $L = 80$ ft, $H = 30$ ft, $F_1 = 20$ kips, and $F_2 = 60$ kips.

6-6 Compute the reactions of the frame shown in Fig. 6-25*c*, if $L = 100$ ft, $H = 40$ ft, $H_1 = 10$ ft, and $w = 4$ kips/ft.

6-7 Compute all reactions of the structure shown in Fig. 6-23*a*, if only the beam *AB* is loaded by a uniformly distributed load $w = 10$ kips/ ft, while the rest of the structure (between *B* and *F*) is unloaded. Let $L_1 = L_5 = 40$ ft, $L_2 = L_4 = 20$ ft, and $L_3 = 80$ ft.

Roof truss system of the ice hockey arena in Bloomington, Minnesota (*Courtesy of Clark Engineering, Minneapolis*).

Internal Forces in Structures

7

7.1 INTRODUCTION

In transmitting loads to their supports, structural elements develop internal forces, such as the hemp rope used to lift the weight of a pail of water considered in Chap. 1 (p. 4).

The column shown in Fig. 7-1 is loaded at the top by an axial compressive force F, acting at the centroid of its cross section. It must transmit this force to the footing and to the subgrade soil. The reaction R_A of the footing on the column may be com-

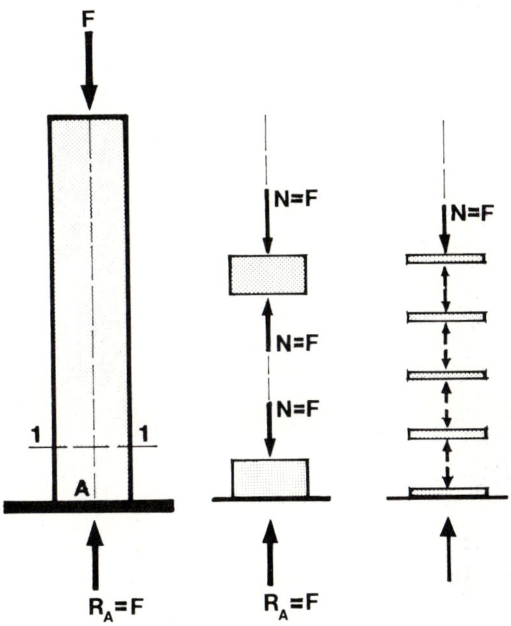

FIGURE 7-1 Internal forces in an axially loaded column.

puted from the equilibrium condition of the vertical forces acting on the column considered as a free body; neglecting the column weight, we obtain

$$R_A - F = 0$$

or

$$R_A = F$$

Let us now make a section 1-1 perpendicular to the column axis. Isolate by means of this section the bottom part of the column from the rest of the column and investigate its equilibrium considering this part as a free body. If only the reaction R_A acted on this free body, it would not be in equilibrium: it would start translating in an upward direction. Since the same condition of equilibrium must apply to the part of the column we have cut off as when it was an integral portion of the column, an internal force N must act on its top section. Such a force N represents the effect of the removed part of the column above section 1-1 on the free body, and for equilibrium of the vertical forces, it must be equal and opposite to the reaction R_A. Under the action of these two forces, the free body is compressed and a *compressive axial internal force* N develops in it. The same situation is found as we examine, as a free body, any other slice cut off from the column by two sections perpendicular to the column axis. Hence compressive axial internal forces $N = F$ exist in *all* cross sections (perpendicular to the column axis) along the length of the column. All parts are compressed by the compressive axial internal force

$$N = F$$

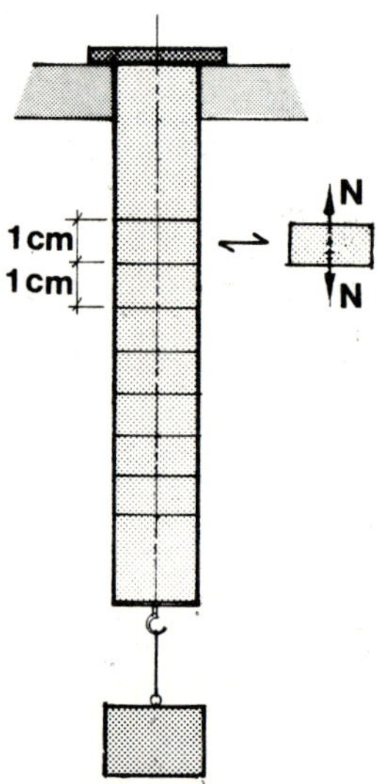

FIGURE 7-2 Internal forces in a hanger.

The presence of internal forces in a structural element can be visualized by experiment (Fig. 7-2). Take a piece of elastic material, say, a rubber band and draw on its surface a set of lines perpendicular to its longitudinal axis. Measure the distance between any two lines. Suspend the rubber band from a fixed point and apply a downward axial force F to its free end. Observe that the distance between the two lines

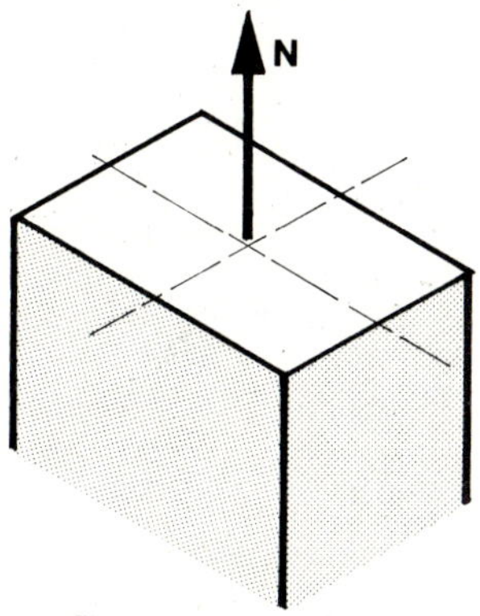

FIGURE 7-3 Tensile axial internal force.

increases. From this observation we can deduce that a tensile axial internal force N exists in the portion between the two lines (Fig. 7-3). If all the lines are equally spaced, the distances increase equally. Hence the axial internal force N is equal over the entire length of the element.

7.2 DETERMINATION OF AXIAL INTERNAL FORCES

If we isolate a free body from the rest of a structural element and introduce internal forces on its sections, all the forces, external and internal, acting on the free body must be in equilibrium. The internal forces at a *section* are then found by investigating the equilibrium of the free body isolated by that *section* from the rest of the element.

As an example, find the axial internal forces in a column of a two-story building which is loaded by a force $F_1 = 10$ kN from the roof and a force $F_2 = 15$ kN from the second floor (Fig. 7-4), neglecting the weight of the column. The reaction of the footing on the column is vertical and equal to

$$R_A = 10 + 15 = 25 \text{ kN} \qquad \text{(upward)}$$

Let us determine the axial internal force in sections 1-1 and 2-2. Isolate a free body by means of a section 1-1, as shown in the upper part of Fig. 7-4*b*. To keep the free body in equilibrium, an upward axial internal force N_1 must act on section 1-1 such that

$$F_1 - N_1 = 0$$

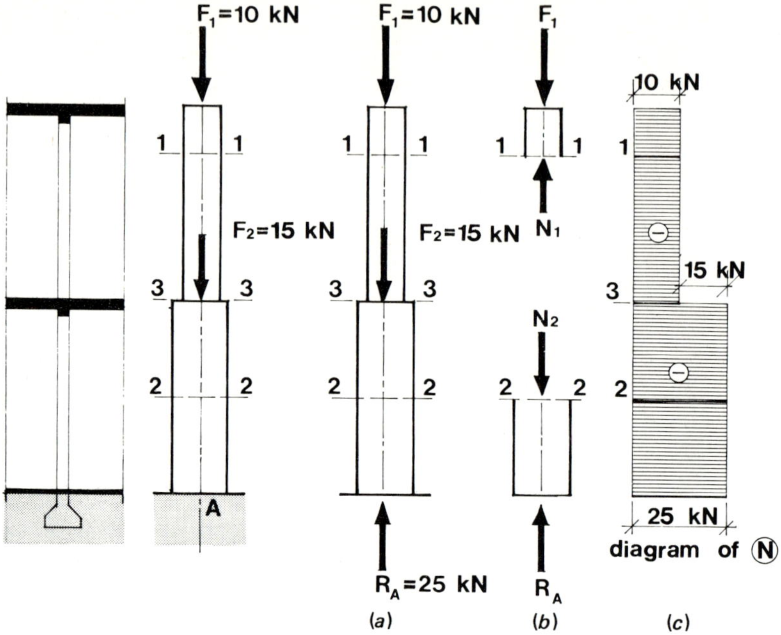

FIGURE 7-4 Internal forces in a column of a two-story building.

so that

$$N_1 = F_1 = 10 \text{ kN} \qquad \text{(upward)}$$

The axial force N_1 pushes on section 1-1; it is a compressive axial force. By this process, a compressive axial internal force $N = 10$ kN may be found to act on *all* the sections of the column in the upper story. It is present in all sections from the top down to section 3-3, where a force $F_2 = 15$ kN starts to act.

To find the axial force in section 2-2, isolate a free body by means of this section, as shown in the bottom part of Fig. 7-4*b*. The reaction $R_A = 25$ kN acts on the bottom section of the free body. For equilibrium of the vertical forces, the downward axial internal force N_2 must be such that

$$R_A - N_2 = 0$$

or

$$N_2 = R_A = 25 \text{ kN} \qquad \text{(downward)}$$

The axial force N_2 is again compressive and can be found to act on *all* the sections of the column in the lower story.

Compressive axial forces push on bodies, squeezing them, while tensile axial forces tend to elongate them. To indicate these actions, we conventionally use a minus (-) sign before the magnitude of a compressive axial force and a plus (+) sign before that of a tensile axial force (Fig. 7-5).

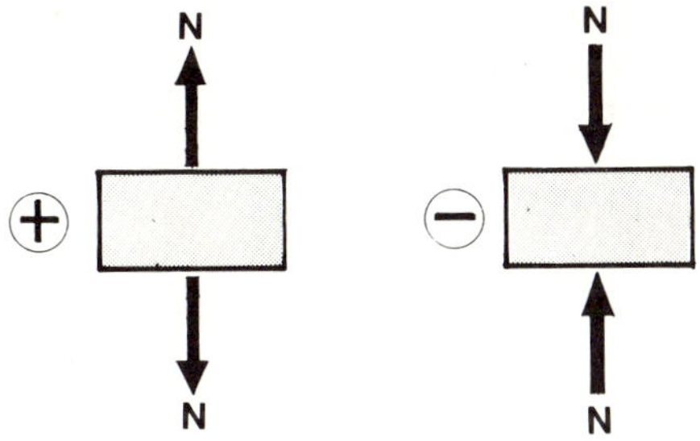

FIGURE 7-5 Tensile and compressive internal axial forces.

We may now plot a diagram of axial forces for both portions of the column, indicating their sign and, in a chosen scale, their magnitude in a direction perpendicular to the longitudinal axis of the column (Fig. 7-4*c*). Notice that in the section where the external force F_2 = 15 kN starts to act, the axial force changes abruptly from N_1 = 10 kN to N_2 = 25 kN. This change is exactly equal to the magnitude of the externally applied force F_2 = 15 kN.

Illustrative Problem 7-1

Some of the structural systems used to build high-rise construction feature a central concrete "core" tower (Fig. 7-6). A grid of beams with overhanging ends is built at the top of the "core." Hangers are then suspended from the overhanging beam ends all around the core to support the floor slabs. Determine the internal forces in one of such hangers pictured in Fig. 7-7 which supports only two floors at points M and N. Each hanger supports 10 m^2 of each floor, and each square meter of floor weighs 20 kN.

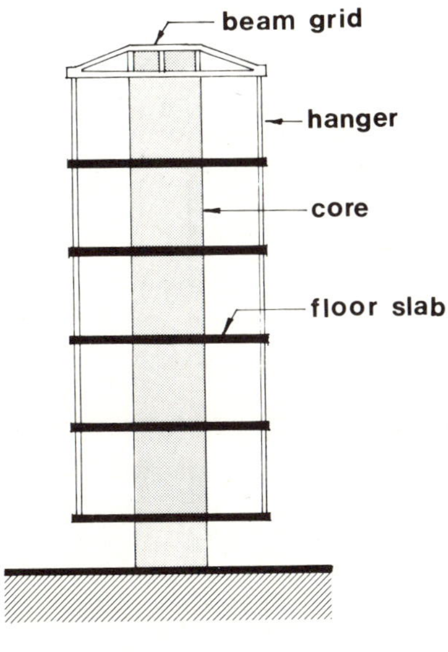

FIGURE 7-6 Building with floors suspended on hangers.

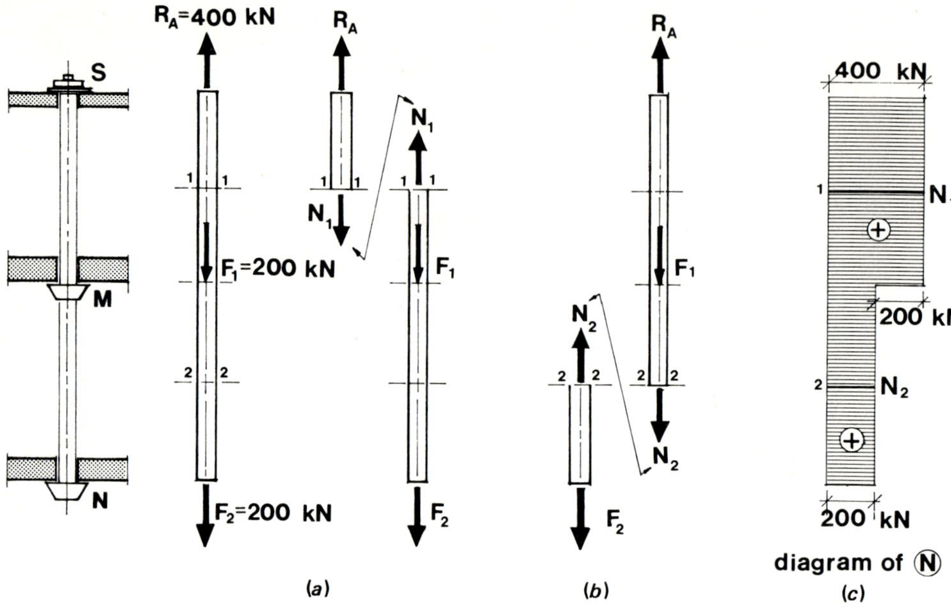

FIGURE 7-7 Internal forces in one hanger of the building in Fig. 7-6.

SOLUTION Each floor contributes a force of 10 × 20 = 200 kN to a hanger. The reaction R_A at the point of suspension S is found from the equilibrium condition of the vertical forces on the hanger. If we assume that R_A is an upward reaction and consider upward forces to be positive, we have (neglecting the small weight of the hanger)

$$R_A - F_1 - F_2 = 0$$

$$R_A = F_1 + F_2 = 200 + 200 = 400 \text{ kN}$$

The axial force N_1 at section 1-1 is found from the equilibrium of either of the free bodies in Fig. 7-7*a*. From the upper free body

$$R_A - N_1 = 0$$

$$N_1 = R_A = 400 \text{ kN} \qquad \text{(tension)}$$

where N_1 is a tensile axial (internal) force, since it pulls away from section 1-1; it elongates the hanger.

From the bottom free body in Fig. 7-7*a*, we obtain

$$N_1 - F_1 - F_2 = 0$$

and again

$$N_1 = F_1 + F_2 = 200 + 200 = 400 \text{ kN} \qquad \text{(tension)}$$

Building in Vancouver, B.C., with floors suspended on hangers fixed to the top of the core (*Photo by L. Cerny*).

The axial force N_2 at section 2-2 may be determined from either of the free bodies shown in Fig. 7-7*b*. From the upper free body we obtain

$$R_A - F_1 - N_2 = 0$$

and

$$N_2 = R_A - F_1 = 400 - 200 = 200 \text{ kN} \qquad \text{(tension)}$$

From the bottom free body we have

$$N_2 - F_2 = 0$$

$$N_2 = F_2 = 200 \text{ kN} \qquad \text{(tension)}$$

All the axial forces in the hanger are tensile. Their diagram is shown in Fig. 7-7*c*.

Note: When determining the internal forces in a structure, the points of application of the external forces (points M and N in the previous example) cannot be moved along the lines of action of the forces F_1 and F_2, since this would change the diagram of the internal forces along the length of the hanger, whose shape depends on the actual points of application of these forces.

7.3 AXIAL FORCES IN TRUSS MEMBERS

Trusses are structures consisting of relatively slender bars connected at *joints*, to form triangular panels. Trusses are made of steel, aluminum, timber, or concrete reinforced with steel bars (reinforced concrete). Trusses are usually so built that the loads acting on them are applied mainly at their joints (Fig. 7-8). A truss may have been developed from a solid beam by cutting large openings in it (Fig. 7-9). The structural action of

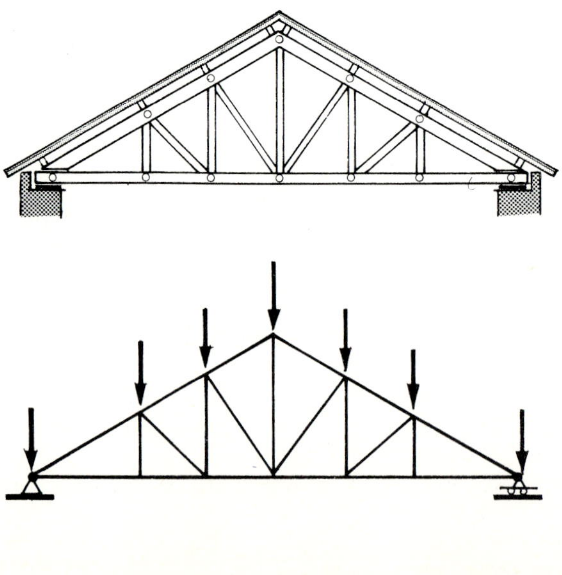

FIGURE 7-8 A roof truss with loads acting at the joints.

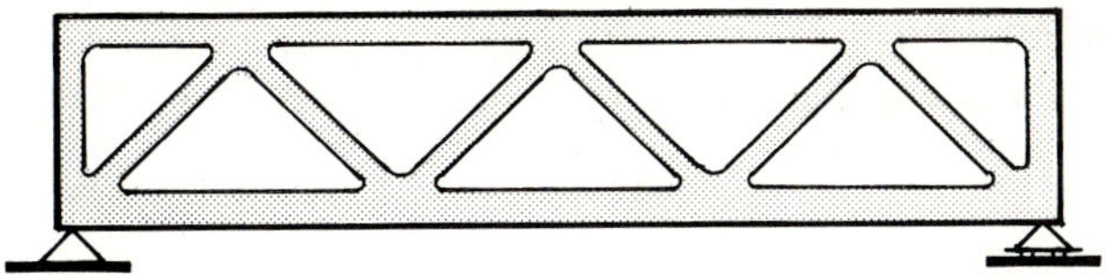

FIGURE 7-9

trusses is efficient, and a considerable amount of material may be saved by using trusses, particularly when large spans are involved.

A simplified but fairly accurate analysis of trusses can be obtained on the basis of two assumptions: (1) that the truss members are pin connected at the joints and (2) that all the loads act on the joints only. Under these assumptions only axial forces can develop in the truss members because each member is acted on by only one force from each endjoint, and for equilibrium of the member, when the small weight of the member is neglected, these two forces must be equal and opposite. Moreover, they must act along the axis of the member, or they would constitute a couple (Fig. 7-10*a*, *b*) and the member would not be in rotational equilibrium. To satisfy assumption 2 more exactly, even the weight of the truss members is approximated by two equivalent forces acting at the endjoints of the members.

In this chapter we will deal exclusively with simply supported, statically determinate trusses, all of whose members are contained in the same vertical plane. Since the same combination of external supports will be used as in simply supported beams, the external reactions of these trusses will be computed by the three static equilibrium conditions that we learned to use in Chap. 6.

Let us return, for the moment, to Illustrative Prob. 3-10 in Chap. 3, where we determined the forces at joint J of a roof truss. Since two of the four forces acting at the joint were known, the remaining two unknown forces were easily determined by means of a force polygon. Similarly, in the general analysis of trusses, we will always start our solution at a joint, where there are only two unknown forces. The following example will show that the knowledge acquired in the preceding chapters is sufficient to perform truss analysis.

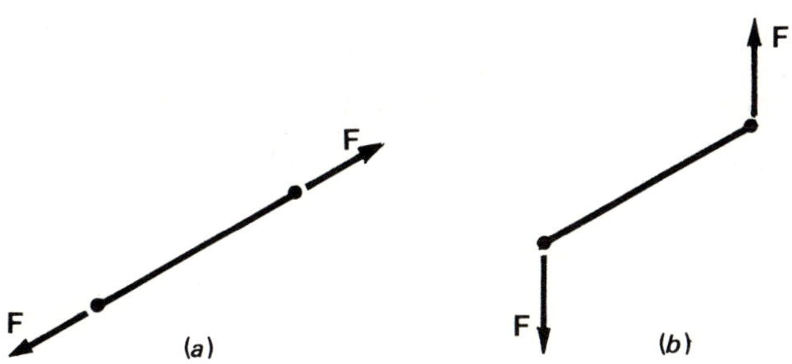

FIGURE 7-10

Illustrative Problem 7-2

Find the axial forces in all the members of the truss shown in Fig. 7-11. The truss consists of seven bars (1 to 7), which are connected at five joints (A to E).

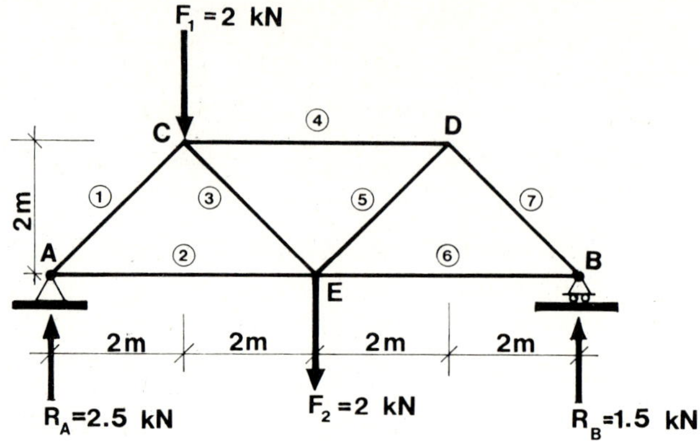

FIGURE 7-11 A diagram of the truss in Illustrative Prob. 7-2.

The entire truss is simply supported by a pin support at joint A and a roller support at joint B.

SOLUTION We must compute, first, the external reactions R_A and R_B of the truss, which must be in equilibrium with the external forces F_1 and F_2 acting at joints C and E, respectively. To determine the reaction R_A, we write a moment equilibrium condition about B, taking clockwise moments as positive,

$$R_A \times 8 - F_1 \times 6 - F_2 \times 4 = 0$$

$$R_A \times 8 - 2 \times 6 - 2 \times 4 = 0$$

$$8R_A - 12 - 8 = 0$$

$$8R_A = 20$$

so that

$$R_A = \tfrac{20}{8} = 2.5 \text{ kN}$$

Then, from an equilibrium condition for the vertical forces we obtain the reaction R_B, assuming again upward forces to be positive,

$$R_A + R_B - F_1 - F_2 = 0$$

so that

$$R_B = F_1 + F_2 - R_A = 2 + 2 - 2.5 = 1.5 \text{ kN}$$

Before determining internal axial forces, we must find a joint where only two unknown axial forces act. This cannot be joint C, since the only known force at this joint is the externally applied force $F_1 = 2$ kN, while the three axial forces in the

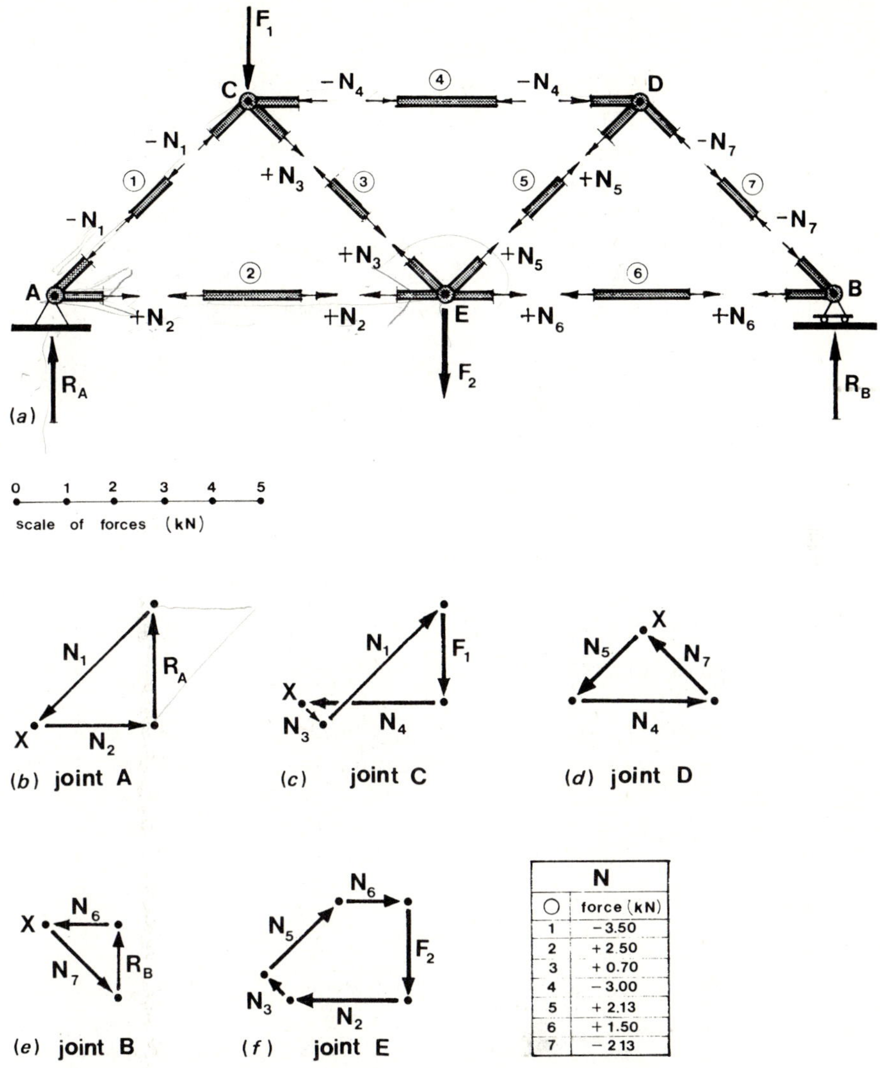

○	N force (kN)
1	−3.50
2	+2.50
3	+0.70
4	−3.00
5	+2.13
6	+1.50
7	−2 13

FIGURE 7-12 Free-body diagrams and equilibrium of truss joints.

members 1, 3, and 4 are unknown. However, we can start at joint A (Fig. 7-12*a*), where, besides the known reaction R_A, only two unknown axial forces, N_1 and N_2, act in members 1 and 2.

Consider the equilibrium of joint A, as a free body (Fig. 7-12*a*), under the action of three forces: the known reaction R_A and the two unknown axial forces N_1 and N_2, which are the actions of members 1 and 2 on joint A. We have already solved this problem a number of times by means of the force triangle (see Chap. 3). Using a convenient force scale, plot the vector R_A (Fig. 7-12*b*). From its endpoints, draw parallels to the lines of action of N_1 and N_2, that is, to the lines of members 1 and 2. Their intersection point X defines the magnitude of N_1 and N_2. For equilibrium of joint A, all arrows in the force triangle must follow the same direction (tip to tail), starting with the arrow of R_A which governs their flow. If you transfer the arrows from the force triangle (Fig. 7-12*b*) to the free-body diagram of joint A (Fig. 7-12*a*), you find that N_1 pushes into the joint while N_2 pulls away from it. Consequently, N_1 is a compressive (−) axial force and N_2 a tensile (+) axial force acting on joint A. Their magnitude is found from the chosen force scale. The forces N_1 and N_2 just determined are the actions of members 1 and 2 on joint A. According to the principle of action and reaction, the same forces N_1 and N_2, but with opposite sense, must be exerted from joint A on members 1 and 2. Thus member 1 is in compression as a result of the compressive axial force

$$N_1 = -3.50 \text{ kN}$$

and member 2 is in tension as a result of the tensile axial force

$$N_2 = +2.50 \text{ kN}$$

We may now proceed to joint C (Fig. 7-12*a*). Since there is a compressive axial force in member 1, this member exerts the same axial compressive force on joint C; this must be shown by the arrow of N_1 at joint C. The external force F_1 and the compressive force N_1 of member 1 are the known forces at joint C. The unknown forces are the actions N_3 and N_4 of members 3 and 4, respectively, on C. Let us find them by means of a separate force polygon for joint C (Fig. 7-12*c*), which we start with the known vector of N_1 (the arrow is that found for joint C). We then add the known vector of F_1. Through the tail of N_1, we draw the parallel to the line of action of N_3 and through the tip of F_1 the parallel to the line of action of N_4. The intersection point X of these two lines defines the magnitude of N_3 and N_4. All the arrows in the force polygon must run in the same direction (tip to tail) with the arrows of N_1 and F_1 governing their flow. By transferring the arrows of N_3 and N_4 from the force polygon (Fig. 7-12*c*) to the free-body diagram of joint C (Fig. 7-12*a*), we find that N_3 pulls away from the joint while N_4 pushes into it. Therefore, N_3 is a tensile and N_4 a compressive force. From the force scale, the axial force in member 3 is

$$N_3 = +0.7 \text{ kN} \qquad \text{(tension)}$$

and that in member 4 is

$$N_4 = -3.0 \text{ kN} \qquad \text{(compression)}$$

We can now proceed to joint D or joint E. Let us choose joint D (Fig. 7-12*a*). The compressive action from member 4 on joint D, $N_4 = -3.0$ kN, is a known force. Its arrow goes into joint D. The unknown forces at D are N_5 and N_7, the actions of members 5 and 7 on joint D. We find them from a force triangle (Fig. 7-12d), which we start with the vector of the known force N_4 (its arrow goes to the right). Through its endpoints we draw the parallels to the lines of action of N_5 and N_7. Their intersection point X defines the magnitudes of N_5 and N_7. Using the tip to tail rule, we find the arrows of N_5 and N_7 in the force triangle. When we transfer them to the free-body diagram of joint D, we find that N_5 pulls away from, while N_7 pushes into, joint D. Thus the axial forces in members 5 and 7 are

$$N_5 = +2.13 \text{ kN} \qquad \text{(tension)}$$

$$N_7 = -2.13 \text{ kN} \qquad \text{(compression)}$$

The last unknown axial force N_6 is found from the equilibrium of joint B (Fig. 7-12e) to be

$$N_6 = +1.5 \text{ kN} \qquad \text{(tension)}$$

Hinge support of the truss over the ice-hockey arena in Bloomington, Minnesota (*Courtesy of Clark Engineering, Minneapolis*).

The equilibrium of joint E can then be used as a check since all the forces acting on it are known. The force polygon of the forces N_2, N_3, N_5, N_6, and F_2 must close ($R = 0$) when all the arrows follow the same direction (Fig. 7-12*f*).

After all the axial forces in the truss have been determined, it is convenient to arrange the results in a table as shown in Fig. 7-12*g*.

The sequence from joint to joint used in this example was somewhat arbitrary. We could have used instead the sequence *A-C-E-D* with a final equilibrium check at joint B or the sequence *A-C-E-B* with the final check at joint D.

A great variety of truss types can be developed by connecting tensile and compressive bars in triangles and by properly supporting the resulting structure. Some types of commonly used trusses are shown in Fig. 7-13.

There are special cases of joints and their loading, where the axial force in some joint members can be found independently of the solution for the rest of the truss. One of such joints is joint A (or B) shown in Fig. 7-13*a*. Isolate joint A from the rest of the structure; its free-body diagram is shown in Fig. 7-14. Consider that one known force acting on the joint is the reaction R_A. The two unknown forces are the axial forces N_1 and N_2, the actions from members 1 and 2. From an equilibrium condition for the horizontal forces acting on the joint

$$N_2 = 0$$

since no other horizontal force or horizontal force component acts at joint A. From an equilibrium condition for the vertical forces, *assuming* N_1 *to push into the joint*, that is, to be compressive, we obtain

$$R_A - N_1 = 0$$

$$N_1 = R_A$$

The positive sign confirms the correctness of our sense assumption for the axial force N_1. It does not mean that the axial force N_1 is tensile; it is compressive because it pushes into joint A, as we have assumed. In tabulating the axial forces, N_1 would then be recorded with a negative sign, thus indicating compression in the member. The equation from which we obtained N_1 expresses, again, the principle of action and reaction: The two forces, N_1 and R_A, which have a common line of action, can be in equilibrium only if they are equal in magnitude and of opposite sense.

Similarly, let us investigate the equilibrium of joint I in the truss of Fig. 7-13*a*. Its free-body diagram is shown in Fig. 7-15. From an equilibrium condition for the vertical forces we find that

$$N_9 = F$$

where N_9 is a compressive force, since it pushes into the joint. From an equilibrium condition for the horizontal forces we obtain

$$N_8 = N_{10}$$

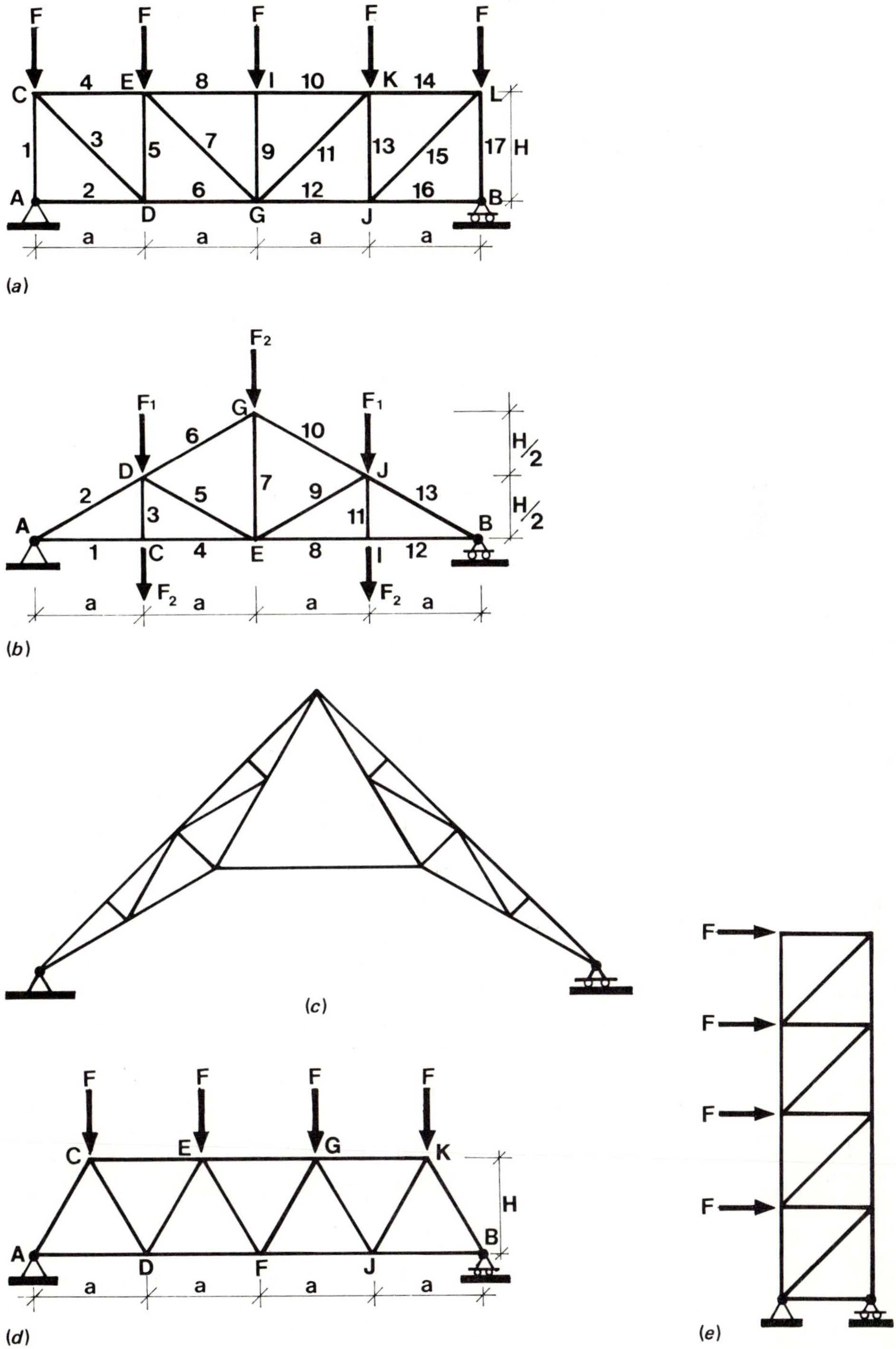

F
F
F
F
F
C
4
E
8
I
10
K
14
L
1
3
5
7
9
11
13
15
17
H
A
2
D
6
G
12
J
16
B
a
a
a
a
(a)
F2
F1
G
F1
6
10
D
J
2
5
7
9
13
H/2
H/2
3
11
A
B
1
C
4
E
8
I
12
F2
F2
a
a
a
a
(b)
(c)
F
F
F
F
C
E
G
K
H
A
D
F
J
B
a
a
a
a
(d)
F
F
F
F
(e)

FIGURE 7-13

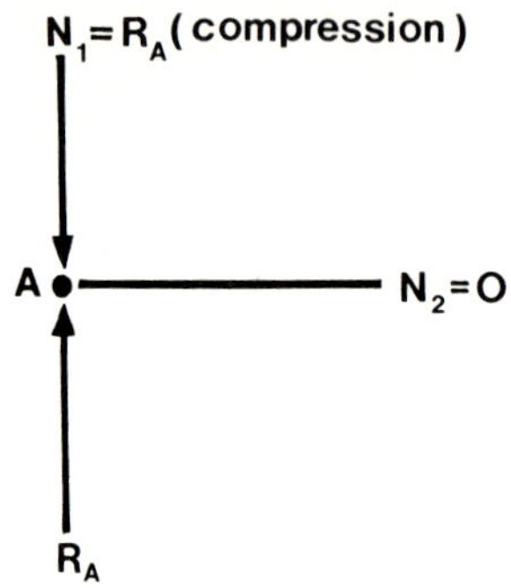

FIGURE 7-14

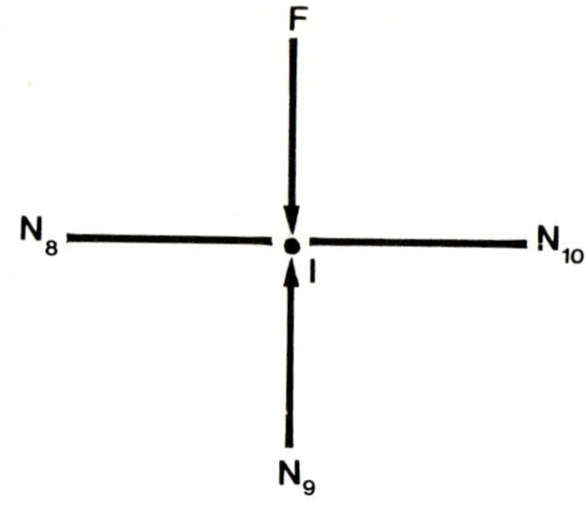

FIGURE 7-15

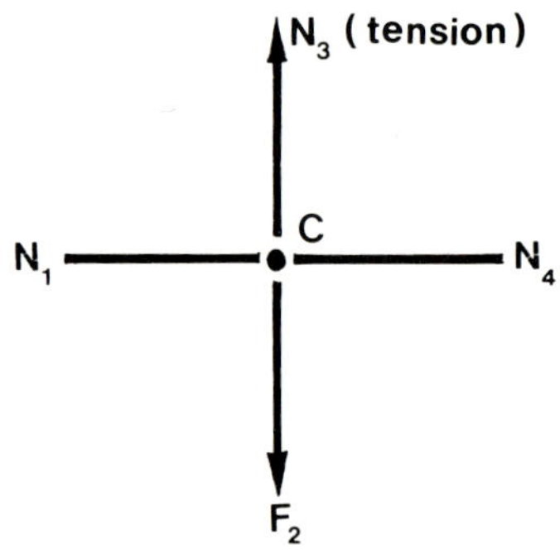

FIGURE 7-16

This equation holds even if this symmetrical truss were not symmetrically loaded, but in symmetrical trusses with symmetrically located loads, symmetrical members must have identical axial forces. Thus, for the truss shown in Fig. 7-13*a*

$$N_1 = N_{17} \qquad N_4 = N_{14} \qquad N_3 = N_{15}$$

and so on.

A similar situation appears at joints C and I of the truss shown in Fig. 7-13*b*. From an equilibrium condition for the vertical forces at joint C (Fig. 7-16) we obtain

$$N_3 = F_2$$

where N_3 is a tensile axial force, since it pulls away from the joint. From the equilibrium of the horizontal forces at C

$$N_1 = N_4$$

7.4 INTERNAL FORCES IN BEAMS

In Sec. 7.3 we investigated axial forces in truss members. We found that tensile and compressive axial forces develop in truss members as a result of the loads acting on the truss. Because of axial forces, truss members either elongate or shorten.

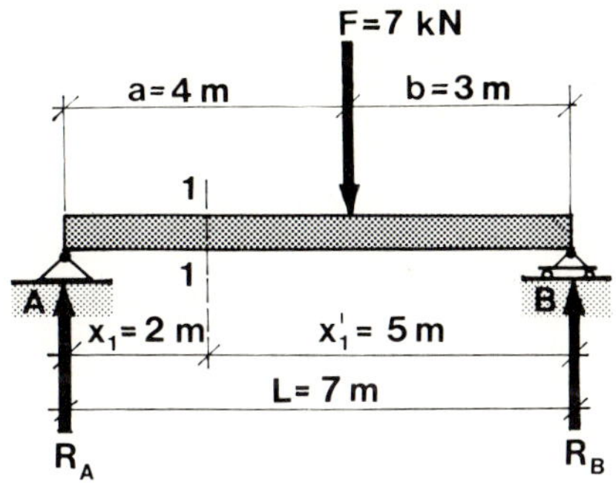

FIGURE 7-17

Let us now investigate what internal forces develop in beams. Under a general coplanar force system, a beam may elongate or shorten, but it also bends. To determine the internal forces that produce such deformations in beams, we will use the same general principles used in the analysis of trusses. The internal forces at a cross section of a beam are found by using this section to isolate a free body from the rest of the beam and by investigating its equilibrium. The cross sections are cut perpendicular to the beam's longitudinal axis.

Let us determine the internal forces at section 1-1, located 2.0 m from the left support of the simply supported beam shown in Fig. 7-17. The beam has a span of 7.0 m and is loaded by a vertical concentrated force $F = 7$ kN at a point 3.0 m from the right support. The dead load of the beam is assumed to be negligible.

The beam reactions must be computed first. From the moment equilibrium equation about B and the equilibrium equation of vertical forces, we obtain

$$R_A = 3 \text{ kN} \qquad R_B = 4 \text{ kN}$$

Let us now isolate the part of the beam to the left of section 1-1 and investigate the equilibrium of the free body thus created (Fig. 7-18*a*).

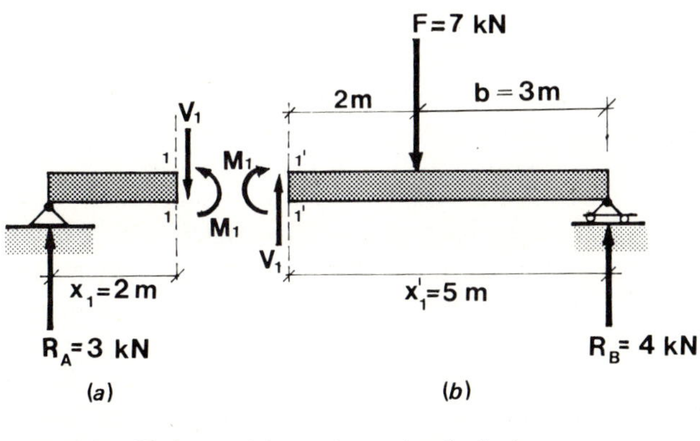

FIGURE 7-18 Internal forces in section 1 of a beam.

1. From the equilibrium of the horizontal forces, we find by inspection that no axial internal force can develop in this free body, since no horizontal forces act on it. Therefore, the axial force at section 1-1

$$N_1 = 0$$

2. Further, it is evident that if there were no vertical internal force acting on section 1-1, the reaction R_A would not be equilibrated. To satisfy the equilibrium condition for the vertical forces acting on the free body, a vertical (transverse) force V_1 must act at section 1-1. If its sense is assumed downward,

$$R_A - V_1 = 0$$

$$V_1 = R_A = 3 \text{ kN}$$

The positive sign of V_1 shows that, as assumed, V_1 acts downward on the free body at section 1-1.

3. Considering now the free body shown in Fig. 7-18*a*, we realize that the two equal, parallel and opposite forces V_1 and R_A, a distance $x_1 = 2.0$ m apart, constitute a clockwise couple. The moment of this couple is

$$M = R_A x_1 = V_1 x_1 = 3 \times 2 = 6 \text{ kN}\cdot\text{m}$$

If there were no equilibrating moment acting on section 1-1, the free body would start to rotate clockwise because of the unbalanced moment M of the couple. Therefore, an internal moment M_1 must develop at section 1-1, capable of equilibrating the moment $M = 6.0$ kN·m. Assuming the sense of the unknown internal moment M_1 to be counterclockwise and writing a moment equilibrium condition about any point of section 1-1, we obtain

$$-M_1 + R_A x_1 = 0$$

$$M_1 = +R_A x_1 = 3 \text{ kN} \times 2 \text{ m} = +6.0 \text{ kN}\cdot\text{m}$$

The positive sign of the result confirms that our assumption about the sense of the moment M_1 was correct.

M **V**

When the internal force V at a cross section of a beam is perpendicular to the longitudinal axis of the beam, it is called the *shear force* at that section.

The internal moment M at a cross section of a beam is called the *bending moment* at that section.

Thus, under the action of a general coplanar force system, an *axial force*, a *shear force*, and a *bending moment* may develop at a cross section of a beam. All these three internal quantities will generally be called the *internal forces* at the section, although one of them is a bending moment. The problem of finding the internal forces in a beam consists of determining axial forces, shear forces, and bending moments at different cross sections of the beam.

Let us now investigate the equilibrium of the free-body diagram in Fig. 7-18*b*, which consists of the right portion of the beam, separated by section 1′-1′ from the rest of the beam. Sections 1-1 and 1′-1′ are actually the left and right faces of the same section or cut. By the principle of action and reaction, the internal forces on section 1′-1′ must be equal and opposite to those on section 1-1.

1. From the equilibrium condition for the horizontal forces, again

$$N_1 = 0$$

2. From the equilibrium condition for the vertical forces (do not forget that the force $F = 7$ kN acts on this free body)

$$R_B + V_1 - F = 0$$

$$V_1 = F - R_B = 7 - 4 = +3.0 \text{ kN}$$

The force V_1 has an upward sense as it must, by the principle of action and reaction.

3. From the moment equilibrium condition about any point of section 1′-1′, assuming the bending moment M_1 to be clockwise, we obtain

$$M_1 + F(x_1' - b) - R_B x_1' = 0$$

$$M_1 + 7(5 - 3) - 4 \times 5 = 0$$

$$M_1 + 14 - 20 = 0$$

$$M_1 - 6 = 0$$

$$M_1 = +6.0 \text{ kN}$$

The moment M_1 acting on section 1′-1′ has a clockwise sense as it must, by the principle of action and reaction.

Inspecting the preceding results, we may conclude that the shear force V_1 acting on section 1-1 of the left part of the beam is equal and opposite to the shear force V_1

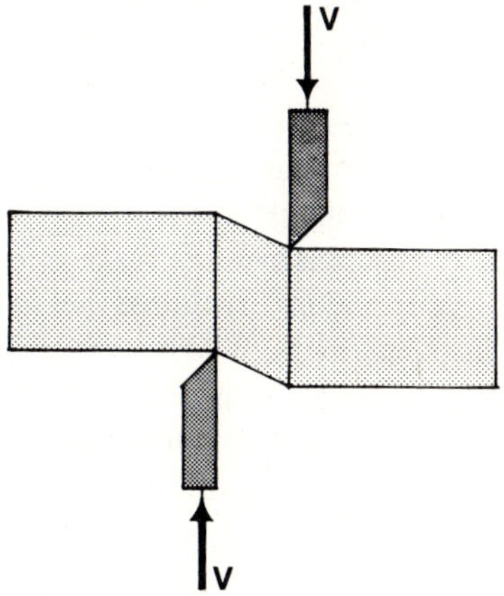

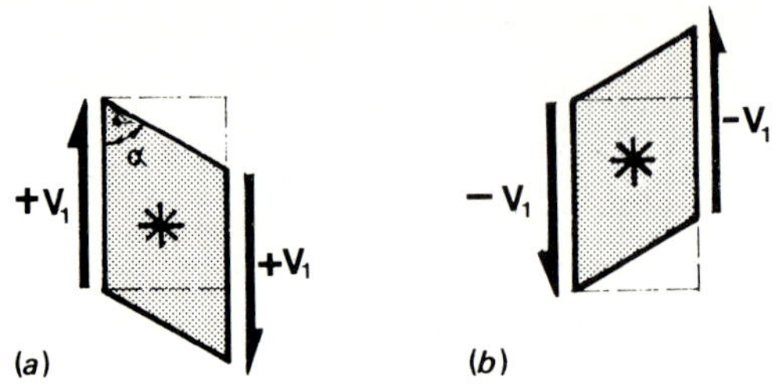

FIGURE 7-19 (*a*) Positive shear forces on a beam slice; (*b*) negative shear forces on a beam slice.

acting on section 1′-1′ of the right part of the beam. The downward shear force V_1 on section 1-1 represents the effect of the removed right portion of the beam on the left portion. The upward force V_1 on section 1′-1′ is the action of the removed left portion of the beam on the right portion. If we isolate a thin slice of the beam under consideration at the section, the effect of the shear forces on such a beam slice is shown in Fig. 7-19. The shear force V_1 from the left is equal and opposite to the shear force V_1 from the right. The action of the shear forces on the slice is to shear it off as shears act when cutting sheet metal; the originally rectangular slice is changed by this action into a rhomboidal slice. The original right angle shown in Fig. 7-19 has been changed into an acute angle α.

If we imagine our beam to consist of many thin slices (Fig. 7-20), we notice that the shear forces tend to make the slices slide vertically one with respect to the other. A similar situation occurs when we press a deck of cards between our hands and then try to slide one of our hands with respect to the other.

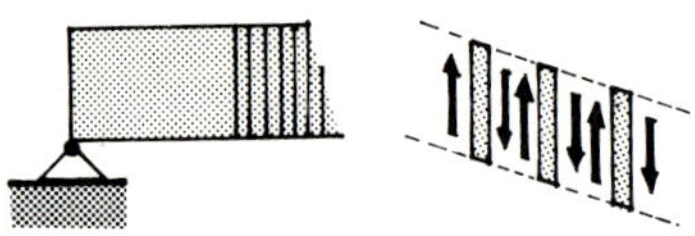

FIGURE 7-20 The action of shear forces in beams.

The effect of the shear forces on the beam slice shown in Fig. 7-19*a* may be assumed as positive, and we may agree to call shear forces causing such an effect *positive* shear forces at that section. In other words, each shear force, whether we approach the slice from the left or from the right, is considered positive if their couple is *clockwise* about the center of the slice (see the asterisk in Fig. 7-19). In Fig. 7-19*b* we show the negative shear forces which are upward on the right side and downward on the left side of the slice; their couple is counterclockwise. Thus the shear force on section 1-1 (or 1′-1′) of the beam shown in Fig. 7-18 is positive.

In determining the effects of the axial forces in a structure, we have assigned them a positive or negative sign depending on whether they were tensile or compressive. A similar convention is adopted for the bending moments of the beams. Figure 7-21*a* shows an isolated beam slice with *positive* moments M_1 acting on it. These moments bend the beam slice in such a way that its *bottom* fibers are in *tension* and lengthen, while the *top* fibers are in *compression* and shorten. Hence a bending moment is called "positive" when it is *clockwise* on the *left* side and *counterclockwise* on the *right* side of the slice. In Fig. 7-21*b* we see the negative bending moments acting on a slice. Negative bending moments are clockwise on the right side and counterclockwise

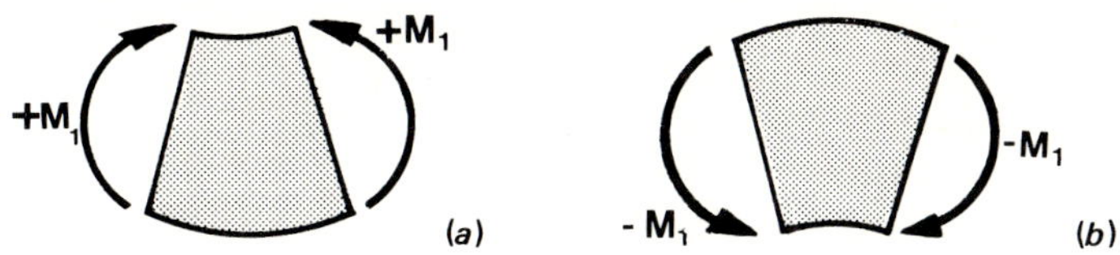

FIGURE 7-21 (*a*) Positive bending moments on a beam slice; (*b*) negative bending moments on a beam slice.

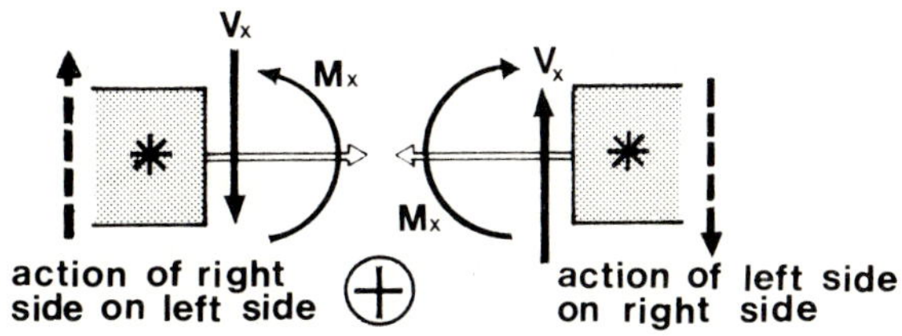

FIGURE 7-22 Positive shear forces and positive bending moments on the two sides of a beam section.

on the left side of the slice. They lengthen the top fibers and shorten the bottom fibers of the beam at the section.

By these conventions Fig. 7-22 shows positive shear forces and positive bending moments on the two sides of a beam section. Compare their senses with those of Figs. 7-19 and 7-21.

Note: The reader is advised to draw Figs. 7-19 and 7-21 on a separate card and consult it while studying the illustrative problems and solving the exercises on bending moments and shear forces in beams.

Illustrative Problem 7-3

Determine the internal forces at section 2-2 of the beam discussed in Illustrative Prob. 7-2 (Fig. 7-23). The reactions have already been determined to be R_A = 3 kN and R_B = 4 kN.

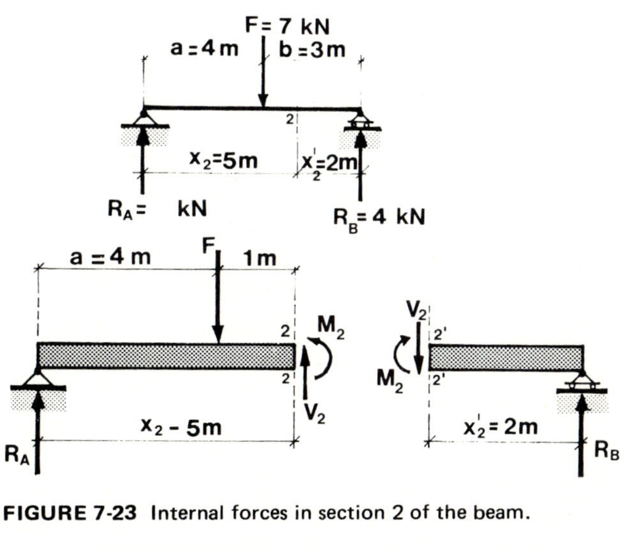

FIGURE 7-23 Internal forces in section 2 of the beam.

SOLUTION Isolate by means of section $2'$-$2'$ a right portion of the beam and assume that the sense of the shear force V_2 on it is downward. From the equilibrium condition for the vertical forces acting on this free body

$$R_B - V_2 = 0$$

$$V_2 = R_B = 4.0 \text{ kN}$$

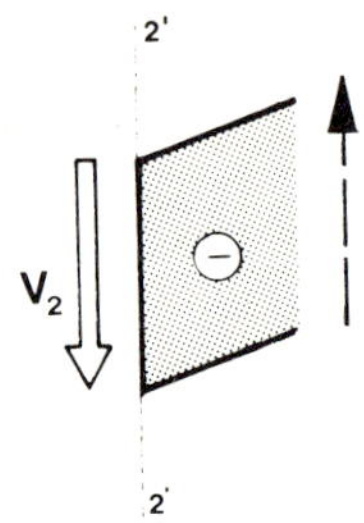

The positive sign of the result confirms *only* our assumption about the sense of V_2. However, the shear effect of V_2 is negative since the couple of the forces of V_2 is counterclockwise (see Figs. 7-19*b* and 7-22).

Assume now a clockwise sense for the bending moment M_2 acting on the right portion of the beam, and write the moment equilibrium condition about any point of section $2'$-$2'$:

$$M_2 - R_B x_2' = 0$$

$$M_2 = R_B x_2' = 4 \times 2 = 8 \text{ kN}\cdot\text{m}$$

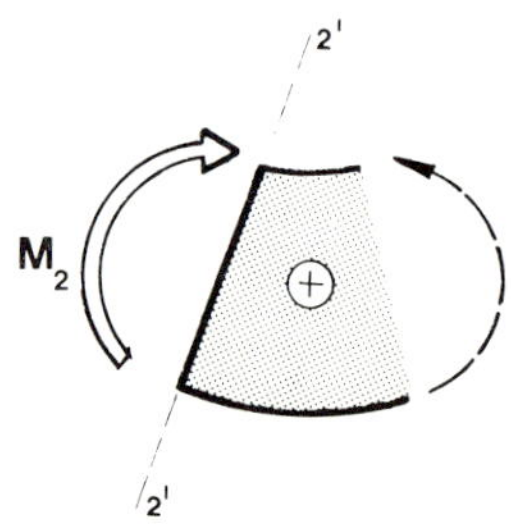

The positive result confirms *only* our assumption about the sense of M_2. The bending effect of M_2 is positive, as can be found by consulting with Fig. 7-21*a* or the right portion of Fig. 7-22. Also, M_2 is positive because, by another accepted rule, it lengthens the bottom fibers and shortens the top fibers of the beam at section $2'$-$2'$.

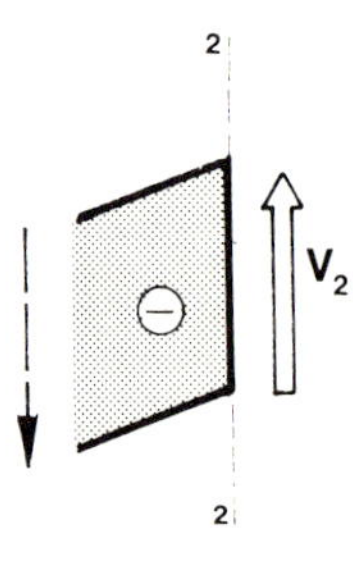

Identical results can be obtained from the investigation of the equilibrium of the left portion of the beam. From the condition of equilibrium for the vertical forces, we obtain

$$R_A - F + V_2 = 0$$

$$V_2 = F - R_A = 7 - 3 = 4 \text{ kN} \qquad \text{(negative—see Fig. 7-19}b\text{)}$$

From the equilibrium condition for the moments about any point on section 2-2, we obtain

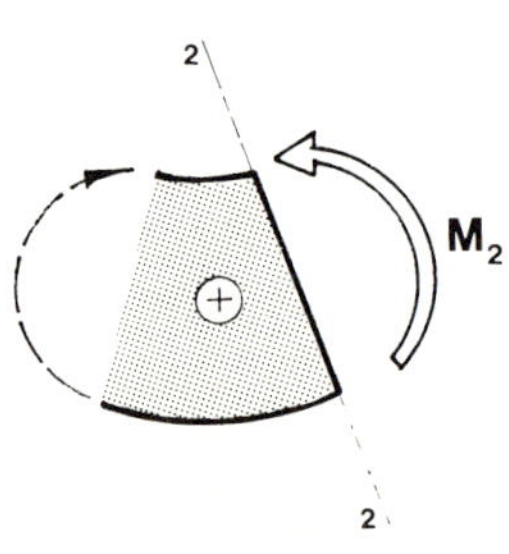

$$R_A x_2 - F(x_2 - a) - M_2 = 0$$

$$M_2 = R_A x_2 - F(x_2 - a) = 3 \times 5 - 7(5 - 4)$$

$$= 15 - 7 = 8 \text{ kN}\cdot\text{m} \qquad \text{(positive—see Fig. 7-21}a\text{)}$$

Let us now evaluate the internal forces at section 3-3 (or 3′-3′), which is located just to the right of the force F (Fig. 7-24). The vertical equilibrium condition for all the forces acting on the right portion of the beam is

$$R_B - V_3 = 0$$

$$V_3 = R_B = 4.0 \text{ kN} \qquad \text{(negative—see Fig. 7-19}b\text{)}$$

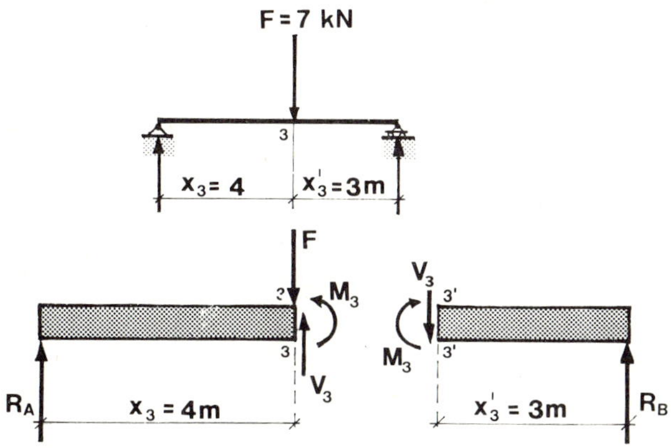

FIGURE 7-24 Internal forces in section 3 of the beam.

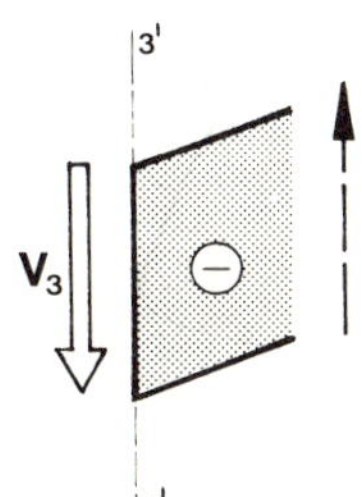

The shear force in section 3′-3′, immediately to the right of the force F, is negative; it has a negative shearing effect, as shown in Fig. 7-19b.

We may check this result by writing the equilibrium condition for the forces acting on the left portion of the beam. Here, the force F must be included in the equilibrium equations.

Thus

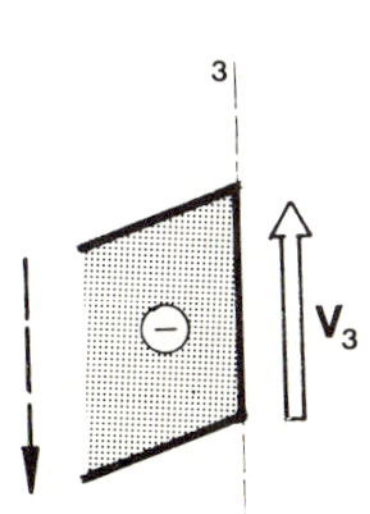

$$R_A + V_3 - F = 0$$

$$V_3 = -R_A + F = -3 + 7 = 4.0 \text{ kN} \qquad \text{(negative—see Fig. 7-19}b\text{)}$$

The bending moment M_3 may be computed from the equilibrium condition for the moments about section 3′-3′ of the forces acting on the right portion of the beam. Assuming a clockwise sense for M_3, we obtain

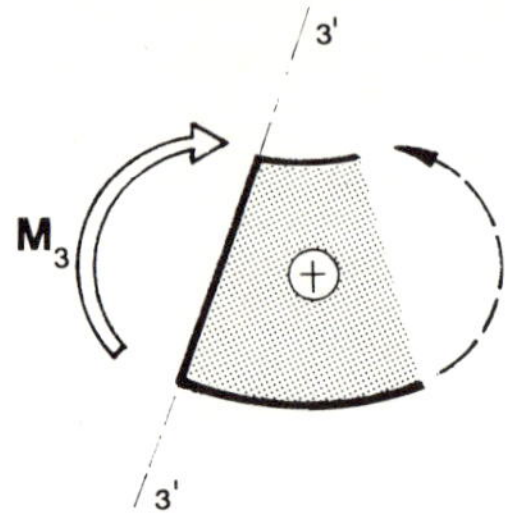

$$-R_B \times 3 + M_3 = 0$$

$$M_3 = 3R_B = 3 \times 4 = 12 \text{ kN·m} \qquad \text{(positive—see Fig. 7-21}a\text{)}$$

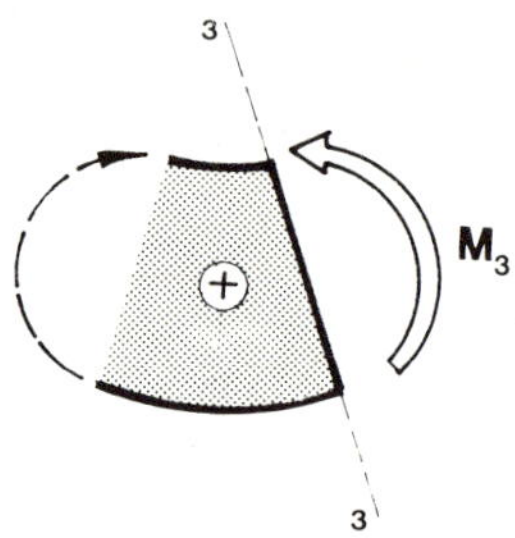

Using the left portion of the beam, from the condition of equilibrium for the moments about any point on section 3-3, we have

$$R_A \times 4 - M_3 = 0$$

$$M_3 = 4R_A = 4 \times 3 = 12 \text{ kN·m} \qquad \text{(positive)}$$

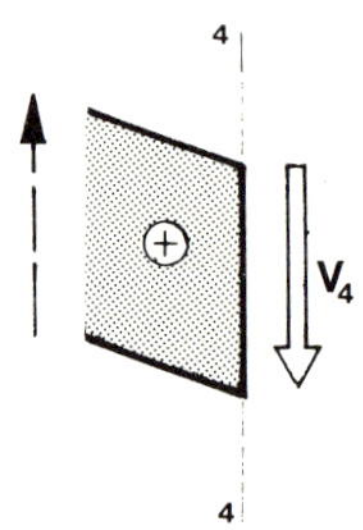

Finally, let us compute the internal forces in section 4-4, immediately to the left of the concentrated force F (Fig. 7-25). From the equilibrium condition for the forces acting on the left part of the beam (the force F does not act on this part), we obtain

$$R_A - V_4 = 0$$

$$V_4 = R_A = 3.0 \text{ kN} \qquad \text{(positive)}$$

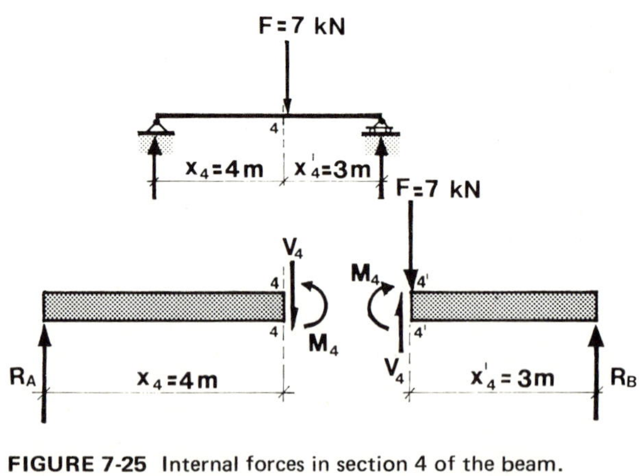

FIGURE 7-25 Internal forces in section 4 of the beam.

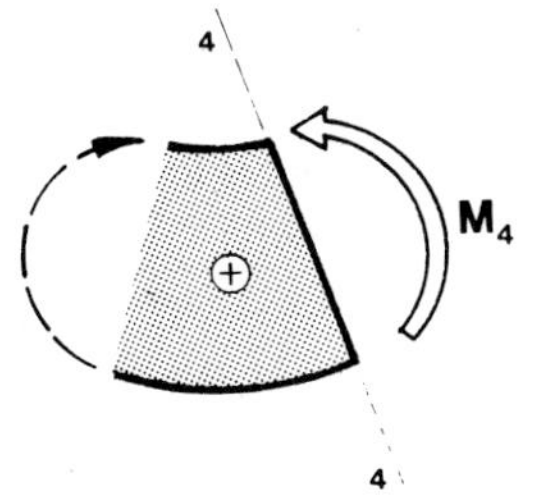

The bending moment M_4 may be computed from the equilibrium condition for the moments about section 4-4 of the forces acting on the left part of the beam:

$$R_A \times 4 - M_4 = 0$$

$$M_4 = R_A \times 4 = 3 \times 4 = 12 \text{ kN·m} \qquad \text{(positive)}$$

After the values of the internal forces have been computed in different sections of the beam, they can be represented graphically by plotting their values along the length of the beam. Choose a scale for the shear forces and plot their values vertically, with positive values above and negative values below the axis (Fig. 7-26). The shear-force diagram shows that the shear-force value changes abruptly at the section where the concentrated force F is applied. Immediately to the left of this section it is still positive, $V_4 = +3.0$ kN, but immediately to the right $V_3 = -4.0$ kN is negative. The total change, or jump, from +3 to -4 is equal to the magnitude of the force F (that is, 7.0 kN).

Similarly, we can plot the bending-moment diagram, which shows the change of the bending-moment values along the beam. The bending moments are positive all along the beam; thus we conclude that our beam curves downward, the con-

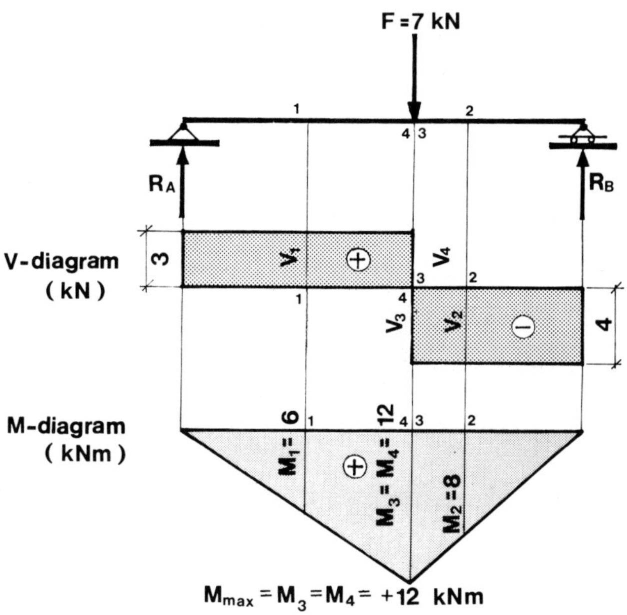

FIGURE 7-26 Shear force diagram and bending moment diagram of the beam.

sequence of tension on the bottom side of the beam due to positive moments. Let us plot positive moments below the horizontal line to emphasize that under positive moments the bottom fibers of the beam are in tension.[†] The largest bending moment on the beam occurs at the section where the concentrated force F is applied and is equal to

$$M_{max} = M_3 = M_4 = +12 \text{ kN·m}$$

Illustrative Problem 7-4

Determine the shear force and bending-moment diagrams for a simply supported beam loaded by a uniformly distributed load $w = 1.0$ kN/m over its entire length (Fig. 7-27).

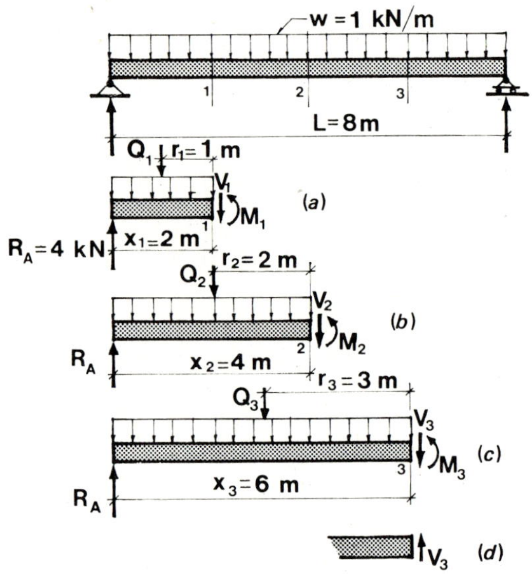

FIGURE 7-27 Internal forces in a simply supported beam loaded by a uniformly distributed load.

SOLUTION The equivalent concentrated load is

$$Q = w \times L = 1.0 \times 8 = 8.0 \text{ kN}$$

and, by symmetry, the reactions are

$$R_A = R_B = \frac{Q}{2} = \frac{8}{2} = 4.0 \text{ kN}$$

[†]As an alternative, it is customary to plot the positive moments above the horizontal line. This convention is based on the differential relationship (from calculus) between the shear force and bending moment, which is beyond the scope of this book. The reader can easily adopt either of the two conventions.

Let us compute first the internal forces at section 1-1, 2 m from the left support (Fig. 7-27*a*). Isolate by means of section 1-1 the left portion of the beam, and investigate the equilibrium of the free body thus obtained. To satisfy the equilibrium conditions, internal forces V_1 and M_1 must be introduced at the section 1-1. Assume their sense as shown in Fig. 7-26*a*, and write the equilibrium condition for the vertical forces:

$$R_A - V_1 - Q_1 = 0$$

where Q_1 is the equivalent concentrated force representing the uniformly distributed load between the support A and section 1-1:

$$Q_1 = wx_1 = 1.0 \times 2 = 2.0 \text{ kN}$$

Then

$$R_A - V_1 - wx_1 = 0$$

$$V_1 = R_A - wx_1 = 4 - 1 \times 2 = 4 - 2 = 2.0 \text{ kN}$$ (positive, since the sense of V_1 is the same as shown on the right side of the slice in Fig. 7-19*a*)

From the equilibrium condition for the moments about any point on section 1-1, we obtain

$$R_A x_1 - Q_1 \times r_1 - M_1 = 0$$

$$M_1 = R_A x_1 - Q_1 \times r_1 = 4 \times 2 - 2 \times 1 = 8 - 2$$

$$= 6.0 \text{ kN}\cdot\text{m} \qquad \text{(positive)}$$

Let us proceed now to section 2-2 at midspan of the beam (Fig. 7-27*b*). The equilibrium condition for all forces acting on the left part of the beam gives

$$R_A - Q_2 - V_2 = 0$$

or

$$V_2 = R_A - Q_2$$

where Q_2 is the equivalent concentrated force representing the uniformly distributed load between A and section 2-2:

$$Q_2 = w \times 4 = 1 \times 4 = 4 \text{ kN}$$

Thus

$$V_2 = 4 - 4 = 0$$

The shear force at midspan is zero. For moment equilibrium about section 2-2 of the forces acting on the left part of the beam, we obtain

$$R_A x_2 - Q_2 \times r_2 - M_2 = 0$$

$$M_2 = R_A x_2 - Q_2 \times r_2 = 4 \times 4 - 4 \times 2 = 16 - 8$$

$$= 8.0 \text{ kN}\cdot\text{m} \qquad \text{(positive)}$$

Finally, let us proceed to section 3-3, located 2 m to the left of the right support. Assume a downward sense of V_3. From the equilibrium of forces, we have (Fig. 7-27*c*)

$$R_A - Q_3 - V_3 = 0$$

$$V_3 = R_A - Q_3 = 4 - 6 = -2.0 \text{ kN}$$

The negative sign of V_3 shows that our assumption of its sense was incorrect. The shear force V_3 acts upward, as shown in Fig. 7-27*d*. According to our sign convention, such shear force is negative

$V_3 = -2.0$ kN (negative, since V_3 has the same sense as shown on the right side of the slice in Fig. 7-19*b*)

From the equilibrium of moments about section 3-3, we obtain

$$R_A x_3 - Q_3 \times r_3 - M_3 = 0$$

which yields

$$M_3 = R_A x_3 - Q_3 \times r_3 = 4 \times 6 - 1 \times 6 \times 3 = 24 - 18$$

$$= 6.0 \text{ kN}\cdot\text{m} \quad \text{(positive)}$$

Let us now plot the computed values of V and M in a shear-force diagram and a bending-moment diagram (Fig. 7-28). The shear-force diagram is an inclined straight line changing from positive to negative values at midspan. We have evaluated only three values of the bending-moment diagram, but if more values were obtained at nearer and nearer points, this diagram would be seen to be a type of curve called a *parabola*, with the value of M maximum at midspan and equal to

$$M_{\max} = \frac{wL^2}{8}$$

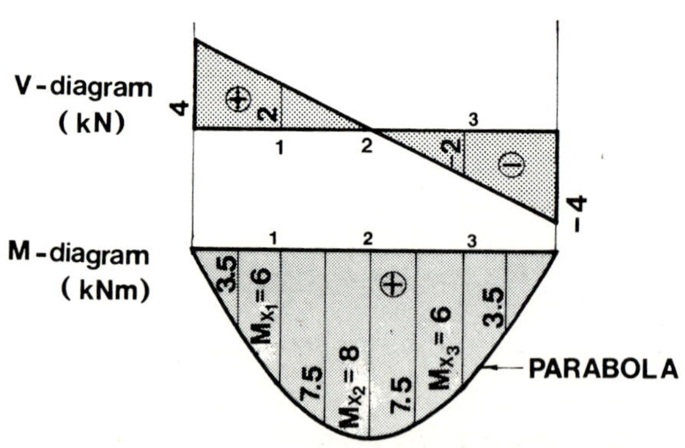

FIGURE 7-28 Shear-force diagram and bending-moment diagram of a simply supported beam loaded by a uniformly distributed load.

In our case

$$M_{\max} = \frac{1 \times 8^2}{8} = 8 \text{ kN}\cdot\text{m}$$

Shear-force diagrams and bending-moment diagrams for beams with different loadings are shown in Figs. 7-29 through 7-31. Check whether the results shown are correct.

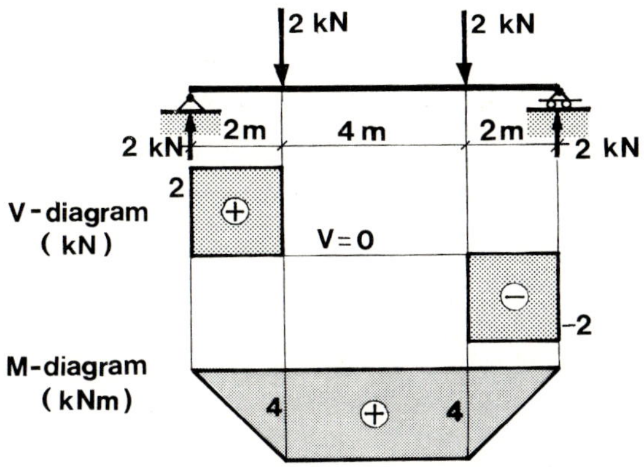

FIGURE 7-29

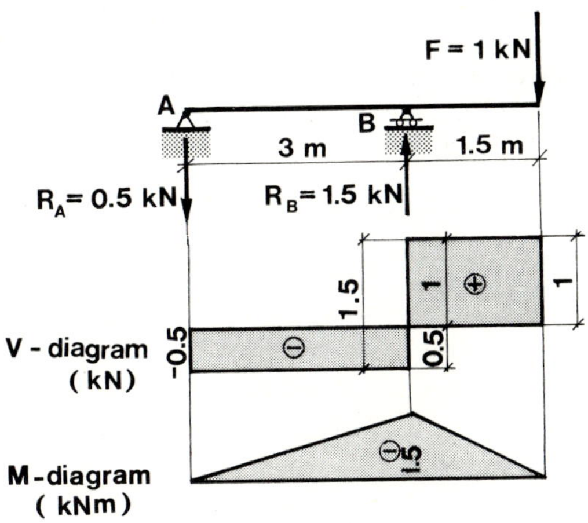

FIGURE 7-30

Some important relationships can be derived from the preceding examples:

1. The largest value of the bending moment always occurs at the section, where the shear force is either zero or abruptly changes sign. It is very important to locate such section, since the beam curves the most sharply there. Such a section is sometimes called the *critical section.* The shearing of a beam slice at the critical section changes sign, as shown in Fig. 7-32.

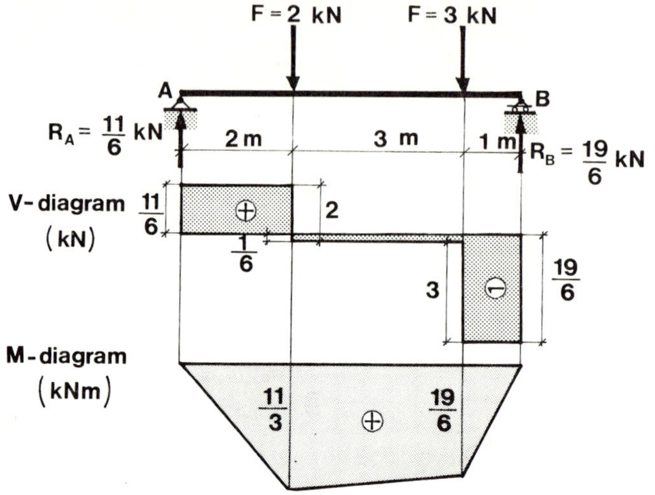

FIGURE 7-31

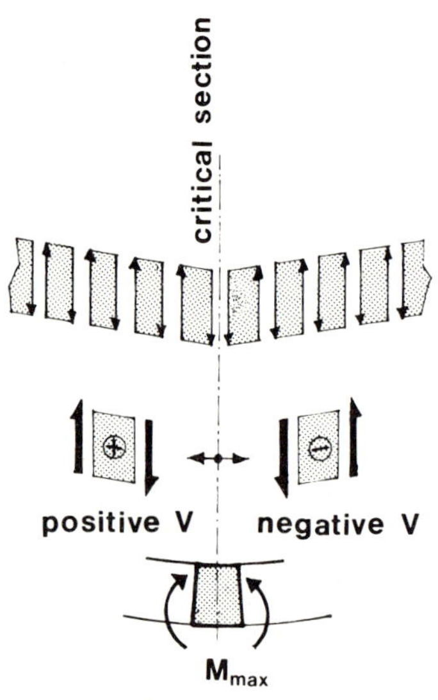

FIGURE 7-32 At the critical section, shear forces change their sign.

2. The shear force changes abruptly at sections where concentrated forces are applied, but bending moments do not change value at such sections.
3. The shear force diagram is constant between sections where the beam is unloaded; the bending moment diagram is a straight line between such sections.
4. Along the portion of a beam uniformly loaded the shear diagram is a straight line and the bending moment diagram is curved (the curve is an arc of parabola).

The shear-force diagram and the bending-moment diagram of a cantilever beam loaded by vertical loads is shown in Fig. 7-33. The section where the maximum

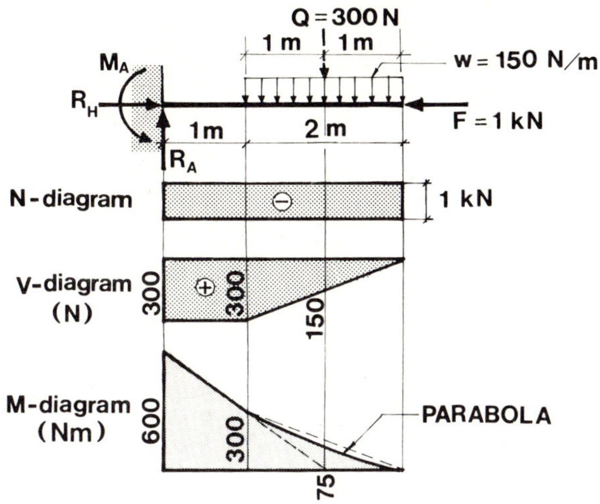

FIGURE 7-33

moment occurs is at the fixed end. There the beam bends most sharply; therefore, this is the critical section.

The principles used to compute shear forces and bending moments in the preceding examples can be used to analyze simply supported beams with overhangs, as shown by the following illustrative problem.

Illustrative Problem 7-5

Find the maximum shear force and the maximum bending moment in the beam with overhangs shown in Fig. 7-34.

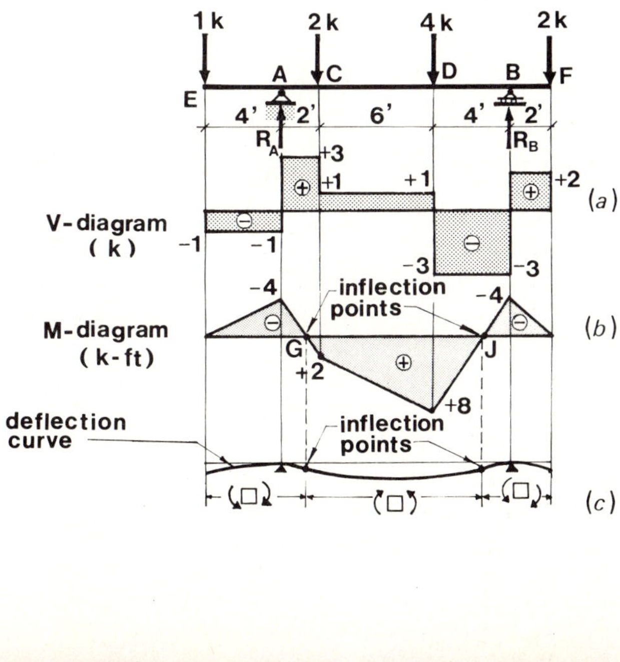

FIGURE 7-34

SOLUTION

1. *Reactions R_A and R_B*. Assuming the sense of these reactions to be upward and from the moment equilibrium condition about B, we have

$$12R_A + 2 \times 2 - 4 \times 4 - 2 \times 10 - 1 \times 16 = 0$$

$$12R_A = 48$$

$$R_A = \tfrac{48}{12} = 4.0 \text{ kips}$$

From the moment equilibrium condition about A, we obtain

$$12R_B + 1 \times 4 - 2 \times 2 - 4 \times 8 - 2 \times 14 = 0$$

$$12R_B = 60$$

$$R_B = \tfrac{60}{12} = 5.0 \text{ kips}$$

As a check of the equilibrium of vertical forces, we have

$$R_A + R_B - 1 - 2 - 4 - 2 = 0$$

$$4 + 5 - 9 = 0$$

2. *Shear forces.* Isolating the corresponding free bodies (Figs. 7-35 through 7-38) and assigning the sign to the shear forces according to our convention, we have

Section E	$V_E = -1.0$ kip	
To the left of A	$V_A = -1.0$ kip	$V'_A - V_A = 4$ kips
To the right of A	$V'_A = -1.0 + 4 = +3.0$ kips	
To the left of C	$V_C = +3.0$ kips	$V'_C - V_C = 2.0$ kips
To the right of C	$V'_C = +3.0 - 2.0 = +1.0$ kip	
To the left of D	$V_D = +1.0$ kip	$V'_D - V_D = 4.0$ kips
To the right of D	$V'_D = +1.0 - 4.0 = -3.0$ kips	
To the left of B	$V_B = -3.0$ kips	$V'_B - V_B = 5.0$ kips
To the right of B	$V'_B = -3.0 + 5.0 - +2.0$ kips	
At F	$V_F = +2.0$ kips	

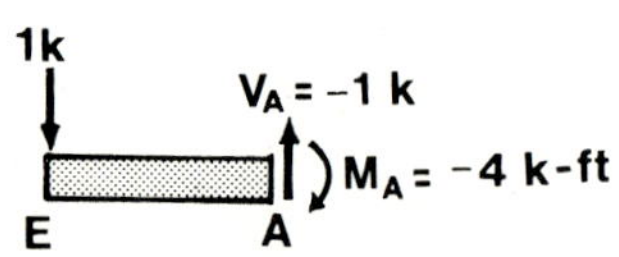

FIGURE 7-35

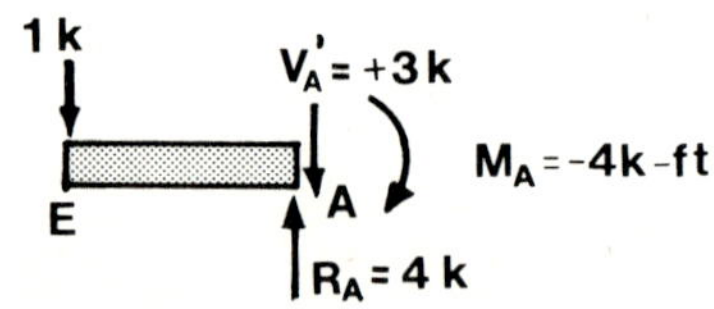

FIGURE 7-36

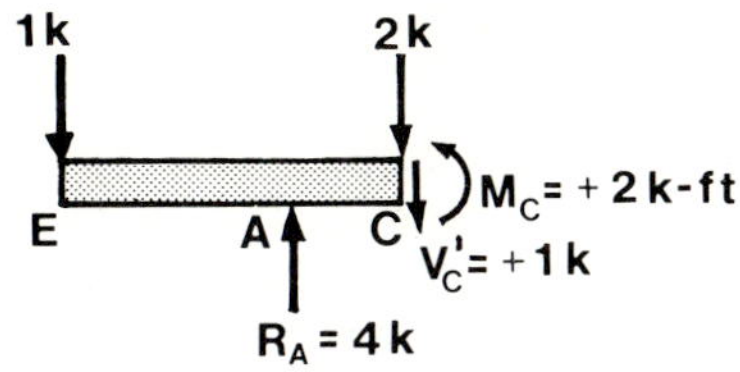

FIGURE 7-37

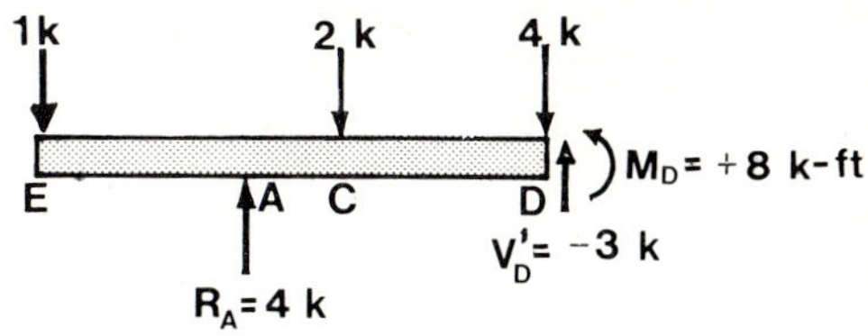

FIGURE 7-38

3. *Bending moments.*

$$M_E = M_F = 0$$

$$M_A = -1 \times 4 = -4.0 \text{ kip}\cdot\text{ft}$$

$$M_C = -1 \times 6 + 4 \times 2 = -6 + 8 = +2.0 \text{ kip}\cdot\text{ft}$$

$$M_D = -1 \times 12 + 4 \times 8 - 2 \times 6 = -12 + 32 - 12 = +8.0 \text{ kip}\cdot\text{ft}$$

$$M_B = -2 \times 2 = -4.0 \text{ kip}\cdot\text{ft} \quad \text{(take the segment } B\text{-}F \text{ as a free body)}$$

$$V_{max} = V_A' = -V_D' = +3.0 \text{ kips} \qquad M_{max} = M_D = +8 \text{ kip}\cdot\text{ft}$$

There are three critical sections in this beam, at A, B, and D. In each of them the shear force abruptly changes its sign. The largest of the three peak moments, M_D, will govern the design of the beam. You may also notice that the outer portions of the beam are under the action of negative bending moments; they bend with tension in the top fibers (Fig. 7-34*c*). At points G and J (Fig. 7-34*b*) the moments change sign (from negative to positive and vice versa). In the central portion of the beam between G and J the bottom fibers are in tension. The points where the moment is zero (i.e., changes sign) are called *inflection points.* Since points A and B of the beam must remain in contact with the supports, we may now draw the deformed shape of the beam (Fig. 7-34*c*) or its *deflection curve.*

New Terms Appearing in Chapter 7

Internal forces
Axial (internal) force
Compressive axial force
Tensile axial force
Diagram of axial forces
Shear force, V_x
Shear-force diagram
Bending moment, M_x
Bending-moment diagram
Positive and negative V_x and M_x
Parabola
Critical section
Maximum shear force
Maximum bending moment
Inflection point
Deflection curve

Quiz Number 6

A. TRUE OR FALSE

1. When a column is loaded by an external compressive axial force, internal tensile axial forces develop in it.
2. The tensile internal axial force in a member of a truss tends to elongate the member.
3. The principal assumption in the approximate evaluation of axial forces in a truss is that its members are rigidly fixed into the joints.
4. We may start to solve for the axial forces in a truss at a joint where only three unknown axial forces act.
5. In the approximate evaluation of axial forces in the members of a truss we assume that the members are not loaded by external forces between the endjoints.
6. Since we assume that all the members of a truss are pin connected at the joints and that all the external loads act at the joints, we conclude that only axial internal forces can develop in the truss members.
7. If the reactions of a truss can be determined by the three conditions of equilibrium, the truss is statically determinate.
8. If all the forces acting at a joint of a truss are in equilibrium, the force polygon for the joint must be closed ($R = 0$) when all the arrows flow tip to tail.
9. In the approximate analysis of the axial forces in the members of a truss, their weight is considered to be a uniformly distributed load between the endjoints.
10. In the approximate analysis of the internal forces in a truss, we evaluate the axial forces, the shear forces, and the bending moments in each truss member.
11. In the analysis of a truss, we always start the solution at a joint where only two unknown forces act.
12. We usually denote a compressive axial force in a truss member as positive.
13. Axial forces, shear forces, and bending moments develop in a beam under a general coplanar force system.
14. The bending moment tends to shear off adjacent sections of a beam.
15. A positive bending moment tends to bend the beam so that tension develops in its top fibers.
16. Bending moments change sign at points of inflection.
17. The points of inflection of the deflection curve of a beam coincide with the critical sections, where the bending moments reach their maximum values.
18. At a critical section the shear force is either zero or changes its sign.
19. The critical section of a cantilever beam under downward loads is at its free end.
20. Once the bending-moment diagram of a beam is known, its deflection curve can be approximately sketched.

B. MULTIPLE CHOICE

1. Which of these assumptions is inapplicable to the approximate determination of the internal forces in a truss?

(*a*) moments develop in its members
(*b*) its members are pin connected at the joints
(*c*) all loads are applied to the joints only
(*d*) the weight of each member is replaced by two equivalent forces acting at the endjoints of the member

2. In the analysis of a plane truss under a coplanar force system the solution is started at a joint where no more than ________ unknown forces act:
(*a*) two (*b*) three (*c*) four
3. What kind of internal forces may develop in a beam under a general coplanar force system?
(*a*) axial forces only (*b*) shear forces only
(*c*) bending moments only (*d*) all of the above
4. An internal axial force in a truss member, which tends to shorten the length of the member, is called a
(*a*) negative axial force (*b*) reaction
(*c*) shear force (*d*) positive axial force
5. The sections of a beam at which the bending moments acquire their greatest values are called
(*a*) equivalent sections (*b*) points of inflection
(*c*) critical sections (*d*) support points
6. The sections of a beam at which the bending moments change sign are called
(*a*) equivalent sections (*b*) points of inflection
(*c*) critical sections (*d*) support points
7. A cantilever beam loaded by a concentrated force at its free end has a critical section
(*a*) at its free end (*b*) at its fixed end
(*c*) at the midpoint of its length
8. A simply supported beam loaded by a uniformly distributed load over its entire length has a critical section
(*a*) at its free end (*b*) at its fixed end
(*c*) at midspan (*d*) at its supports
9. The bending-moment diagram of a simply supported beam loaded by a uniformly distributed load over its entire length has the shape of a
(*a*) triangle (*b*) parabola
(*c*) trapezoid (*d*) rectangle
10. The bending-moment diagram of a simply supported beam loaded by a concentrated force at midspan has the shape of a
(*a*) triangle (*b*) parabola
(*c*) trapezoid (*d*) rectangle

C. SHORT ANSWERS

1. The ________________ sign of an internal axial force in a truss member designates compression.
2. A downward shear force acts on the extreme left section of a free body of a beam. According to our sign convention, such shear force is ________________.

3. According to our sign convention, a bending moment which causes tension in the top fibers of a beam is ______________.
4. A clockwise bending moment acts on the extreme left section of a free body of a beam. According to our sign convention, such a bending moment is ________ ________.
5. A cantilever beam is loaded by a concentrated force at its free end (Fig. 7-39). What is the shape and the sign of its shear-force diagram?

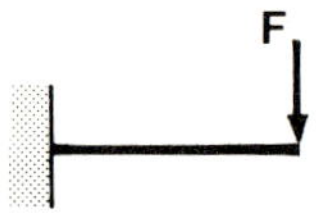

FIGURE 7-39

6. What is the sign of the shear-force diagram for the cantilever shown in Fig. 7-40?

FIGURE 7-40

7. What are the shape and the sign of the bending-moment diagram for the cantilevers shown in Figs. 7-39 and 7-40?
8. Where is the critical section in the cantilever shown in Fig. 7-39?
9. What is the value of the axial force in truss member 5 in Fig. 7-41?

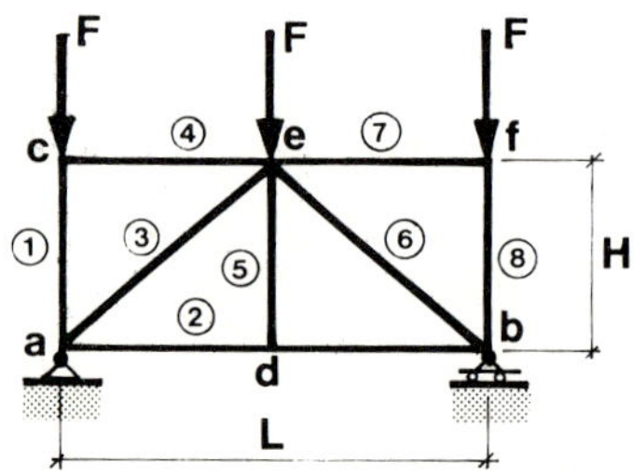

FIGURE 7-41

10. What is the value and sign of the axial force in truss member 4 in Fig. 7-41?

EXERCISES

7-1 Find the axial forces in all the members of the truss shown in Fig. 7-41, if its span $L = 40$ m, its depth $H = 10$ m, and each force $F = 2.0$ kN.

7-2 Find the axial forces in all the members of the truss shown in Fig. 7-42.

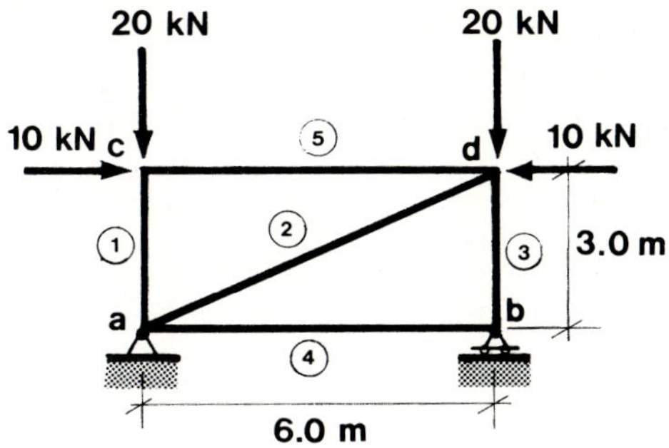

FIGURE 7-42

7-3 Find the axial forces in all the members of the truss in Fig. 7-13*a*, if $a = H = 10$ ft and each force $F = 4$ kips.

7-4 Find the axial forces in all the members of the truss shown in Fig. 7-13*b*, if $a = 10$ ft, $H = 8$ ft, each force $F_1 = 4$ kips, and $F_2 = 2$ kips.

7-5 Find the shear and bending-moment diagrams of the beams shown in Figs. 7-43*a* through 7-43*e*.

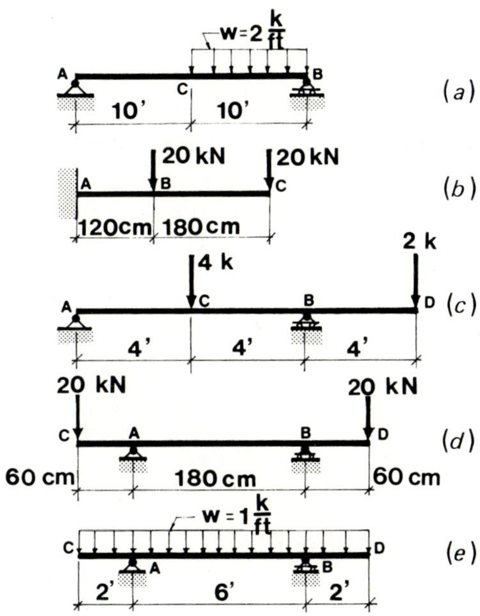

FIGURE 7-43

A plastic tensile-test specimen in the testing machine (*Courtesy of MTS Systems Corporation, Minneapolis*).

Fundamentals of Strength of Materials

8

8.1 STRESSES AND DEFORMATIONS IN STRUCTURES

In the preceding chapters we learned how to find the resultants or equilibrants of force systems. We learned how to evaluate reactions at the supports of simple, statically determinate structures and, finally, how to compute internal forces in structural elements, that is, axial forces, shear forces, and bending moments on their cross sections. In doing so, we noticed that under the action of external and internal forces all structural elements deform: columns shorten, hangers elongate, and beams deflect. Because of these deformations, structural elements change their shape and geometrical dimensions. For practical reasons, it is usually desirable for structures to develop deformations *very* small compared to their dimensions. Thus the shapes and the dimensions of structural members must be so designed as to make their deformations small. When this is done, we can perform all force calculations on structural systems, neglecting their deformations, that is, as if they were perfectly rigid (undeformable).

It is true that a column shortens under the action of a compressive axial force. However, the change of its length is actually so small that we can neglect it and analyze the column as if it were undeformed, that is, with its original length unchanged. Similarly, under the action of loads, beams bend and deflect, thus causing small horizontal displacements of their supports and small changes in their spans, as shown in Fig. 6-7. We disregard these changes because they actually are (and *must* be) small. Thus we compute the reactions and the internal forces of beams by assuming that they are rigid, undeformable bodies, with their original spans and lengths unchanged.

However, in reality, perfectly or ideally rigid materials do not exist. Structural materials may be assumed to be composed of microscopic particles connected by "springs" in all directions. According to the "toughness" of these springs, we speak about more or less deformable materials. (This is why strength of materials is considered a part of deformable body mechanics.) In rubber the springs are considerably softer than in steel: rubber is more deformable than steel.

The springs can be imagined as an internal mechanism resisting the deformation of the individual fibers of a body. The internal resistance to deformation of a body is called *stress.*

Internal forces, which we have determined by statics, do not act at one point of a cross section. They are distributed over the entire cross section and resisted by a very large number of individual "springs." Rather than dealing with such complicated set of elementary forces, we prefer to work with the force acting on a *unit* area of the cross section, such as a square inch or a square centimeter. Thus the stress p as an internal resistance of the material to deformation is conveniently expressed as a force F per unit area a and, consequently, is sometimes called a *unit stress*:

$$p = \frac{F}{a}$$

Therefore, the units of stress are pounds per square inch (lb/in^2), pounds per square foot (lb/ft^2), newtons per square millimeter (N/mm^2), newtons per square centimeter (N/cm^2), and so on.

The *internal* force acting on a unit area of a cross section might act, generally, in any direction (Fig. 8-1). For engineering purposes two particular components of a unit internal force are of importance: the component perpendicular or *normal* to the cross section, and that acting (as the shear force does) in the plane of the cross section, that is, *tangent* to the cross section. Analogically, speaking in terms of stress, the ratio of a (small or elementary) normal component n of the internal force to the small or elementary area a of the cross section on which it acts is called a *normal stress*, since it acts in a direction normal to the area. It is denoted by the Greek lowercase letter

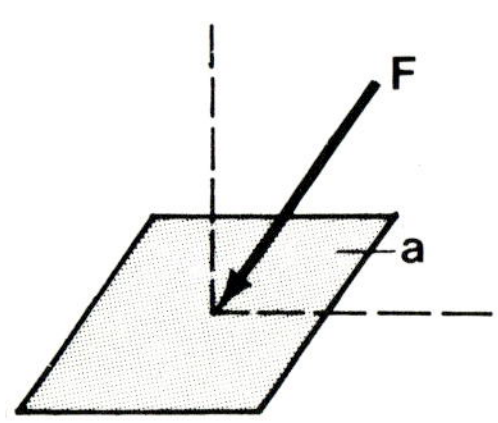

FIGURE 8-1

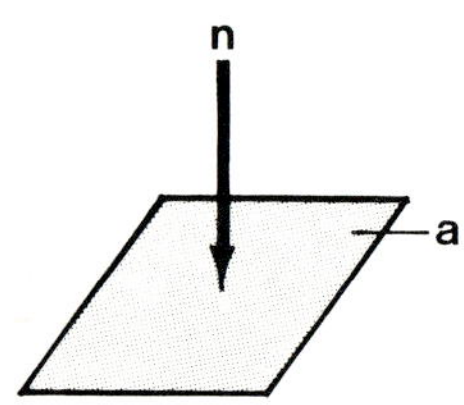

FIGURE 8-2

sigma (σ), so that (Fig. 8-2)

$$\sigma = \frac{n}{a} \qquad \left(\frac{\text{lb}}{\text{in}^2}, \frac{N}{\text{cm}^2}, \text{and so on}\right)$$

The ratio of a (small or elementary) shear component v to the small area a on which it acts is called a *shear stress* or a *tangential stress*. It is denoted by the Greek lowercase letter tau (τ), so that (Fig. 8-3)

$$\tau = \frac{v}{a} \qquad \left(\frac{\text{lb}}{\text{in}^2}, \frac{N}{\text{cm}^2}, \text{and so on}\right)$$

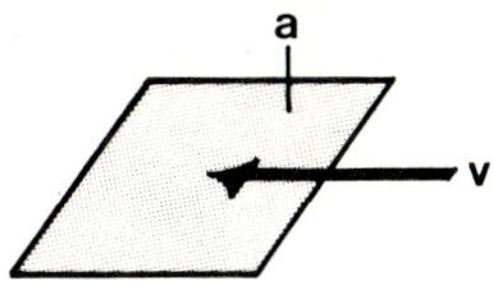

FIGURE 8-3

The elementary internal forces (of components n and v) are distributed among individual springs and deform them. Consequently, in reality a structural element changes its shape and geometrical dimensions; it deforms. If the element is made of an elastic material and the load is not excessive, these deformations disappear after the load has been removed from the element. Deformations which disappear after the removal of loads are called *elastic deformations.* Structures must be designed to behave elastically; after the removal of loads, they should not remain deformed, but assume their original shapes and dimensions. Otherwise, after a few loadings they would be so deformed as to become unusable.

After this preliminary discussion, we can specify the two types of problems which we shall consider in strength of materials:

1. How is the total internal force (determined by statics) distributed over the area of the cross section? The determination of the corresponding stress and its distribution over the cross section is our first problem.
2. How does a structural element deform when stressed? The evaluation of the deformations of a structure, to ensure that they are not excessive, is our second problem.

Once the first problem is solved, we must choose the proper structural material, which will withstand safely the computed stresses and will have small elastic deformations under these stresses. Once the structural material has been selected, the shape and the size of the structural members must be determined; the proportioning of structural members, that is, the determination of their dimensions, is called *structural design*.

Sometimes the two above-mentioned problems are reversed. The structure has already been designed or even built, and we must check whether its stresses and deformations are within allowable limits. These checks are called *structural analysis*. Structural analysis may also consist of determining allowable loads of such magnitude that stresses and deformations do not exceed allowable limits.

8.2 DISTRIBUTION OF NORMAL STRESSES IN MEMBERS UNDER AXIAL TENSION OR COMPRESSION

Under the action of axial loads, compressive or tensile internal axial forces develop in the members of trusses as well as in columns, hangers, cables, and so on. These internal

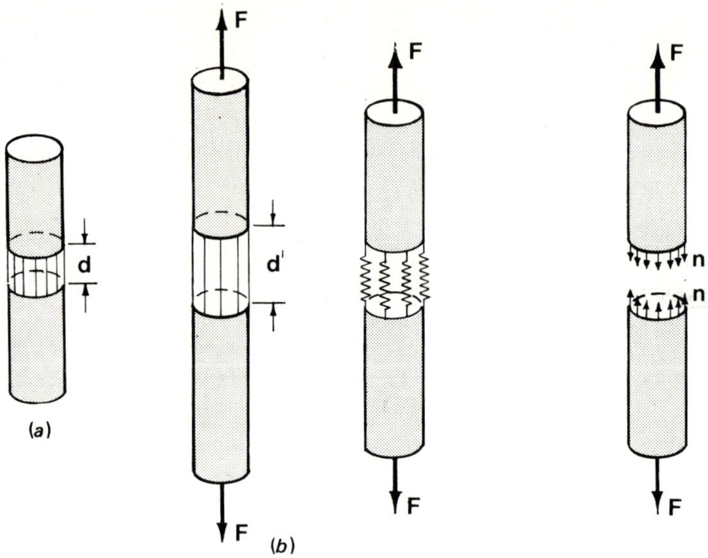

FIGURE 8-4 (*a*) A bar loaded by axial forces elongates; (*b*) normal stresses in the bar are uniformly distributed over the whole cross section of the bar.

forces do not act at one point of the cross section, but are distributed over the entire cross section. This is accompanied by the development of stresses in the cross section. Our first problem is to determine how the force is distributed, that is, the corresponding distribution of stresses over the cross section. To solve this problem, consider a rubber model of a bar with circular cross section. Draw two parallel circles a distance d apart on the surface of the bar and then load the bar at its ends with two tensile axial forces F (Fig. 8-4*a*). An internal tensile force $N = F$ develops in each cross section of the bar, as can be shown by equilibrium in the axial direction of parts of the bar. As a result of this load the bar elongates and the distance d between the two circles increases to d'. By measuring the distance d', we find that it is constant around the bar and we surmise that it is also constant all over the cross section (an assumption tests prove, in general, to be correct) (Fig. 8-4*b*).

Now imagine that at the section under investigation the upper part of the bar is connected to the lower part of the bar by a system of springs. Since we assume that the distance d between corresponding points on the two circles increases by the same amount *all over the cross section*, all the elementary springs elongate equally. This means that they are all tensioned by equal elementary forces. Thus the total internal axial force N is distributed *uniformly* over the whole cross section of the bar, and equal elementary forces act on each unit area of the cross section. The corresponding normal stress σ is uniformly distributed (i.e., constant), over the cross section. Its magnitude is found simply by dividing the total internal axial force N by the total area of the cross section A:

$$\sigma = \frac{N}{A} = \frac{F}{A}$$

This normal stress is tensile; it is responsible for the resistance of the bar against its elongation. It is associated with the tensile internal axial force $N = F$. In agreement with our sign convention for axial forces, *tensile* normal stresses will be called *positive* and have a plus sign in front of their numerical value, while *compressive* normal stresses will be called *negative* and have a minus sign in front of their numerical value.

Illustrative Problem 8-1

Compute the normal stress in sections 1-1 and 2-2 of the column shown in Fig. 7-4. The column supports two stories. The top portion of the column has a square cross section 20 X 20 cm. The square cross section of the bottom portion is 25 X 25 cm (Fig. 8-5).

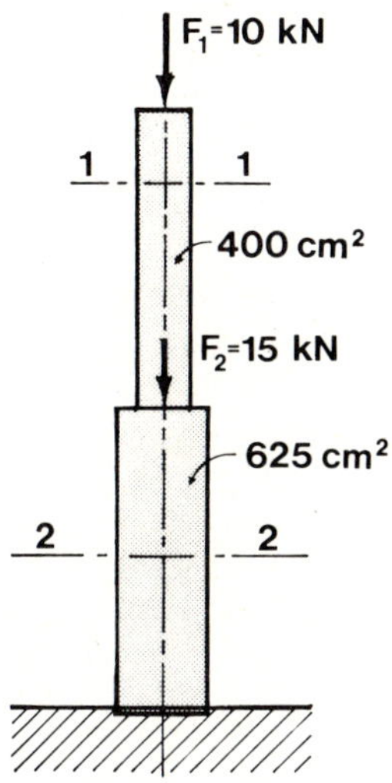

FIGURE 8-5

SOLUTION The area of cross section 1-1

$$A_1 = 20 \times 20 = 400 \text{ cm}^2 \qquad (62 \text{ in}^2)$$

and that of cross section 2-2:

$$A_2 = 25 \times 25 = 625 \text{ cm}^2 \qquad (97 \text{ in}^2)$$

From the analysis of the internal forces in the column, we find the following compressive axial forces acting on sections 1-1 and 2-2:

$$N_1 = -10 \text{ kN} \qquad (2\,250 \text{ lb}) \qquad \text{(compression)}$$

$$N_2 = -10 - 15 = -25 \text{ kN} \qquad (5\,620 \text{ lb}) \qquad \text{(compression)}$$

Thus the normal stress in section 1-1 is compressive and is

$$\sigma_1 = \frac{N_1}{A_1} = -\frac{10\,000 \text{ N}}{400 \text{ cm}^2} = -25 \frac{N}{\text{cm}^2} \qquad \left(36.3 \frac{\text{lb}}{\text{in}^2}\right)$$

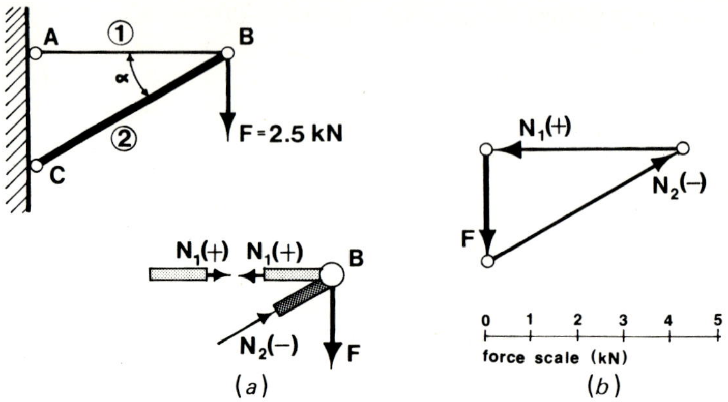

FIGURE 8-6 The bracket in Illustrative Prob. 8-2. (*a*) Free-body diagram of joint *B*. (*b*) Force triangle. For equilibrium at joint *B*, all arrows follow the same direction.

The compressive normal stress in section 2-2 is

$$\sigma_2 = \frac{N_2}{A_2} = -\frac{25\ 000\ \text{N}}{625\ \text{cm}^2} = -40\ \frac{N}{\text{cm}^2} \qquad \left(58\ \frac{\text{lb}}{\text{in}^2}\right)$$

Illustrative Problem 8-2

Evaluate the normal stress in the horizontal member *AB* of the bracket shown in Fig. 8-6. The member has a circular cross section with a diameter of 2.5 cm (1.0 in). The concentrated force acting at joint *B* is $F = 2.5$ kN (562 lb). The angle $\alpha = 30°$.

SOLUTION We must determine, first, the axial force in the member *AB*. This can be found from the equilibrium of a free body of joint *B*. Forces acting on the joint are: the action N_1 from the member *AB*, the action N_2 from the member *BC*, and the external concentrated force $F = 2.5$ kN. The free-body diagram of joint *B* is shown in Fig. 8-6*a*. The equilibrium condition for all the forces acting on joint *B* is stated by the force triangle shown in Fig. 8-6*b*. To obtain it, we plot in a certain scale the vector of *F* and through its endpoints draw parallels to the lines of action of the axial forces N_1 and N_2. The intersection point of these two lines determines the magnitude of N_1 and N_2. Remember that the arrows of all the forces in the triangle must flow tip to tail, since the forces are in equilibrium. Transfer the arrows to the free-body diagram of joint *B* (Fig. 8-6*a*) to find that N_1 is a tensile force since it pulls away from joint *B*, while N_2 is a compressive force since it pushes on the joint. Using the scale of forces, we find that

$$N_1 = +4.33\ \text{kN} \qquad (973\ \text{lb})$$

and

$$N_2 = -5.0\ \text{kN} \qquad (1\ 124\ \text{lb})$$

The area of the cross section of member AB is

$$A_1 = \frac{\pi d^2}{4} = \frac{\pi(2.5)^2}{4} = 4.9 \text{ cm}^2 \qquad (0.76 \text{ in}^2)$$

Therefore, the tensile normal stress in the member AB is

$$\sigma = \frac{N_1}{A_1} = \frac{4\,330 \text{ N}}{4.9 \text{ cm}^2} = +883.7 \frac{N}{\text{cm}^2} \qquad \left(1\,280 \frac{\text{lb}}{\text{in}^2}\right)$$

8.3 DEFORMATION OF A BAR UNDER AXIAL LOAD AND ITS STRESS-STRAIN DIAGRAM

A bar elongates under tension and shortens under compression. While it is desirable to know the relationship between these deformations and the corresponding stresses, we must realize that such a relationship can be obtained only experimentally, that is, through tests. For this purpose, standard tensile specimens have been developed and specified by the American Society for Testing and Materials (ASTM). Specimens for testing the tensile properties of different materials have different specified shapes and dimensions. A typical specimen for tensile tests on steel is shown in Fig. 8-7. The specimen is screwed in a tensile testing machine and an axial force F is applied to the

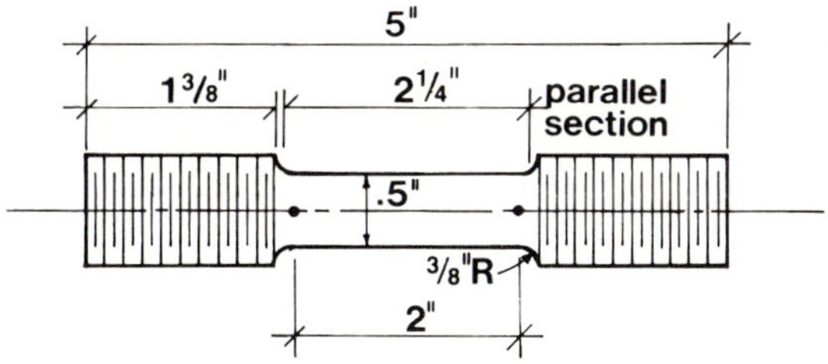

FIGURE 8-7 A typical specimen for tensile tests of steel.

specimen and measured. Since the area A of the cross section of the specimen is known, we can calculate the tensile normal stress for each value of the force F as

$$\sigma = \frac{F}{A}$$

Since the original area A of the undeformed specimen is used in computing the stress σ, such a stress is called the *nominal* tensile stress in the specimen. (The actual stress is slightly greater because under tension the area of the specimen becomes slightly smaller.)

To measure the axial deformation of the specimen, an instrument called an *extensometer* is attached to it, which measures precisely the change of its original or base length L due to the applied stress. If for a certain stress increment, the base length L changes by a length ΔL (Fig. 8-8), the quantity ΔL is called the *total elongation* of

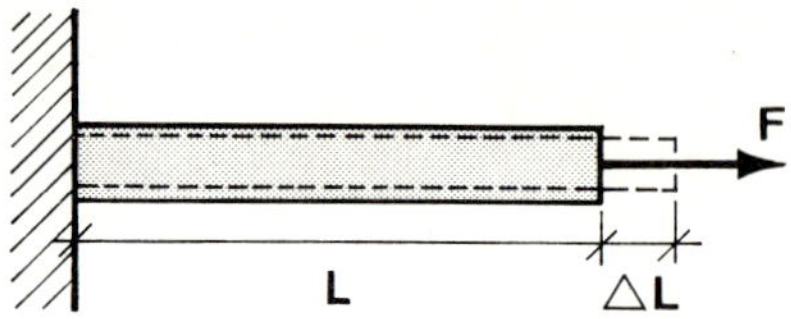

FIGURE 8-8

the length L. The elongation per the unit of length ϵ (Greek lowercase letter "epsilon") is obtained by dividing ΔL by the base length L:

$$\epsilon = \frac{\Delta L}{L}$$

The quantity ϵ is called the *unit strain* or simply the *strain.* The strain ϵ is the elongation of a unit length of the specimen, and it is a *dimensionless* number since it is a ratio of two lengths (e.g., in/in, ft/ft, or cm/cm).

Let us consider a tensile specimen of steel with a circular cross section of diameter $d = 0.504$ in (1.28 cm). The specimen is mounted in a testing machine, and an axial force is applied to it. The length of the specimen increases as the load is increased. The change of length of the specimen is measured by an extensometer which is attached to the specimen and has a base length of 2.0 in (5.08 cm). The data obtained from the tension test are shown in Table 8-1. In this table the normal stress due to a particular value of the axial load F has been computed by dividing the load by the original cross sectional area of the specimen, which is

$$A = \frac{\pi d^2}{4} = \frac{\pi \times 0.504^2}{4} = 0.20 \text{ in}^2 \qquad (1.29 \text{ cm}^2)$$

For example, the normal stress at $F = 2\,000$ lb (8 890 N) is

$$\sigma = \frac{2\,000 \text{ lb}}{0.20 \text{ in}^2} = 10\,000 \frac{\text{lb}}{\text{in}^2} \qquad \left(6\,900 \frac{N}{\text{cm}^2}\right)$$

TABLE 8-1 Tension Test Data (Steel)

TEST DATA		COMPUTED DATA	
LOAD ON THE SPECIMEN, lb	ΔL, in	STRESS σ, lb/in^2	STRAIN $\epsilon = \Delta L/L$
2 000	0.0007	10 000	0.00035
4 000	0.0013	20 000	0.00065
6 000	0.0020	30 000	0.00100
8 000	0.0027	40 000	0.00135
10 000	0.0034	50 000	0.00170
12 000	0.0046	60 000	0.00230

If the measured elongation for this stress is $\Delta L = 0.0007$ in, and the base length is $L = 2.0$ in, the corresponding strain is

$$\epsilon = \frac{\Delta L}{L} = \frac{0.0007}{2} = 0.00035$$

The rest of the data in Table 8-1 have been computed in the same way.

Inspection of the data in Table 8-1 shows that up to a certain level of the stress σ, the elongation ΔL is almost directly proportional to the stress σ. For example, if the stress increases three times, the elongation ΔL, and therefore the strain $\epsilon = \Delta L/L$, also increases three times. Thus there is a linear relationship between stress and strain that can be expressed by the simple (straight-line) equation

$$\sigma = E \times \epsilon \tag{8-1}$$

where E is a constant of proportionality between the stress σ and the corresponding strain ϵ. The quantity E can be computed by taking any value of σ in the table and the corresponding value of ϵ and substituting them in Eq. (8-1). If, for example, σ = 40 000 lb/in^2 and the corresponding $\epsilon = 0.00135$, then

$$E = \frac{\sigma}{\epsilon} = \frac{40\,000}{0.00135} = 29\,620\,000\ \frac{\text{lb}}{\text{in}^2} \qquad \left(20\,400\ \frac{\text{kN}}{\text{cm}^2}\right)$$

The constant E has the units of the stress σ (lb/in^2) since ϵ is a dimensionless quantity. The constant E is called the *tensile modulus of elasticity* of the specimen's steel, and Eq. (8-1) is known as *Hooke's law* after its discoverer, the English scientist Robert Hooke, who formulated it on the basis of experimental results in the year 1660. Hooke's law states that *within given limits, the stress in certain materials is linearly proportional to the strain. The constant of proportionality in this relationship is the modulus of elasticity E.*

The relationship in Eq. (8-1) may be plotted from the results of our steel tensile test, in a cartesian coordinate plane (Fig. 8-9). The magnitude of stress σ is plotted in a given scale on the vertical axis, while ϵ is plotted on the horizontal axis. Such a diagram is called a *stress-strain diagram*. In the stress-strain diagram, for our steel specimen, linear proportionality is shown by a straight line, which in our test extends approximately to point A, where σ = 50 000 lb/in^2. From that point on, the graph curves down, the strain increases more and more rapidly as the stress increases, and linear proportionality is not valid any more. The value of the stress at point A on the graph is called the limit of *proportionality* of the material. Hooke's law for steel material holds up to the limit of proportionality.

Another stress value called the *elastic limit* usually lies very close to the limit of proportionality. If we stress the specimen up to the elastic limit and then remove the load, the strain practically disappears. However, if we stress the specimen beyond its elastic limit and then remove the load, the deformation of the specimen does not vanish completely: a certain elongation remains, the so-called permanent elongation.

Since it is essential that structures should not exhibit permanent deformations

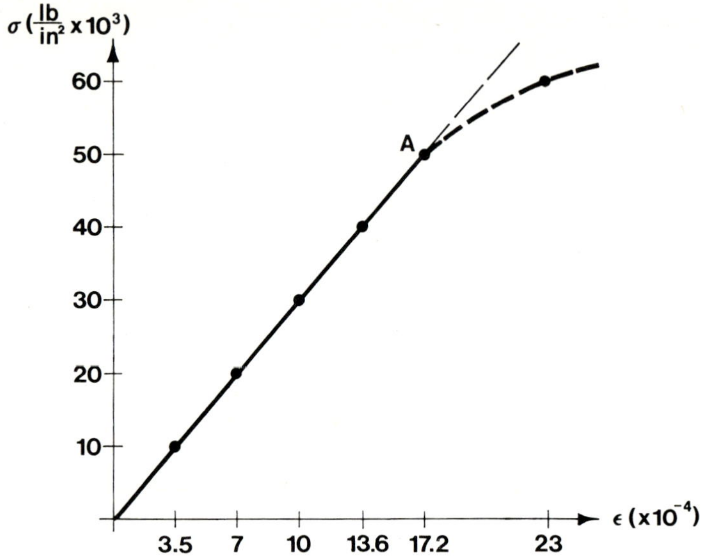

FIGURE 8-9 Stress-strain diagram for steel tensile specimen.

after the removal of loads, structural elements must never be stressed above the elastic limit or the almost coincident proportionality limit.

Identical tests on compression specimen can be used to define a *compressive strain* and a *compressive elastic modulus*. Some materials, like steel, have the same elastic modulus in tension and compression. Others, like concrete, do not.

The modulus of elasticity is an important property of a material. Usually, the higher the modulus of elasticity, the higher the allowable stresses in a material within the elastic limit. The modulus of elasticity E of the most commonly used structural materials are given in Table A-11. From this table we find, for example, that steel has approximately an $E_s = 30\ 000$ kips/in^2 = 30 000 000 lb/in^2 (20 700 kN/cm^2), while aluminum has, approximately, an $E_a = 10\ 000$ kips/in^2 = 10 000 000 lb/in^2 (7 000 kN/cm^2) or one third that of steel:

$$E_a = \frac{E_s}{3}$$

This indicates that for the same stress σ, the strain ϵ in the aluminum specimen is three times as large as that in the steel specimen. For example, if the normal stress in both steel and aluminum specimens is

$$\sigma = 30\ 000\ \frac{\text{lb}}{\text{in}^2} \quad \left(20.70\ \frac{\text{kN}}{\text{cm}^2}\right)$$

the strain in the steel specimen is (see Eq. 8-1)

$$\epsilon_s = \frac{\sigma}{E_s} = \frac{30\ 000}{30\ 000\ 000} = 0.001$$

while the strain in the aluminum specimen is

$$\epsilon_a = \frac{\sigma}{E_a} = \frac{30\,000}{10\,000\,000} = 0.003$$

Assume now that the length of both specimens is

$$L = 100 \text{ in}$$

the absolute elongation ΔL is found from the definition of strain

$$\epsilon = \frac{\Delta L}{L}$$

from which for the steel specimen

$$\Delta L_s = \epsilon_s \times L = 0.001 \times 100 = 0.1 \text{ in}$$

while the absolute elongation of the aluminum specimen is

$$\Delta L_a = \epsilon_a \times L = 0.003 \times 100 = 0.3 \text{ in}$$

If we continue to increase the load on the steel specimen used in Table 8-1 and plot additional points of the stress-strain diagram (Fig. 8-10), we might find that at a certain level of stress (point B) the strain increases without any appreciable

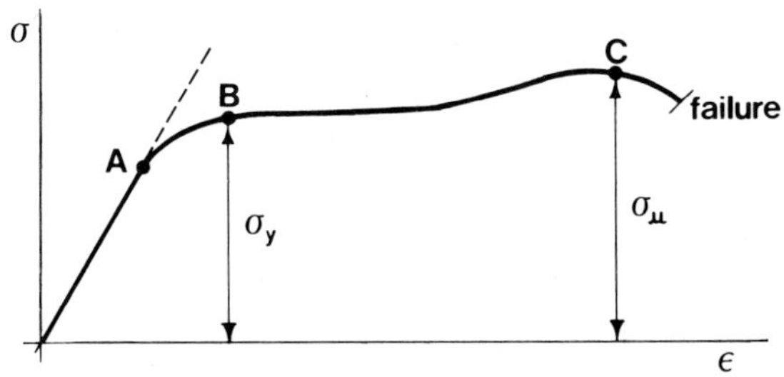

FIGURE 8-10 Full stress-strain diagram for steel.

increase in the stress. At this point the material starts to behave plastically and flows or yields. The stress at which this happens is called the *yield stress* or *yield point* σ_y (a most dangerous stress level for structures). After a certain amount of yielding some steels exhibit what is known as *strain hardening*, a rearrangement of the molecular grid which allows an increase in stress capacity. Thus we may increase the stress in a steel specimen beyond the yield point, with an accompanying increase in strain, until we reach point C, which defines the maximum value of the stress σ in the stress-strain diagram. At this point the cross section of the specimen rapidly narrows, a neck appears in it (Fig. 8-11), and failure follows. The stress at point C is called the *ultimate tensile stress* or *tensile strength* σ_u of the specimen material. (Remember that the stress values plotted in the stress-strain diagram are computed by dividing the applied force by the *original* area of the specimen before deformation and that the stress computed in this way is called the *nominal stress*.)

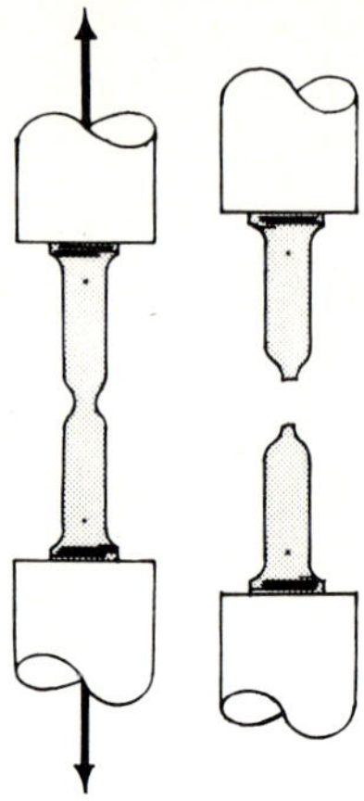

FIGURE 8-11 Necking of a steel tensile specimen shortly before failure.

Some materials, such as those whose stress-strain diagram is shown in Fig. 8-10, exhibit considerable yielding and necking before failure. Such materials are said to be *ductile*, and the amount of their yielding is a measure of their *ductility*. Wrought iron and steel are typical ductile materials. On the other hand, some materials do not experience any yielding: their stress-strain diagram is of the type shown in Fig. 8-12, that is, a straight line up to failure. Materials with this type of stress-strain diagram are called *brittle*. Cast iron, glass, and ice are brittle materials. The values of σ_u and σ_y for the most commonly used structural metals are given in Table A-12.

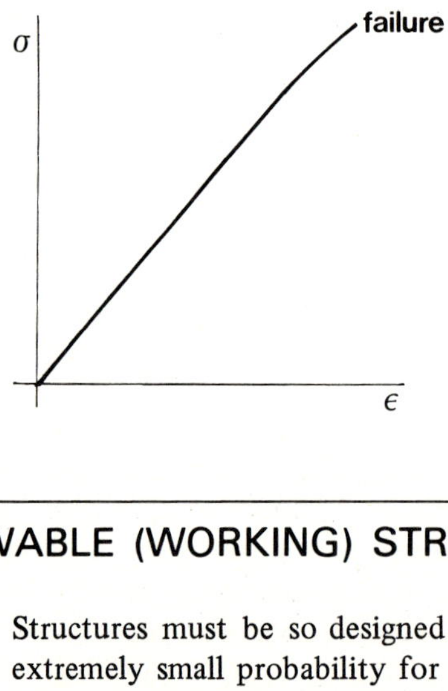

FIGURE 8-12 Stress-strain diagram of a brittle material.

8.4 ALLOWABLE (WORKING) STRESSES

Structures must be so designed that under any load combination there should be an extremely small probability for the stress σ in any cross section to reach σ_y or σ_u. In other words, we must make sure that the stresses developed in structures under service conditions are far below the values of σ_u or σ_y. The service stresses, thus limited to be a fraction of σ_u or σ_y, are called *allowable stresses* or *working stresses*. We usually determine the allowable stresses by the equation

$$\sigma_{all}^{\dagger} = \frac{\sigma_u}{(SF)_u}$$

when the allowable stress is based on the ultimate strength of the material, or by the equation

$$\sigma_{all} = \frac{\sigma_y}{(SF)_y}$$

when the allowable stress is based on the yield stress of the material. The coefficients SF are called *safety factors.* They are given by codes and industrial specifications and vary according to the types of structures, materials used, types of stresses, and loading conditions. Factors of safety are always larger than unity: They indicate how many times the load would have to be increased to reach the levels of stresses σ_u or σ_y. In this way, the largest stresses which can develop under the most severe loading conditions will be far below the critical stresses σ_u or σ_y. Typical values of allowable stresses for the most commonly used structural materials are listed in Table A-13. In the detailed design of a structure, the designer must follow the requirements of codes and industry specifications.

8.5 DESIGN OF MEMBERS UNDER AXIAL TENSION OR COMPRESSION

In Chapter 7 we learned how to find the internal axial forces in columns and truss members. The dimensions of these elements must now be determined so that the largest axial stress developed in them by the loads is smaller or, at most, equal to the allowable stress for the given material and type of loading. Thus we write

$$\sigma_{max} \leqq \sigma_{all} \tag{8-2}$$

For elements with constant cross section, the largest axial stress σ_{max} develops in the cross section with the largest axial force N_{max}

$$\sigma_{max} = \frac{N_{max}}{A} \tag{8-3}$$

where A is area of the cross section.

If we substitute in Eq. (8-2) for σ_{max} its expression from Eq. (8-3), we obtain

$$\frac{N_{max}}{A} \leqq \sigma_{all} \tag{8-4}$$

†Although this notation is correct from the standpoint of the strength of materials (physics), in texts on design of steel, wooden and other structures the allowable stresses are denoted by the capital letter F with the subscript *a* for axial, *b* for bending, *v* for shear allowable stresses. Consequently, the scientific notation F for force had to be replaced by *P*.

From Eq. (8-4) the required cross-sectional area

$$A_{req} \geqq \frac{N_{max}}{\sigma_{all}} \tag{8-5}$$

or, in words:

> The required cross-sectional area must be larger or at least equal to the ratio of the largest axial force in the member divided by the allowable stress of the given material.

Since we use allowable or working stresses to design members, this method is often called the *working-stress design method*.

Illustrative Problem 8-3

Find the lateral dimension and the elongation of a steel hanger 20 ft (6 m) long, with a square cross section loaded by a tensile axial force $F = 88$ kips (391 kN) (Fig. 8-13). The allowable stress in tension of the steel used is 22 000 lb/in² (15 000 N/cm²).

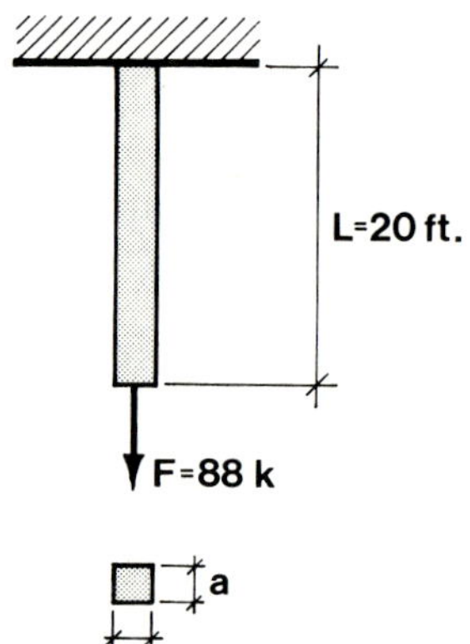

FIGURE 8-13

SOLUTION Neglecting the small weight of the bar itself, the internal axial force is constant over the whole length of the bar and equals

$$N_{max} = +88\,000 \text{ lb} \qquad \text{(tension)}$$

The required area of the cross section is computed from Eq. (8-5)

$$A_{req} = \frac{N_{max}}{\sigma_{all}} = \frac{88\,000}{22\,000} = 4.0 \text{ in}^2 \qquad (25.8 \text{ cm}^2)$$

The area of the square cross section of side a is

$$a^2 = 4.0 \text{ in}^2$$

Hence

$$a = \sqrt{4} = 2.0 \text{ in} \qquad (5.08 \text{ cm})$$

To find the total elongation ΔL of the bar of length $L = 20$ ft, we use Hooke's law [Eq. (8-1)]:

$$\sigma = E \times \epsilon$$

where

$$E = 30\,000\,000 \frac{\text{lb}}{\text{in}^2}$$

$$\sigma = \frac{F}{A} = \frac{88\,000 \text{ lb}}{4.0 \text{ in}^2} = 22\,000 \frac{\text{lb}}{\text{in}^2}$$

and

$$\epsilon = \frac{\Delta L}{L}$$

with

$$L = 20 \times 12 = 240 \text{ in}$$

Thus

$$22\,000 \frac{\text{lb}}{\text{in}^2} = 30\,000\,000 \frac{\text{lb}}{\text{in}^2} \times \frac{\Delta L}{240 \text{ in}}$$

$$\Delta L = \frac{22\,000 \times 240}{30\,000\,000} = 0.176 \text{ in} \qquad (0.447 \text{ cm} = 4.47 \text{ mm})$$

The length of the bar after deformation is increased by 0.176 in (4.47 mm).

Note: The weight of the bar is obtained by multiplying its volume $V = L \times A = 240 \times 4 = 960 \text{ in}^3$ by the unit weight of steel $\gamma = 0.30 \text{ lb/in}^3$. Thus the weight of the bar is $W = 960 \times 0.30 = 288$ lb, which is much less than the force $F = 88$ kips = 88 000 lb.

Illustrative Problem 8-4

Determine the cross section of bar 2 in the truss described in Illustrative Prob. 7-2, if the allowable tensile stress of the steel used is 10 kN/cm² (10 000 N/cm²).

SOLUTION In Illustrative Prob. 7-2 we found that the tensile axial force in member 2 is

$$N_2 = +2.5 \text{ kN} = 2\,500 \text{ N}$$

The required area of the cross section is then

$$A_{\text{req}} = \frac{N_2}{\sigma_{\text{all}}} = \frac{2\,500}{10\,000} = 0.25 \text{ cm}^2$$

Let us choose a rod of circular cross section. The diameter d of the rod is then computed by equating the area of a circular cross section of diameter d to 0.25 cm^2, or

$$\frac{\pi d^2}{4} = 0.25$$

from which

$$d^2 = \frac{0.25 \times 4}{\pi}$$

so that

$$d = \sqrt{\frac{0.25 \times 4}{\pi}} = 0.564 \text{ cm} = 5.64 \text{ mm}$$

Since a rod with such a diameter is not available on the market, the practical solution consists in selecting a rod with the rounded diameter of 0.6 cm = 6 mm.

Let us now *analyze* the bar we have just chosen. The actual area of the selected cross section is

$$A = \frac{\pi (0.6)^2}{4} = 0.283 \text{ cm}^2$$

instead of the required minimum of 0.25 cm^2. The actual stress in the rod is then

$$\sigma = \frac{2\,500}{0.283} = 8\,800 \frac{N}{\text{cm}^2} = 8.8 \frac{\text{kN}}{\text{cm}^2}$$

This stress is lower than the given tensile allowable stress of the materials (10 kN/cm^2). Thus we write

$$8.8 \frac{\text{kN}}{\text{cm}^2} < 10 \frac{\text{kN}}{\text{cm}^2} \text{ allowable}$$

and the design is on the safe side.

Illustrative Problem 8-5

The roof structure shown in Fig. 8-14 contains a tie-rod T which transfers a tensile axial force of 300 kN (67.4 kips). Design the tie-rod with a cross section

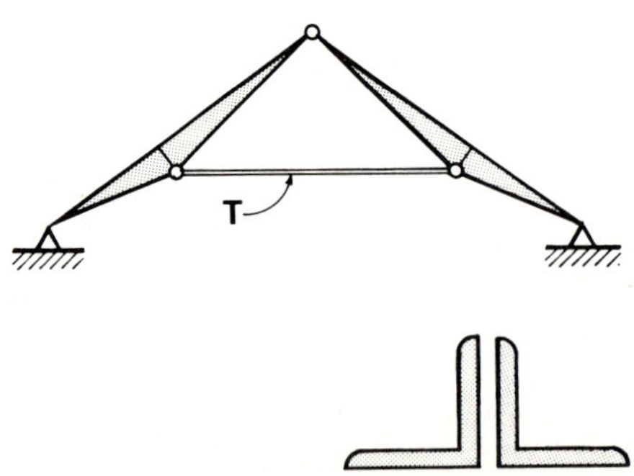

FIGURE 8-14

composed of two equal-leg angles (Fig. 8-14), assuming an allowable tensile stress in the steel of 15 kN/cm².

SOLUTION The required area of the tie-rod is

$$A_{req} = \frac{N_{max}}{\sigma_{all}} = \frac{300}{15} = 20 \text{ cm}^2$$

Thus each angle must have an area

$$A_1 = \frac{20}{2} = 10 \text{ cm}^2$$

We look in Table A-9 in the Appendix for an angle with an area close to the required 10 cm². This is an angle **L** 75 × 75 × 8, with an area

$$A_1 = 11.0 \text{ cm}^2$$

It has a leg length of 75 mm and a thickness of 8 mm. Thus, for the design of the tie-rod, we use

2 **L** 75 × 75 × 8

with a total cross-sectional area

$$A = 2 \times 11 = 22 \text{ cm}^2$$

ANALYSIS The actual stress in the designed tie-rod is

$$\sigma = \frac{N}{A} = \frac{300}{22} = 13.64 \frac{\text{kN}}{\text{cm}^2} < 15 \frac{\text{kN}}{\text{cm}^2}$$

The actual stress in the designed tie-rod is smaller than the allowable tensile stress of the material; the tie-rod is safely designed.

What is the total elongation of the tie-rod if its length $L = 70$ m = 7 000 cm and the modulus of elasticity of the steel is 20 700 kN/cm²? With the actual stress in the rod $\sigma = 13.64$ kN/cm², $E = 20\ 700$ kN/cm², and

$$\epsilon = \frac{\Delta L}{L} = \frac{\Delta L}{7\ 000},$$

Hooke's law $\sigma = E \times \epsilon$ gives

$$13.64 = 20\ 700 \times \frac{\Delta L}{7\ 000}$$

from which

$$\Delta L = \frac{13.64 \times 7\ 000}{20\ 700} = 4.6 \text{ cm} \qquad (1.8 \text{ in})$$

Illustrative Problem 8-6

A cast-iron column runs through two stories of a building and is loaded by the axial forces as shown in Fig. 8-15. Compute the normal stress in the upper and

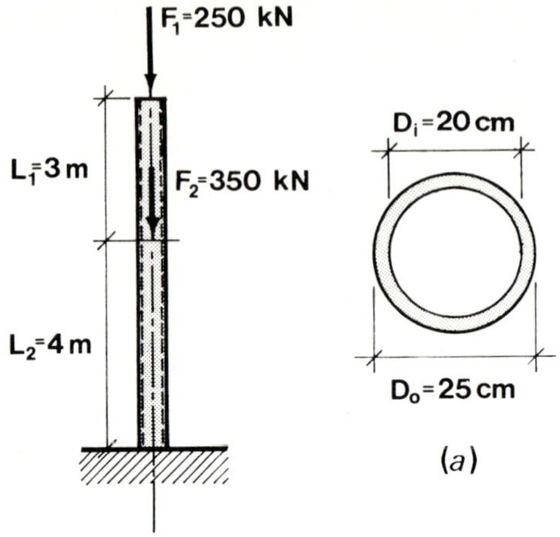

FIGURE 8-15 Cast-iron column in Illustrative Prob. 8-6: (*a*) cross-section of the column.

lower parts of the column and the total shortening ΔL of the column, if the modulus of elasticity of cast iron is $E = 7\,500\ \text{kN/cm}^2$.

SOLUTION The constant compressive axial force acting in the top portion of the column is

$$N_1 = -250\ \text{kN}$$

while that acting in the lower portion of the column is

$$N_2 = -250 - 350 = -600\ \text{kN}$$

The area of the cross section of the column shown in Fig. 8-15*a* is computed as the difference between the area of the outer circle of diameter $D_o = 25$ cm and the area of the inner circle (hole) of diameter $D_i = 20$ cm, or

$$A = \frac{\pi D_o^2}{4} - \frac{\pi D_i^2}{4} = \frac{\pi}{4}(D_o^2 - D_i^2) = \frac{\pi}{4}(25^2 - 20^2)$$

$$= \frac{\pi}{4}(625 - 400) = 177\ \text{cm}^2$$

The compressive normal stress in the upper part of the column is thus

$$\sigma_1 = \frac{N_1}{A} = \frac{-250}{177} = -1.41\ \frac{\text{kN}}{\text{cm}^2} \qquad \left(-2\,045\ \frac{\text{lb}}{\text{in}^2}\right)$$

and that in the lower part

$$\sigma_2 = \frac{N_2}{A} = \frac{-600}{177} = -3.40\ \frac{\text{kN}}{\text{cm}^2} \qquad \left(-4\,900\ \frac{\text{lb}}{\text{cm}^2}\right)$$

The total shortening of the column is the sum of the shortenings of its two portions. For the upper part, we have

$$\sigma_1 = 1.41 \frac{\text{kN}}{\text{cm}^2}$$

$$L_1 = 3.0 \text{ m} = 300 \text{ cm}$$

$$\epsilon = \frac{\Delta L_1}{L_1} = \frac{\Delta L_1}{300}$$

$$E = 7\,500 \frac{\text{kN}}{\text{cm}^2}$$

and by Hooke's law, $\sigma = E \times \epsilon$

$$1.41 = 7\,500 \times \frac{\Delta L_1}{300}$$

from which

$$\Delta L_1 = \frac{1.41 \times 300}{7\,500} = 0.0564 \text{ cm} = 0.564 \text{ mm} \qquad \text{(shortening)}$$

For the lower part of the column, we have

$$\sigma_2 = 3.40 \frac{\text{kN}}{\text{cm}^2}$$

$$L_2 = 4.0 \text{ m} = 400 \text{ cm}$$

$$\epsilon = \frac{\Delta L_2}{L_2} = \frac{\Delta L_2}{400}$$

and by Hooke's law, $\sigma = E \times \epsilon$

$$3.40 = 7\,500 \times \frac{\Delta L_2}{400}$$

from which

$$\Delta L_2 = \frac{3.40 \times 400}{7\,500} = 0.18 \text{ cm} = 1.8 \text{ mm} \qquad \text{(shortening)}$$

The total shortening of the column is

$$\Delta L = \Delta L_1 + \Delta L_2 = 0.564 + 1.8 = 2.364 \text{ mm} \qquad (0.1 \text{ in})$$

New Terms Appearing in Chapter 8

Strength of materials
Rigid, undeformable bodies
Elastic materials
Deformable bodies
Unit stress
Stress
Normal stress, σ
Shear (tangential) stress, τ

Elastic deformations
Distribution of stress over a cross section
Design and analysis of structural members
Tensile test
Tensile specimen
Extensometer
Elongation (absolute or total)
Unit strain, ϵ
Strain, ϵ
Modulus of elasticity in tension, E
Stress-strain diagram, $\sigma - \epsilon$
Hooke's law, $\sigma = E \times \epsilon$
Proportionality limit
Elastic limit
Permanent elongation
Yield stress, σ_y
Strain hardening
Ultimate tensile stress, σ_u (tensile strength)
Nominal stress
Ductile materials
Ductility
Brittle materials
Allowable stress (working stress), σ_{all}
Service stress (at service conditions)

Quiz Number 7

A. TRUE OR FALSE

1. All structural materials are undeformable, that is, absolutely rigid so that no deformations develop in structures under the action of loads.
2. A modulus of elasticity is of no use in determining deformations of structural members.
3. Unit stress, or simply stress, is defined as the ratio of a unit area A to the force F acting on it.
4. A general stress at a point of a cross section may be resolved into a normal stress σ and a shear stress τ at that point.
5. A shear or tangential stress acts at right angles to the cross section.
6. The determination of all the needed dimensions of structural members is called *analysis*.
7. The axial load on a truss element results in a uniformly distributed normal stress on its cross sections.
8. Structures should not have large deformations.
9. Structures should not have permanent deformations.
10. After the removal of loads, the elastic deformations due to the loads do not disappear.
11. An extensometer is a device for measuring the elongation or shortening of a material specimen between two points, which define the base length of the extensometer.
12. A unit strain, or simply strain, is always equal to a unit stress.
13. The strain defines the amount of deformation per unit length of a member.
14. A stress-strain diagram may be obtained from a tensile test.
15. The stress-strain diagram for a linearly elastic material is a curve.
16. Hooke's law expresses a direct (linear) proportionality between the stress and the force causing that stress.

17. In a linearly elastic material, stress is linearly proportional to strain. The constant of proportionality is called the *modulus of elasticity* of the material.
18. The modulus of elasticity has the same units as the strain.
19. The yield point on a stress-strain diagram indicates the nominal stress at which the strain increases without an appreciable increase in the stress.
20. The tensile strength of a material is the value of the maximum nominal tensile stress in the stress-strain diagram for the material.

B. MULTIPLE CHOICE

1. Which of the following is not a ductile material?
   (*a*) glass (*b*) steel (*c*) brass (*d*) wrought iron
2. What usually happens to a tensile specimen of ductile material when it approaches failure?
   (*a*) its cross section increases
   (*b*) it slips out of the grips of the tension testing machine
   (*c*) it necks
   (*d*) it develops an elastic strain
3. Which of the following is a ductile material?
   (*a*) ice (*b*) gypsum (*c*) cast iron (*d*) concrete (*e*) none of these
4. Which of the following stresses should develop in structures under service conditions?
   (*a*) the yield stress (*b*) the elastic limit stress
   (*c*) the ultimate stress (*d*) the allowable stress
5. Which of the following equations expresses Hooke's law?
   (*a*) $\sigma = \frac{F}{A}$ (*b*) $\sigma = E \times \epsilon$ (*c*) $\sigma = \frac{E}{\epsilon}$ (*d*) $\sigma = \frac{\Delta L}{L}$
6. Which material has the smallest modulus of elasticity in tension?
   (*a*) steel (*b*) aluminum (*c*) timber (*d*) wrought iron
7. Which of the following has the highest modulus of elasticity in tension?
   (*a*) steel (*b*) brass (*c*) cast iron (*d*) aluminum
8. Under the action of a tensile axial force a tie-rod 100 ft long develops a strain $\epsilon = 0.002$. What is the total elongation of the tie rod in inches?
   (*a*) 2.4 (*b*) 0.24 (*c*) 2.0 (*d*) 0.2
9. The yield stress of a material is $\sigma_y = 60$ kips/in^2, and the safety factor to be used under working conditions is SF = 2.0. What is the allowable stress in the material?
   (*a*) 120 kips/in^2 (*b*) 30 000 lb/in^2 (*c*) 3 000 kips/in^2 (*d*) 15 kips/in^2
10. If a shear force $V = 10\,000$ lb acts on cross-sectional area $A = 100$ in^2, what is the magnitude of the shear stress uniformly distributed over the entire area?
    (*a*) 100 kips/in^2 (*b*) 1 000 lb/in^2 (*c*) 0.100 kips/in^2 (*d*) 10 lb/in^2

C. SHORT ANSWERS

1. Glass is a ____________ material.
2. The modulus of elasticity of steel is approximately __________ times as large as the modulus of elasticity of aluminum.

3. If $\epsilon = 0.002$ and $E = 10\,000\,000$ lb/in², the corresponding stress is $\sigma =$ ____________________.
4. Draw the shape of the stress-strain diagram for a brittle material.
5. In an elastic material following Hooke's law, the stress is linearly proportional to the strain. If the stress increases three times, the strain increases __________ times.
6. A tensile axial force $F = 200$ kips pulls a truss member made of steel with an allowable tensile stress of 20 000 lb/in². What is the required cross-sectional area of the member?
7. Hooke's law is a relationship between E, σ, and ϵ. What is the expression for ϵ derivable from Hooke's law?

$\epsilon =$ ________

8. Place the letter corresponding to the following values of the moduli of elasticity next to the corresponding material.

(*a*) $E = 30\,000 \dfrac{\text{kips}}{\text{in}^2}$ timber ________

(*b*) $E = 3\,000 \dfrac{\text{kips}}{\text{in}^2}$ concrete ________

(*c*) $E = 1\,760\,000 \dfrac{\text{lb}}{\text{in}^2}$ steel ________

(*d*) $E = 10\,000\,000 \dfrac{\text{lb}}{\text{in}^2}$ aluminum ________

9. The scale of an extensometer shows a reading $\Delta L = 0.03$ in on a base $L = 30$ in; what is the value of ϵ?

$\epsilon =$ __________

10. A tensile strain value $\epsilon = 0.001$ has been determined in a steel tie-rod. What is the tensile stress in the tie-rod?

$\sigma = +$ ____________________

EXERCISES

8-1 A steel column is acted on by an axial force N = 600 kN. Find the required size of a wide-flange section for this column, if the allowable stress is 15 kN/cm².

8-2 Find the magnitude of the tensile force F acting on a steel bar of circular cross section with a diameter $D = 1.0$ in, if the strain in the bar $\epsilon = 0.002$.

8-3 Determine the required thickness t of a column of tubular cross section, if it is loaded by an axial force $F = 2\,000$ kN. The outside diameter D_o of the column is 30 cm, and the allowable stress of its material is 10 kN/cm². Round off the calculated thickness to the nearest 0.5 cm, and then compute the actual stress in the designed column.

8-4 Design a steel tie-rod capable of resisting a tensile axial force of 600 kN. The only available material consists of steel strips with a cross-sectional area of 80 × 10 mm each. How many such strips must be used if the allowable stress of the material is 15 kN/cm²?

8-5 One hanger in a building is composed of four equal-leg angles (Fig. 8-16) and supports 18 floors. The load from each floor on the hanger is 20 kips. What is the required constant cross-sectional area of the hanger, if the allowable stress in the steel is 20 kips/in^2? Select the size of the angles from Table A-9. Ignore the weight of the hanger.

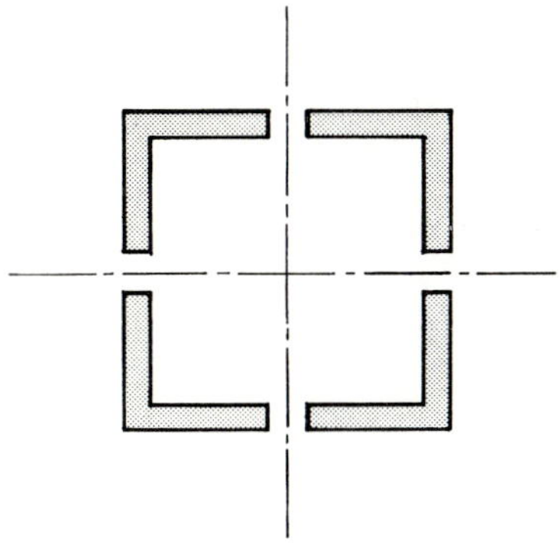

FIGURE 8-16

8-6 The cross section of a steel bar is shown in Fig. 8-17. What maximum safe load F can the bar carry, if the allowable stress of its material is 20 kips/in^2? What is the total elongation of the bar, if its length is 10 ft?

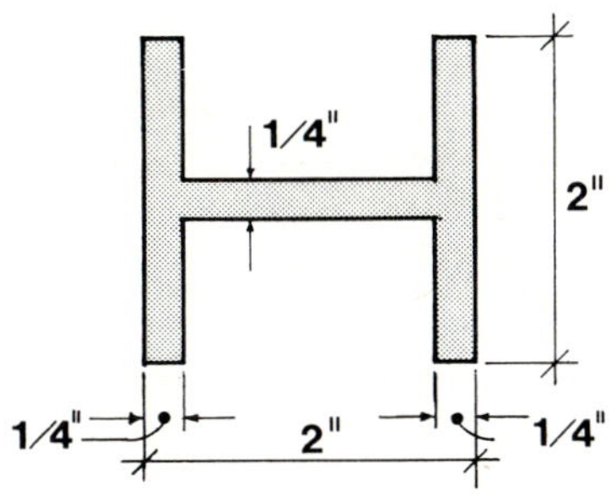

FIGURE 8-17

8-7 A platform 24 × 24 ft is to support a uniformly distributed load of 250 lb/ft^2. How many square timber posts must be used to support the platform if the size of the posts is 4 × 4 in and the allowable stress in compression of the timber is 1 000 lb/in^2? Show how you would arrange these posts so that each one of them carries the load from an equal tributary area.

Failure of a tapered timber beam in the testing machine (*Courtesy of Colorado State University, Fort Collins*).

Bending Stresses in Beams; Design of Beams

9

9.1 MOMENT OF INERTIA AND SECTION MODULUS

Beams bend and deflect under transverse loads. Like any other deformation in structures, bending, or flexure, a deformation typical of beams, is caused by the stresses which develop in their cross sections. Such stresses are called *bending stresses*. We will now determine the distribution of bending stresses over the beam cross section and establish where they are largest.

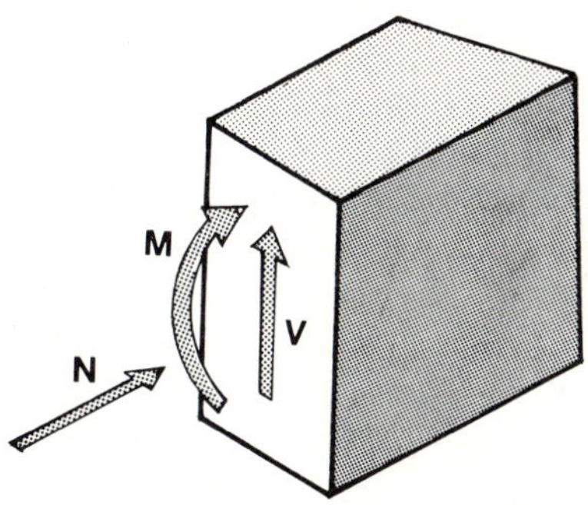

As we learned in Chap. 7, internal forces, that is, bending moments, shears, and axial forces, act on the cross sections of beams. Let us first investigate the state of stress due to bending moments on the cross sections of a beam. For example, in the beam shown in Fig. 9-1, the forces F are symmetrically located with respect to the midspan of the beam. The end reactions are

$$R_A = R_B = F$$

In the middle portion of the beam (between the two forces F) the shear force

$$V = 0$$

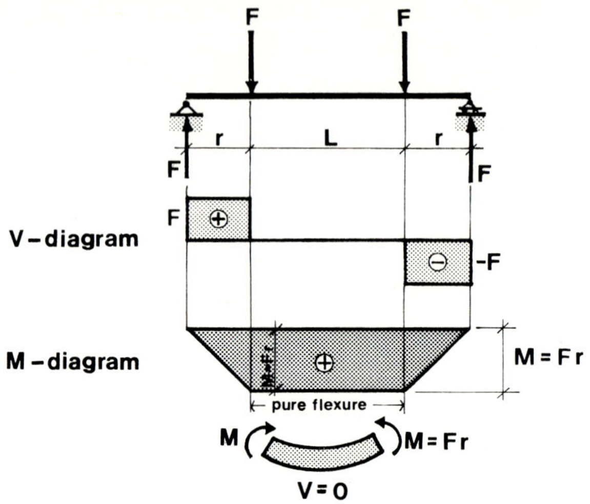

FIGURE 9-1

and the bending moment

$$M = F \times r$$

is constant. The state of stress in a portion of a beam in which $V = 0$ and M = constant is called *pure bending*. The nature of the deformation due to pure bending is best observed on a beam model made of rubber. Let the beam have a rectangular cross section of width b and depth h (Fig. 9-2). Draw fine lines in the middle portion of the model, evenly spaced by d and perpendicular to the longitudinal axis of the beam. (The longitudinal axis is defined as a line connecting the centroids of all cross sections of the beam. In a rectangular section the longitudinal axis is thus located at middepth.)

Now apply such a loading to the model that its middle portion is in pure bending and observe the lines after the model deforms under load (Fig. 9-2): the lines remain straight, but they rotate as the distance between them at the top of the beam is now smaller than d, while at the bottom the distance is larger than d. Along the longitudinal axis of the beam, instead, the distance d remains unchanged.

Imagine now that two adjacent cross sections of the beam, a distance d apart, are connected by elastic springs, representing the elastic fibers of the beam material. If the fibers at the top have shortened, a *compressive* normal stress must act on them. If the fibers at the bottom elongate, a *tensile* normal stress must act on them, and if the length of the fibers on the beam axis remains unchanged, no normal stress acts on them.

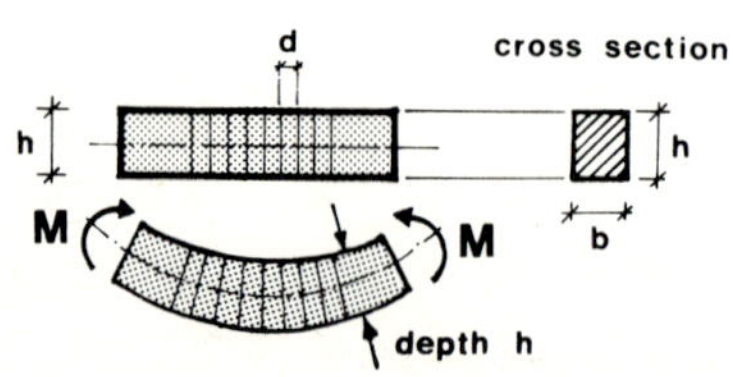

FIGURE 9-2

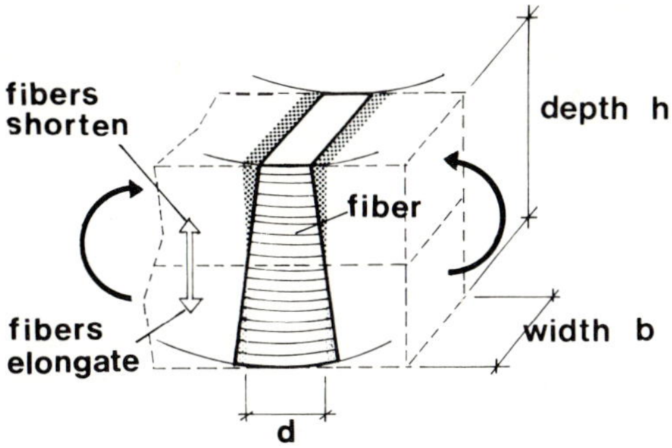

FIGURE 9-3 As a result of moments shown, bottom fibers of the beam elongate and top fibers shorten. Elongation and shortening of the fibers vary linearly along the depth of the beam.

We noticed that the straight lines remain straight after deformation (Fig. 9-3); this means that the elongation and shortening of the fibers varies *linearly* along the depth of the beam. The farther a fiber is from the beam axis, the larger its elongation is and, hence, the larger the stress in it. Thus the distribution of the bending (normal) stresses over the cross section of the beam is as shown in Fig. 9-4. The fibers on the horizontal plane along the longitudinal axis of the beam have zero normal stress and constitute the so-called *neutral plane*. The neutral plane intersects a cross section in what is called the *neutral axis* (NA) of the cross section. Remember that the neutral axis is a centroidal axis of a cross section. The bending stress has the same value at all points of the width b of the cross section, located at the same distance y from the neutral axis, since the fibers at such points have the same elongation or shortening.

For the sense of the bending moment M shown in Fig. 9-3, the largest (positive) tensile stress $\sigma_T(+)$ develops in the bottom fibers of the cross section, a distance $h/2$ from the neutral axis. The largest (negative) compressive stress $\sigma_C(-)$ develops in the top fibers, also a distance $h/2$ from the neutral axis.

The effect of compressive stresses acting on the top half of the cross section may be replaced by a single force C, acting at the centroid of the compression triangle ABC

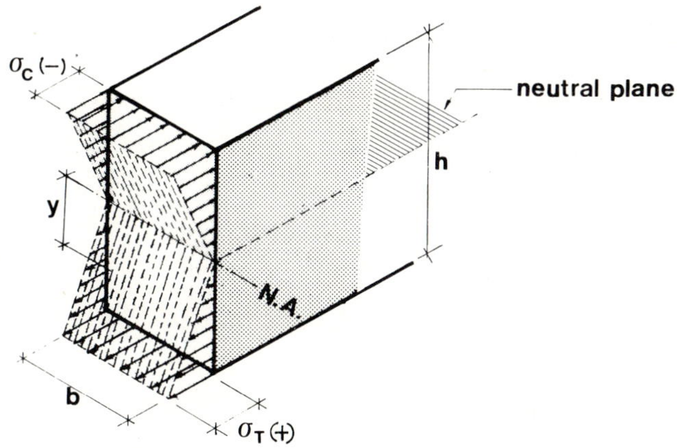

FIGURE 9-4 Distribution of bending stresses over a cross section of the beam.

Connection of steel beams to a column in a steel building (*Photo by L. Cerny*).

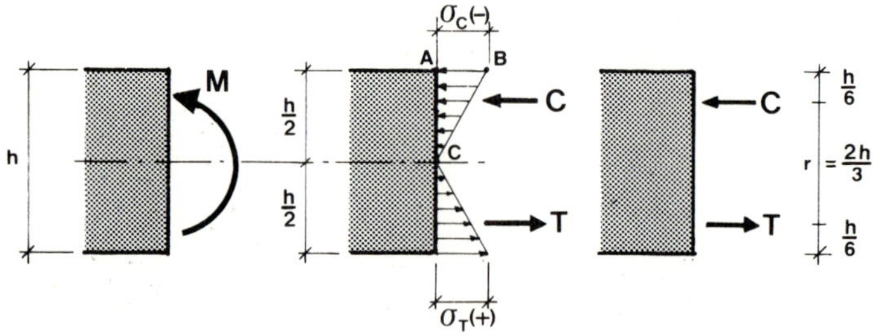

FIGURE 9-5

in Fig. 9-5. The magnitude of the compressive resultant C is equal to the volume of the compression wedge *ABC-DEF*, shown in Fig. 9-6:

$$C = \frac{1}{2}\sigma_C \times \frac{h}{2} \times b = \frac{\sigma_C \times b \times h}{4}$$

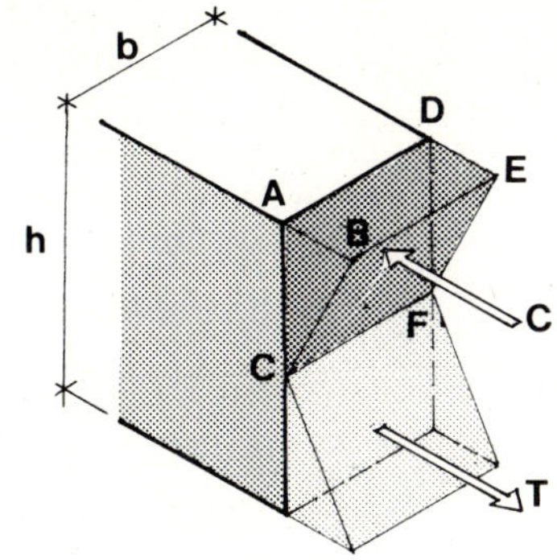

FIGURE 9-6

Similarly, the tensile resultant T on the bottom half of the cross section is

$$T = \frac{\sigma_T \times b \times h}{4}$$

The internal resultants C and T are two equal and opposite forces a distance $\boldsymbol{r}$ apart and constitute a couple. The distance $\boldsymbol{r}$ is called the *lever arm* of the internal couple and is

$$r = \tfrac{2}{3}\, h$$

Thus the moment of the internal couple is

$$M_i = T \times r = \frac{\sigma_T \times b \times h}{4} \times \frac{2}{3} h = \sigma_T \times \frac{bh^2}{6}$$

or

$$M_i = C \times r = \frac{\sigma_C \times b \times h}{4} \times \frac{2}{3} h = \sigma_C \times \frac{bh^2}{6}$$

The moment M_i of the internal couple at a section must be equal to the bending moment M at the same section, as defined in Sec. 7-3.

Thus

$$M_i = M$$

and

$$\sigma_T \times \frac{bh^2}{6} = M$$
$$\sigma_C \times \frac{bh^2}{6} = M \qquad (9\text{-}1)$$

In Eqs. (9-1), σ_T and σ_C are the largest tensile and compressive bending stresses in the cross section due to the bending moment M. Their magnitude may be computed from Eqs. (9-1).

$$\sigma_T = + \frac{M}{bh^2/6} \quad \text{(tension)}$$

$$\sigma_C = - \frac{M}{bh^2/6} \quad \text{(compression)} \tag{9-2}$$

S

The expression $bh^2/6$ is called the *section modulus* S of the rectangular cross section of width b and depth h,

$$S = \frac{bh^2}{6} \quad (\text{in}^3 \text{ or cm}^3) \tag{9-3}$$

Thus, Eqs. (9-2) may also be written as

$$\sigma_T = + \frac{M}{S}$$

$$\sigma_C = - \frac{M}{S} \tag{9-4}$$

and show that for a rectangular section $\sigma_C = - \sigma_T$. The type of stress, tension or compression, acting on a particular half of the cross section depends solely on the sense of the bending moment M.

The section modulus S is an important property of the cross section, which characterizes its *bending capacity* and, indirectly, its *stiffness*. The larger the section modulus S, the smaller the stresses developed in the section for a given value of the bending moment M.

Equations (9-4) may also be rewritten as

$$M = \sigma_T \times S$$

or

$$M = \sigma_C \times S \tag{9-5}$$

to show that the larger the section modulus S, the larger the bending moment M developed at a section for a given value of the stress σ_T or σ_C.

The section moduli S for certain cross-sectional shapes are shown in Table A-10. The section moduli S for steel sections (wide-flange sections, channels, and angles) are found in Tables A-4 through A-9.

I

Another important section property, called the *moment of inertia* I of the section, may be derived from the section modulus. If we indicate by $\bar{y}$ the distance from the neutral axis (which is the centroidal axis of the cross section) to the top or the bottom fibers and S the corresponding section modulus, the

$$I = S \times \bar{y} \tag{9-6}$$

is the moment of inertia I of the cross section about its neutral (centroidal) axis.

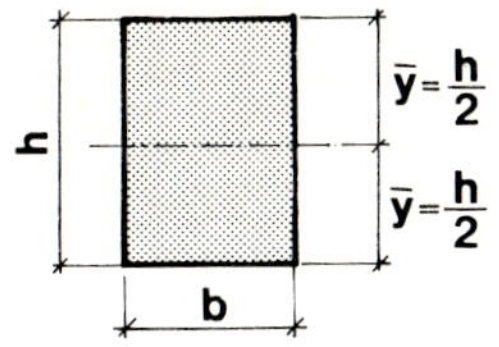

FIGURE 9-7

In a rectangular cross section (Fig. 9-7)

$$\bar{y} = \frac{h}{2}$$

and

$$S = \frac{bh^2}{6}$$

so that

$$I = S \times \bar{y} = \frac{bh^2}{6} \times \frac{h}{2} = \frac{bh^3}{12} \qquad (\text{in}^4 \text{ or cm}^4)$$

The formulas for the evaluation of the moment of inertia I_x about the centroidal axis X–X for some basic cross-sectional shapes are given in Table A-10. For shapes symmetrical about the neutral axis, the distances $\bar{y}$ to the top and to the bottom fibers are equal; hence there is only one value of the section modulus S, and the magnitude of the stresses at the top and bottom fibers is the same. However, for shapes which are not symmetrical about the neutral axis, this is not so. If the distances from the neutral axis to the bottom and to the top fibers are $\bar{y}_1$ and $\bar{y}_2$, respectively, and $\bar{y}_1$ is different from $\bar{y}_2$, the section modulus S_1 for the computation of the stress in the bottom fibers may be proved to be

$$S_1 = \frac{I_x}{\bar{y}_1}$$

and that for the stress in the *top* fibers is

$$S_2 = \frac{I_x}{\bar{y}_2}$$

Thus the stress in the bottom fibers of a triangular section in Fig. 9-8 is

$$\sigma_T = \frac{M}{S_1}$$

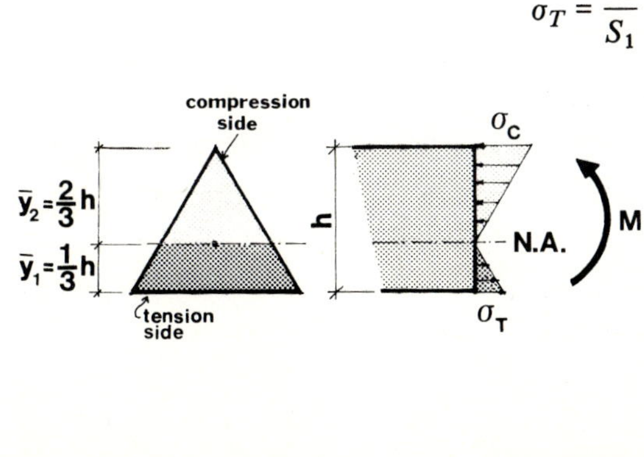

FIGURE 9-8

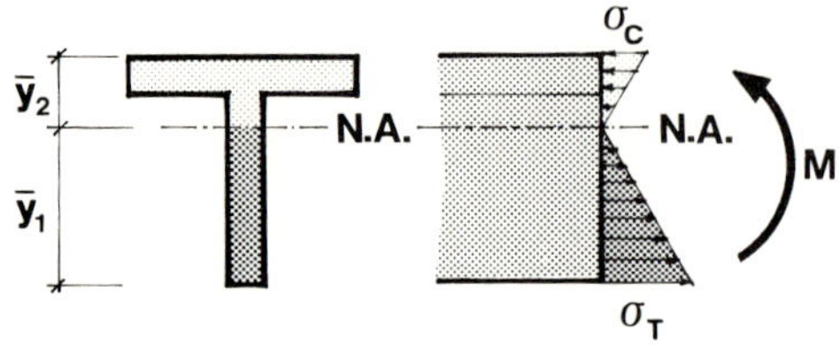

FIGURE 9-9

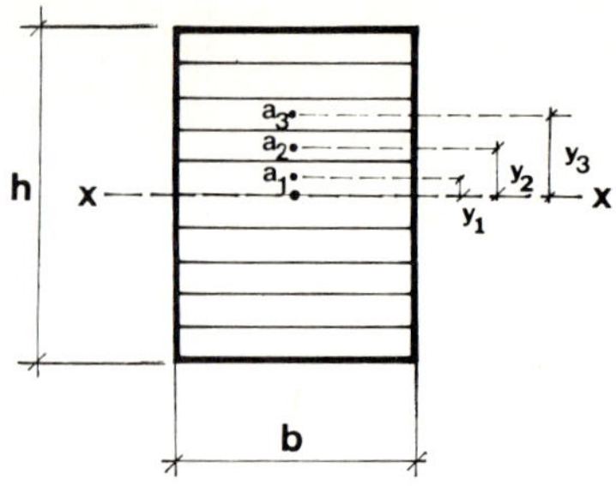

FIGURE 9-10

and that in the top fibers is

$$\sigma_C = \frac{M}{S_2}$$

where S_1 and S_2 are given in Table A-10. The same situation occurs in the unsymmetrical T-shaped section of Fig. 9-9.

To better understand the physical significance of the moment of inertia of a section about its horizontal centroidal axis, let's turn to Fig. 9-10, which shows a rectangle with an X–X axis parallel to the base b, passing through its centroid. Divide the rectangle into a number of thin rectangles or strips of areas a_1, a_2, a_3, and so on, and let the distances from the centroidal X–X axis to the centroid of each strip be y_1, y_2, y_3, etc. If each area a is multiplied by the square of its "lever arm" y, it can be shown that the sum of products

$$a_1 \times y_1^2 + a_2 \times y_2^2 + a_3 \times y_3^2 + \cdots + a_n \times y_n^2$$

is nothing else but the moment of inertia I_x of the rectangle $b \times h$ about the centroidal axis X–X

$$I_x = \Sigma a \times y^2 \tag{9-6a}$$

It is apparent from Eq. (9-6*a*) that the greater the distance y of a strip from the centroidal axis, the larger the contribution of that strip to the moment of inertia of the section. Such a contribution depends on the *square* of the distance y. Thus, if the area a of each strip is equal to 1.0 in^2 and the corresponding "lever arms" y are $y_1 = 1.0$ in, $y_2 = 2$ in, $y_3 = 3$ in, and so on, the contributions of the strips to the moment of inertia of the section are

$$i_1 = 1.0 \times 1.0^2 = 1.0 \text{ in}^4$$

$$i_2 = 1.0 \times 2.0^2 = 4 \text{ in}^4$$

$i_3 = 1.0 \times 3.0^2 =$ ~~27 in^4~~ → 9 in^4

From this sequence, we can appreciate how highly that part of the cross section far away from the neutral axis contributes to the moment of inertia. The most economical

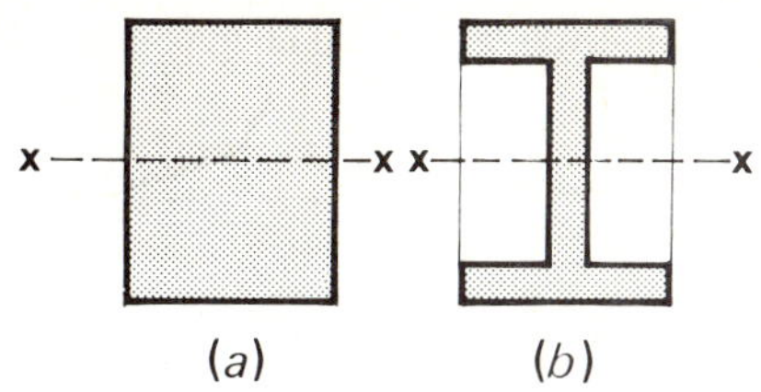

FIGURE 9-11

and efficient cross sections in bending have as little material as possible in the neighborhood of the neutral axis, and as much material as possible located away from the neutral axis (with large "lever arms" y). As far as the amount of material is concerned, the solid rectangle in Fig. 9-11*a* is far less efficient in bending than the I-shaped section in Fig. 9-11*b*. This is the reason for the development of wide-flange sections, which, for a given amount of material, have a high carrying capacity in bending because most of the material is located away from the neutral axis. Similarly, a solid circular section is far less efficient in bending than a hollow one of the same area, which has a lot of material removed from the neighborhood of the centroidal axis.

9.2 DESIGN AND ANALYSIS OF BEAMS

We design a beam when, given its span and loading, we must choose its material and its cross-sectional dimensions so that the beam will safely carry the loading. In designing beams by the *allowable stress method*, the computed stresses at any point of a beam must be smaller than or at most equal to the allowable stress in bending for the material used. The allowable stresses for different material are given in codes and specifications. In designing a beam, we first identify one or more critical sections and then determine the critical section where the largest bending moment M_{max} occurs. The largest bending stresses at this critical section occur at the bottom and top fibers and equal, respectively:

$$\sigma_{max} = \frac{M_{max}}{S_1}$$

$$\sigma_{max} = \frac{M_{max}}{S_2}$$

The sign of these stresses (tension or compression) depends on the sense of the bending moment at the section. Assuming that the material used resists tension and compression equally well, as is the case for steel and other metals, these stresses must be smaller than or, at most, equal to the allowable bending stress σ_{all} for the material used:

$$\sigma_{max} = \frac{M_{max}}{S} \leqq \sigma_{all} \tag{9-7}$$

where S is the *smaller* of the two moduli S_1 and S_2. The necessary value of the section modulus can be now determined from Eq. (9-7):

$$S_{req} \geqq \frac{M_{max}}{\sigma_{all}} \tag{9-8}$$

The dimensions of the cross section are then computed or selected from tables so that the adopted section has at least the required section modulus.

In the analysis of a beam in bending, the loads, span, material, and cross-sectional dimensions of the beam are all known. All we have to do is to prove that the given beam can carry the assigned loads safely, that is, that its computed bending stresses at any point in the beam are smaller than or equal to the allowable bending stress of the beam material. Sometimes, in the analysis of a beam, the largest allowable moment M_{all} is computed from Eq. (9-8):

$$M_{all} = S \times \sigma_{all} \tag{9-9}$$

and compared with the M_{max} due to the loads.

Illustrative Problem 9-1

We want to span a floor with timber beams having a rectangular cross section 6 X 12 cm (Fig. 9-12). These can be laid with the long or with the short side horizontal. Which side must be laid horizontally to achieve the largest bending capacity of the section? We ignore here the fact that timber sections are usually specified by the *nominal* dimensions which are typically $\frac{1}{2}$ in (1.25 cm) larger than the *dressed*, structurally effective dimensions used in the problems of this chapter.

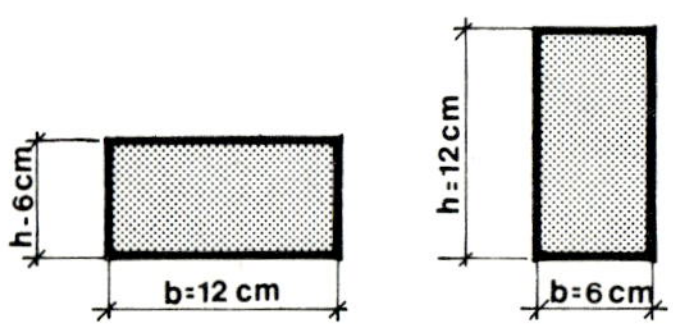

FIGURE 9-12

SOLUTION Let us compare the section moduli for the two cases, noticing that in both cases the beam is symmetrical, so that $S_1 = S_2 = S$.

(*a*) Beam with $b = 6$ cm and $h = 12$ cm, where b is the horizontal dimension or beam width, by Eq. (9-3), has

$$S' = \tfrac{1}{6} \times b \times h^2 = \tfrac{1}{6} \times 6 \times 12^2 = 144 \text{ cm}^3$$

and, by Eq. (9-9),

$$M'_{all} = S' \times \sigma_{all} = 144\sigma_{all}$$

(*b*) For the same section but width $b = 12$ cm and depth $h = 6$ cm we obtain

$$S'' = \tfrac{1}{6} \times 12 \times 6^2 = 72 \text{ cm}^3$$

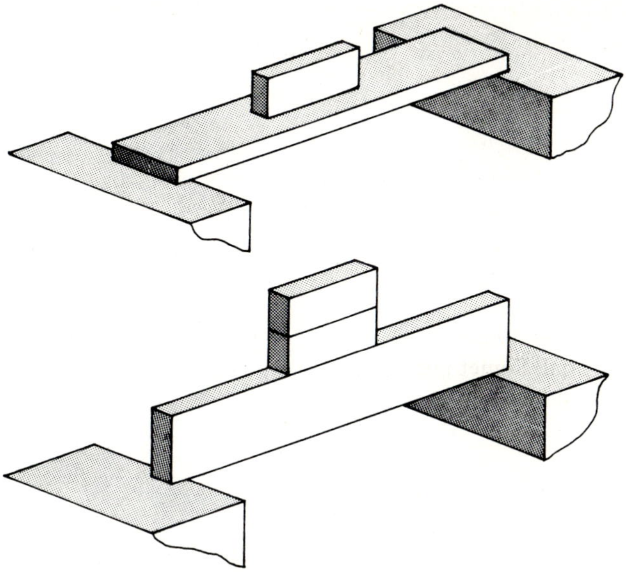

and a maximum allowable moment

$$M''_{\text{all}} = S'' \times \sigma_{\text{all}} = 72\sigma_{\text{all}}$$

For the same allowable bending stress σ_{all}, the allowable moment capacity of the beam in case (*a*) is twice the capacity of the beam in case (*b*).

Illustrative Problem 9-2

Compare the allowable moment capacity of a beam whose cross section is (Fig. 9-13)

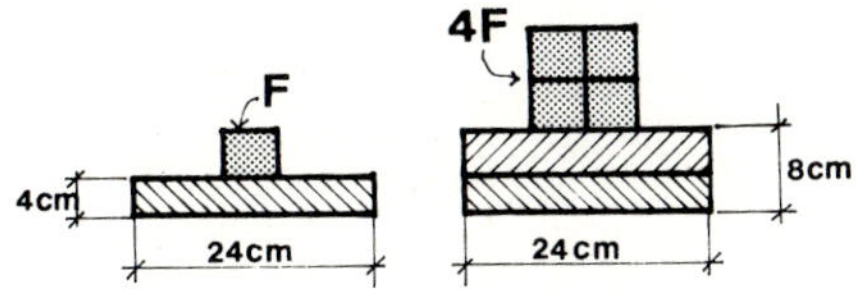

FIGURE 9-13 The beam composed of two planks has four times the capacity of the beam made of one plank.

(*a*) one timber plank with b = 24 cm and h = 4 cm

(*b*) two such planks glued together to create a solid section with b = 24 cm and h = 8 cm

The allowable bending stress

$$\sigma_{\text{all}} = 1.0\ \frac{\text{kN}}{\text{cm}^2}$$

SOLUTION

(*a*) The section modulus of one flat plank is

$$S' = \tfrac{1}{6} \times b \times h^2 = \tfrac{1}{6} \times 24 \times 4^2 = 64 \text{ cm}^3$$

and the corresponding allowable moment capacity is

$$M'_{\text{all}} = S' \times \sigma_{\text{all}} = 64 \times 1.0 = 64 \text{ kN·cm}$$

(*b*) Two planks glued together have $b = 24$ cm and $h = 8$ cm. The section modulus of this cross section is

$$S'' = \tfrac{1}{6} \times 24 \times 8^2 = 256 \text{ cm}^3$$

and the corresponding allowable moment is

$$M''_{\text{all}} = S'' \times \sigma_{\text{all}} = 256 \times 1.0 = 256 \text{ kN·cm}$$

Since

$$M''_{\text{all}} = 256 \text{ kN·cm} = 4 \times 64 \text{ kN·cm} = 4M'_{\text{all}}$$

this cross section has four times the allowable moment capacity of the section consisting of only one plank.

Illustrative Problem 9-3

Determine the rectangular cross section of the timber beam shown in Fig. 7-29, assuming a width-depth ratio (Fig. 9-14) approximately equal to

$$\frac{b}{h} = \frac{5}{7} \qquad (9\text{-}10)$$

and

$$\sigma_{\text{all}} = 0.85 \ \frac{\text{kN}}{\text{cm}^2}$$

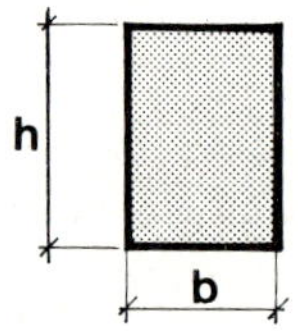

b/h=5/7 FIGURE 9-14

SOLUTION From the moment diagram in Fig. 7-29, we obtain

$$M_{\text{max}} = 1.5 \text{ kN·m} = 150 \text{ kN·cm}$$

From Eq. (9-8), the required section modulus is

$$S_{\text{req}} = \frac{M_{\text{max}}}{\sigma_{\text{all}}} = \frac{150}{0.85} = 176.5 \text{ cm}^3$$

and from Eq. (9-10)

$$b = \tfrac{5}{7} \times h \tag{9-11}$$

Thus

$$S_{req} = \frac{1}{6} \times \frac{5h}{7} \times h^2 = \frac{5h^3}{42} = 176.5 \text{ cm}^3$$

from which

$$h^3 = \frac{42 \times 176.5}{5} = 1\,482.6 \text{ cm}^3$$

and

$$h = \sqrt[3]{1\,482.6} = 11.4 \text{ cm}$$

Once the depth h is known, by Eq. (9-11)

$$b = \tfrac{5}{7} \times h = \tfrac{5}{7} \times 11.4 = 8.14 \text{ cm}$$

For practical reasons, we round up the exact dimensions to the nearest centimeter so that the final dimensions are

$$b = 9 \text{ cm} \qquad \text{and} \qquad h = 12 \text{ cm}$$

Let us now analyze the designed beam. The section modulus of its cross section

$$S = \tfrac{1}{6} \times b \times h^2 = \tfrac{1}{6} \times 9 \times 12^2 = 216 \text{ cm}^3$$

is larger than $S_{req} = 176.5 \text{ cm}^3$; thus the largest bending stress at the critical section will be smaller than the allowable stress. In fact, we have

$$\sigma_{max} = \frac{M_{max}}{S} = \frac{150}{216} = 0.694 \frac{\text{kN}}{\text{cm}^2} < 0.85 \frac{\text{kN}}{\text{cm}^2}$$

Illustrative Problem 9-4

Design the simply supported beam shown in Fig. 7-25 as a wide-flange steel beam for $\sigma_{all} = 15 \text{ kN/cm}^2$.

SOLUTION From Fig. 7-25, we have

$$M_{max} = 12 \text{ kN·m} = 1\,200 \text{ kN·cm}$$

Thus

$$S_{req} = \frac{M_{max}}{\sigma_{all}} = \frac{1\,200}{15} = 80 \text{ cm}^3$$

From Table 5 we select a **W** 100 × 190 with

$$S = S_x = 89 \text{ cm}^3 > 80 \text{ cm}^3 \text{ (required)}$$

The analysis of the designed beam gives

$$\sigma_{max} = \frac{M_{max}}{S} = \frac{1\,200}{89} = 13.5\ \frac{\text{kN}}{\text{cm}^2} < 15\ \frac{\text{kN}}{\text{cm}^2}$$

Illustrative Problem 9-5

Determine the center-to-center spacing c for the simply supported, 2×12-in wood joists for a floor spanning 20 ft (6.0 m), wall to wall. The total uniformly distributed load on the floor is 60 lb/ft² (2.87 kN/cm²), and it includes the weight of the joists. Assume an allowable bending stress for wood of 1 000 lb/in² (690 kN/cm²).

SOLUTION If the center to center distance (Fig. 9-15) between the joists is c ft, the total uniformly distributed load on one lineal foot of joist is

$$w = 60 \times c\ \frac{\text{lb}}{\text{ft}}$$

The largest bending moment in a simply supported, uniformly loaded beam is (see Illustrative Prob. 7-4)

$$M_{max} = \tfrac{1}{8} \times w \times L^2$$

where L is the beam span. In our problem (keeping c in feet)

$$M_{max} = \tfrac{1}{8}\,(60c)\,20^2 = 3\,000c\ \text{lb}\cdot\text{ft} = 3\,000c \times 12 = 36\,000c\ \text{lb}\cdot\text{in}$$

The available section modulus is

$$S = \tfrac{1}{6} \times b \times h^2 = \tfrac{1}{6} \times 2 \times 12^2 = 48\ \text{in}^3$$

From Eq. (9-9) we have

$$M_{max} = S \times \sigma_{all}$$

or

$$36\,000c = 48 \times 1\,000$$

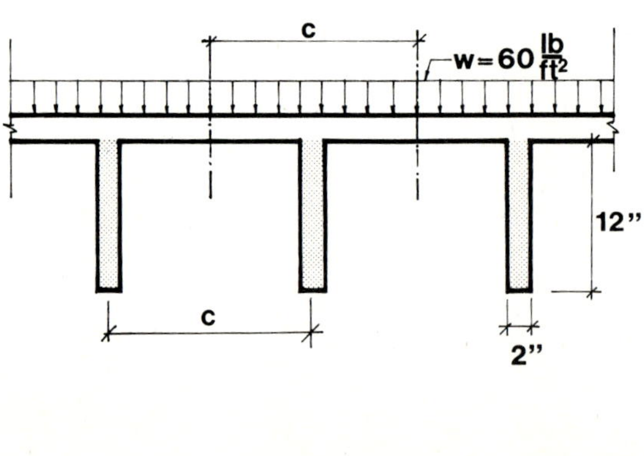

FIGURE 9-15

from which

$$c = \frac{48 \times 1\,000}{36\,000} = 1\frac{1}{3} \text{ ft} = 16 \text{ in} \qquad (40 \text{ cm})$$

You may check the calculations by computing

$$w = 60c = 60 \times 1\frac{1}{3} = 80 \frac{\text{lb}}{\text{ft}} \qquad \left(1.17 \frac{\text{kN}}{\text{cm}^2}\right)$$

$$M_{\text{max}} = \frac{1}{8} \times 80 \times 20^2 = 4\,000 \text{ lb}\cdot\text{ft}$$

$$= 4\,000 \times 12 = 48\,000 \text{ lb}\cdot\text{in}$$

$$\sigma = \frac{M}{S} = \frac{48\,000}{48} = 1\,000 \frac{\text{lb}}{\text{in}^2} = \sigma_{\text{all}}$$

Thus the joists of the floor system should be spaced 16 in on centers.

9.3 PLASTIC OR ULTIMATE STRENGTH DESIGN OF STRUCTURES

In Chap. 8 and in the preceding sections of Chap. 9, structural members were designed by use of the working-stress method. This method is based on the criterion that the largest stress which may develop at any point of a structural member under service conditions should be smaller than or, at most, equal to the allowable stress for the material

$$\sigma_{\text{max}} \leqq \sigma_{\text{all}}$$

Allowable stresses are usually determined by dividing the yield stress σ_y or ultimate strength σ_u by appropriate safety factor

$$\sigma_{\text{all}} = \frac{\sigma_y}{(\text{SF})_y}$$

$$\sigma_{\text{all}} = \frac{\sigma_u}{(\text{SF})_u}$$

When a structural member is loaded by a tensile or compressive axial force, the corresponding normal stresses are uniformly distributed over its entire cross section. Hence, if the working-stress method is used, each fiber has the same safety factor against failure.

In the so-called *plastic* or *ultimate-strength design method*, an ultimate value N_u of the axial force is first determined, which would cause the member to yield (Fig. 9-16). The value N_u is the product of the area A of the member and the yield stress σ_y of the material:

$$N_u = A \times \sigma_y \tag{9-12}$$

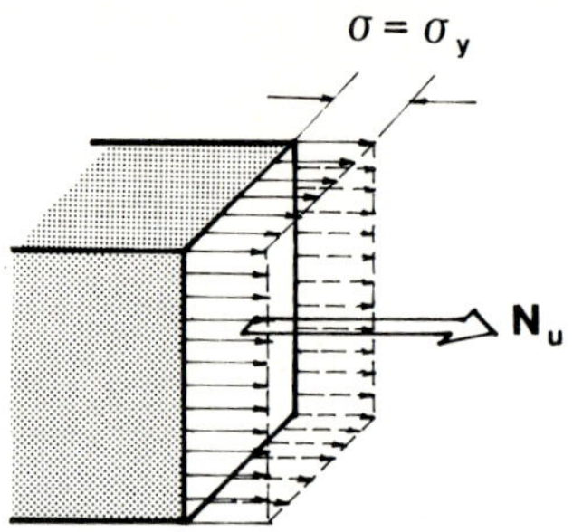

FIGURE 9-16 When the member is loaded with the ultimate axial force N_u, all fibers of its cross section yield at the same time.

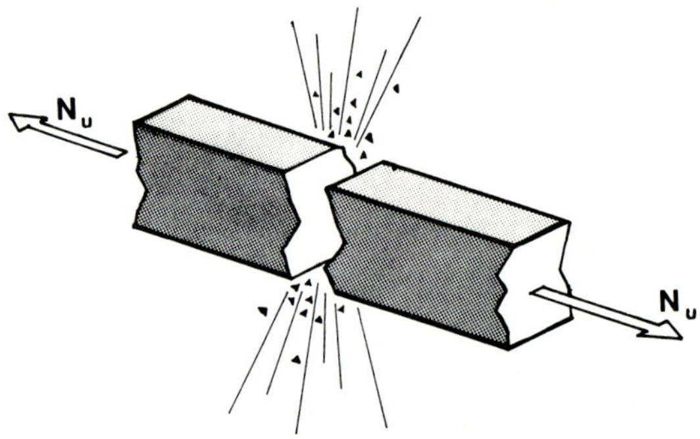

When the member is loaded with the ultimate axial force N_u, all the fibers of its cross section yield at the same time. They are all used to their ultimate stress capacity, which in this case is considered to be given by the yield stress, since after the value of an axial force in a member reaches the ultimate value N_u, the failure of the member is imminent. The allowable value N_{all} of the axial force may be obtained from N_u by dividing it by an appropriate safety factor $(\text{SF})_y$:

$$N_{\text{all}} = \frac{N_u}{(\text{SF})_y} = A\,\frac{\sigma_y}{(\text{SF})_y} \qquad (9\text{-}13)$$

Since

$$\frac{\sigma_y}{(\text{SF})_y} = \sigma_{\text{all}}$$

Eq. (9-13) becomes

$$N_{\text{all}} = A \times \sigma_{\text{all}} \qquad (9\text{-}14)$$

and the required cross section of the member, with $N_{\text{all}} = N_{\text{max}}$, becomes

$$A = \frac{N_{\text{max}}}{\sigma_{\text{all}}} \qquad (9\text{-}15)$$

Comparing Eqs. (9-15) and (8-5), we see that in the case of design for axial loads the working-stress method yields the same result as the ultimate-strength design method. However, this is not so in the case of design based on bending moments.

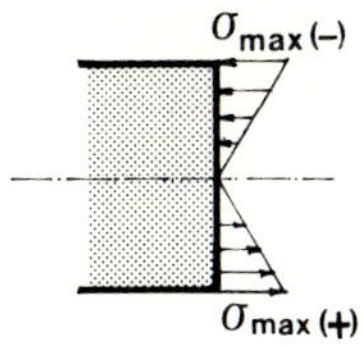

FIGURE 9-17

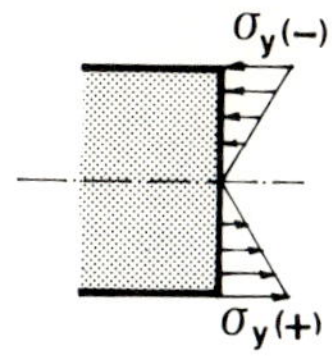

FIGURE 9-18

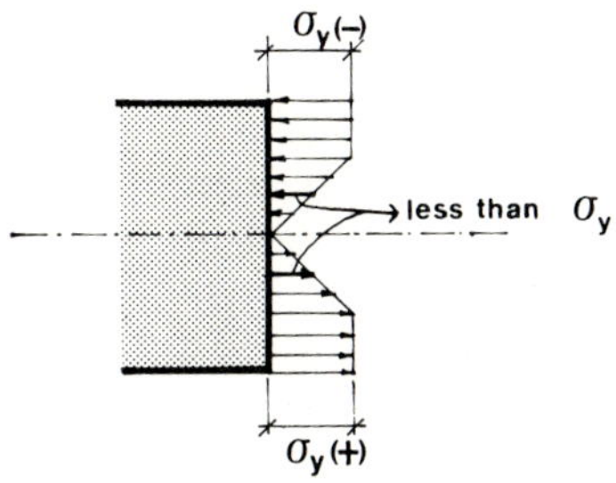

FIGURE 9-19

In the working-stress design of beams we have assumed that under service (working) loads the beam develops over the depth of the cross section the linear distribution of bending stresses shown in Fig. 9-17, in which the stress changes linearly from zero at the neutral axis to maximum values at the top and bottom fibers.

The largest stresses in the extreme fibers, at the critical section with the largest bending moment M_{max}, had to be smaller than or, at most, equal to the allowable bending stress of the beam material. The extreme fibers are thus protected against yielding by a safety factor $(SF)_y$, since

$$\sigma_{all} = \frac{\sigma_y}{(SF)_y}$$

However, at the critical section ***all*** the other fibers develop stresses ***smaller*** than those at the extreme fibers; hence they all have larger safety factors against yielding than that of the extreme fibers. This means that if the loads are increased until the extreme fiber stresses reach the yield point ($\sigma_{max} = \sigma_y$), the rest of the fibers will not be stressed to the yield point yet (Fig. 9-18). At this point further increases in the loads will cause more and more fibers to reach the yield stress, but up to certain values of the loads some fibers in the neighborhood of the neutral axis will not yield (Fig. 9-19).

The loads can grow until all fibers yield (are plasticized) as shown in Fig. 9-20. At this point all the fibers in the cross section flow under the constant tensile and compressive stress σ_y; large relative rotations of the portions of the beam to the right and left of the plasticized section take place. We say that the beam has developed a ***plastic hinge*** at this section. The development of a plastic hinge in a simply supported beam produces its total collapse, since a beam with three hinges cannot support loads; it becomes a "linked mechanism" (Fig. 9-21).

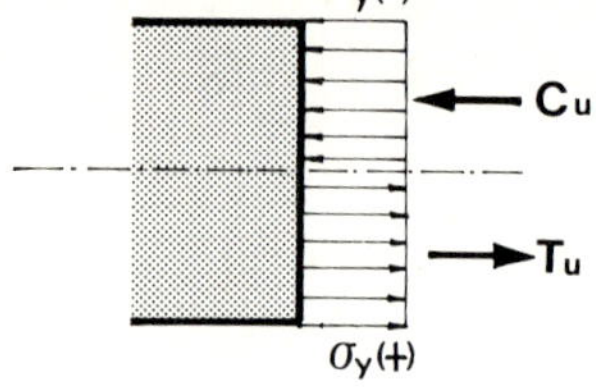

FIGURE 9-20 After all fibers of a cross section reach yield stress, a plastic hinge develops at this section.

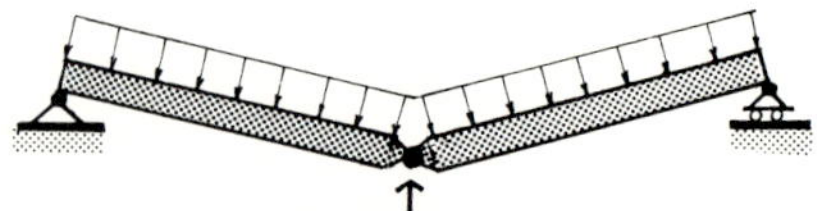

FIGURE 9-21 Development of a plastic hinge in a simply supported beam produces total collapse of the beam.

In following the development of a plastic hinge we have assumed that the strength characteristics of the material are the same in tension and compression, as is true, for example, for steel. The ultimate moment capacity M_u at a plasticized section of a rectangular beam of such a material can be evaluated from the diagram of constant stress distribution in tension and compression of the section just before collapse (Fig. 9-20). The total ultimate compressive force C_u is equal to the volume of the compressive stress block with sides $h/2$, b, and σ_y

$$C_u = \frac{h}{2} \times b \times \sigma_y$$

Similarly, the ultimate tensile force T_u is

$$T_u = \frac{h}{2} \times b \times \sigma_y$$

and is equal to C_u. Hence the forces T_u and C_u constitute a couple. With the lever arm of this couple equal to $h/2$ (Fig. 9-22), the moment of the couple, that is, the ultimate moment M_u, is

$$M_u = T_u \times \frac{h}{2} = C_u \times \frac{h}{2}$$

or (9-16)

$$M_u = \frac{h}{2} \times b \times \sigma_y \times \frac{h}{2} = \sigma_y \frac{bh^2}{4}$$

The allowable moment M_{all} is obtained from the ultimate moment M_u by the use of an appropriate safety factor. In this simplified presentation of the basic ideas of the

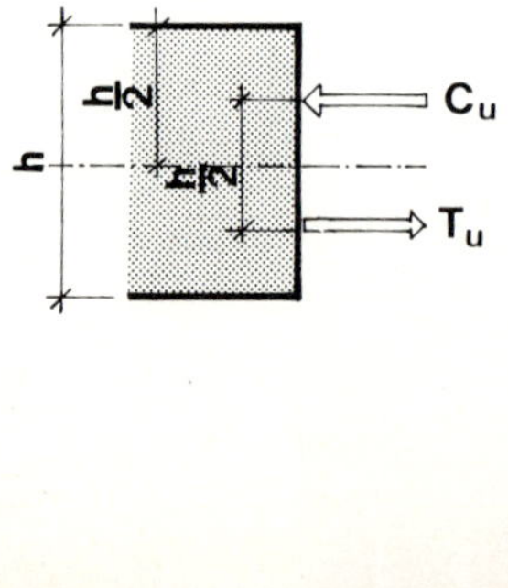

FIGURE 9-22

ultimate-strength design method, the safety factor $(SF)_y$ will have the same value as that used in working-stress design to obtain the allowable stress σ_{all} from the yield stress σ_y. Then

$$M_{all} = \frac{M_u}{(SF)_y} = \frac{\sigma_y}{(SF)_y} \times \frac{bh^2}{4} \tag{9-17}$$

where

$$\frac{\sigma_y}{(SF)_y} = \sigma_{all}$$

and Eq. (9-17) becomes

$$M_{all} = M_{max} = \sigma_{all} \times \frac{bh^2}{4} \tag{9-18}$$

In the working-stress method discussed in Sec. 9-2, we obtained (Eq. 9-9)

$$M_{max} = \sigma_{all} \times S \tag{9-19}$$

where the section modulus S of a rectangular section $b \times h$ was

$$S = \frac{b \times h^2}{6} \tag{9-20}$$

By comparison of Eqs. (9-18) and (9-19) we may define a *plastic section modulus Z* for a rectangular section $b \times h$, to be used in the ultimate-strength design method, as

$$Z = \frac{bh^2}{4} \tag{9-21}$$

which is larger than the working-stress section modulus S of Eq. (9.20). This simplified analysis shows that the allowable moment in the beam computed by the ultimate-strength design method is larger than that computed by the working-stress method, if the same σ_{all} is used for both methods.

The ultimate-strength design method, even with larger safety factors, has met acceptance by *Uniform Building Code* authorities and the design profession because it leads to both material savings and simplified calculations.

Illustrative Problem 9-6

Redesign the beam in Illustrative Prob. 9-3 using the ultimate-strength design method. Assume that the material has equal yield stresses in tension and compression.

SOLUTION In Illustrative Prob. 9-3 we used

$$M_{max} = 150 \text{ kN}\cdot\text{cm}, \sigma_{all} = 0.85 \frac{\text{kN}}{\text{cm}^2}$$

and

$$\frac{b}{h} = \frac{5}{7} \qquad \text{or} \qquad b = \frac{5}{7} \times h$$

Substituting the above values in Eq. (9-18), we obtain

$$M_{\text{max}} = \sigma_{\text{all}} \left(\frac{bh^2}{4} \right)$$

$$150 = (0.85) \frac{bh^2}{4}$$

$$bh^2 = \frac{4 \times 150}{0.85}$$

$$\frac{5}{7} \times h \times h^2 = \frac{4 \times 150}{0.85}$$

from which

$$h^3 = \frac{4 \times 150 \times 7}{0.85 \times 5} = 988 \text{ cm}^3$$

$$h = \sqrt[3]{988} = 9.96 \text{ cm}$$

and

$$b = \tfrac{5}{7}\, 9.96 = 7.1 \text{ cm}$$

(These dimensions are smaller than those obtained in Illustrative Prob. 9-3 by the working-stress method, which resulted in $h = 11.4$ cm and $b = 8.14$ cm.) The overall dimensions obtained by rounding up the above values are $b = 7.5$ cm and $h = 10$ cm.

Since ultimate-strength design often gives smaller cross section, the deflections of beams designed by this method must be checked. (The computation of beam deflections is treated in Sec. 9-5.)

9.4 SHEAR STRESSES IN BEAMS

In the preceding sections of this chapter we evaluated bending stresses in beams, ignoring the existence of shear forces. But shear forces exist and develop shear stresses at most sections of a beam. What is the distribution of these stresses over the cross section? At what points of the section do they assume their maximum values?

To answer these questions, let us turn, as we have done before, to a rectangular beam model and draw a series of rectangles on its sides, as shown in Fig. 9-23*a*. Let us then load the beam model so that both bending moments and shear forces develop in its cross sections. After the load has been applied, the rectangles deform as shown in

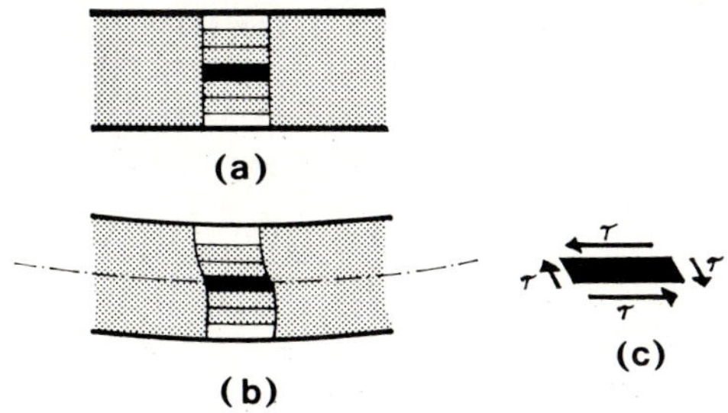

FIGURE 9-23 Presence of shear stresses in beams.

Fig. 9-23*b*. If we isolate the portion of the beam defined by one of the deformed rectangles from the rest of the beam, we notice that it is skewed; such deformation can be caused only by shear stresses on the isolated portion as shown in Fig. 9-23*c*.

We also notice that the elements at the top and bottom of the beam are not skewed, while that on the neutral axis is the most skewed. We may conclude that the shear stress in the top and bottom elements is zero, while it is maximum in the element located at the neutral axis. Finally, the variation of the shear deformation suggests that the distribution of shear stresses over the cross-sectional depth is *curvilinear*. A theoretical investigation (beyond the scope of this book) confirms that for a rectangular cross section, the distribution of shear stresses is *parabolic*, as shown in Fig. 9-24. The total shear force V is the volume of the parabolic cylinder shown in Fig. 9-24, which equals the area of the parabola times the width b of the section.

The area of the quadratic parabola can be shown to be

$$A_p = \tfrac{2}{3} h \times \tau_{max}$$

The volume of the cylinder is

$$V = A_p \times b = \tfrac{2}{3} h \times b \times \tau_{max} \tag{9-22}$$

This volume must be equal to the shear force V at the given cross section. From Eq. (9-22) the largest shear stress in the cross section is

$$\tau_{max} = \frac{3}{2} \times \frac{V}{b \times h} \tag{9-23}$$

The largest value of the shear stress τ_{max} occurs at points of the neutral axis of the critical cross section in shear, where $V = V_{max}$. It is 1.5 times the *average shear stress*, obtained by dividing V by the area $b \times h$ of the cross section.

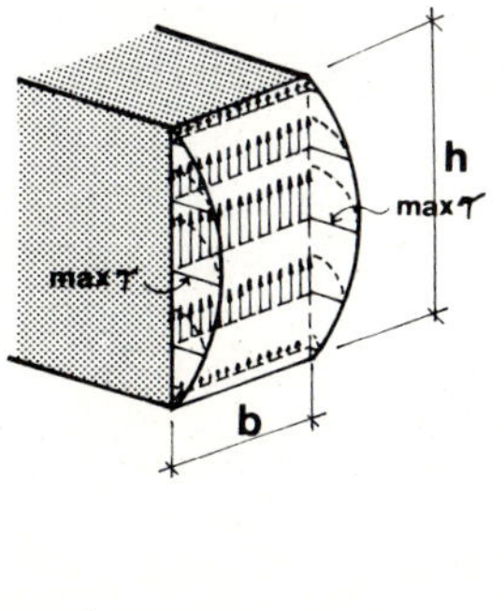

FIGURE 9-24 Distribution of shear stresses in rectangular cross section of a beam is parabolic; max τ is the maximum value of shear stress in the section.

The distribution of shear stresses in a wide-flange section is much more complicated. Specifications of the American Institute of Steel construction permit use of the assumption that in flanged sections only the web (not flanges) is developing shear stresses and that they are uniformly distributed over the area of the web:

$$\tau = \frac{V}{t_w \times d} \tag{9-24}$$

where V = shear force at the section under consideration
t_w = thickness of the web } found in the appropriate
d = total depth of the section } table of the appendix

Illustrative Problem 9-7

Check the shear stress in the beam shown in Fig. 7-29, which was designed in Illustrative Prob. 9-6 by the ultimate-strength design method. Consider the overall dimensions of the cross section to be

$$b = 7.5 \text{ cm} \qquad h = 10.0 \text{ cm}$$

SOLUTION From Fig. 7-29 the largest shear force in the beam is

$$V_{max} = 1.0 \text{ kN}$$

By Eq. (9-23) the maximum value of the shear stress is

$$\tau_{max} = \frac{3}{2} \times \frac{1.0}{7.5 \times 10} = 0.02 \frac{\text{kN}}{\text{cm}^2}$$

and is much smaller than the allowable shear stress τ_{all} given in Table A-12 for Douglas fir as 0.07 kN/cm^2. Thus our beam, which was designed in bending, is safe in shear as well.

Illustrative Problem 9-8

Compute the magnitude of the shear stress in the wide-flange section **W** 100 × 190 made out of A-36 steel when the shear force on that section is $V = 4$ kN. (This is the maximum shear force in the beam of Fig. 7-25, which was designed in Illustrative Prob. 9-4.)

SOLUTION For **W** 100 × 190 from Table A-5, we find that

$$t_w = 7 \text{ mm} = 0.7 \text{ cm}$$

$$d = 106 \text{ mm} = 10.6 \text{ cm}$$

Thus by Eq. (9-24), we have

$$\tau = \frac{V}{t_w \times d} = \frac{4 \text{ kN}}{0.7 \times 10.6 \text{ cm}} = 0.54 \frac{\text{kN}}{\text{cm}^2}$$

which is much smaller than the allowable shear stress in A-36 steel given by Table A-13 as

$$\tau_{all} = 10 \frac{\text{kN}}{\text{cm}^2}$$

9.5 COMPUTATION OF BEAM DEFLECTIONS

The deflections of a beam due to given loads are defined as the vertical displacements of its neutral plane (Fig. 9-25).

The largest deflection in a symmetrically loaded, symmetrical beam occurs at mid-span point. For example, the largest bending deflection y_{max} of a beam, simply sup-

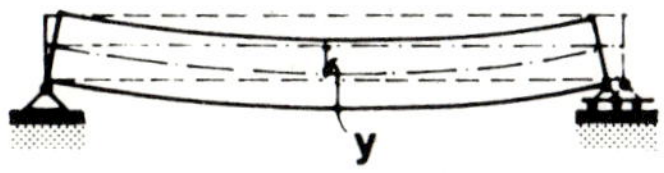

FIGURE 9-25 Deflection of a beam.

ported at its ends and loaded by a uniformly distributed load w (Fig. 9-26), can be shown to be given by

$$y_{max} = \frac{5}{384} \frac{wL^4}{EI} \tag{9-25}$$

where w = uniformly distributed load lb/in
L = span of the beam, in
E = modulus of elasticity of the beam material, lb/in^2
I = moment of inertia of the beam cross section about its neutral axis, in^4

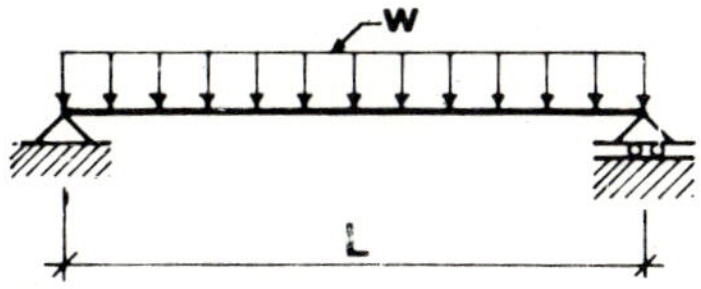

FIGURE 9-26

Table A-14 gives formulas for the maximum bending deflection of commonly used, statically determinate beams under simple loading conditions. They show that the larger the product of the moment of inertia I of the cross section and the modulus of elasticity E, the smaller the deflections, that is, the stiffer the beam. The product EI is an important indicator of the bending rigidity of a beam and is called its ***flexural rigidity*** (stiffness).

Illustrative Problem 9-9

Find the maximum deflection of the 2- X 12-in timber joist in Illustrative Prob. 9-5. The joists are simply supported and have a span of 20 ft (6.0 m). They are loaded by a total uniformly distributed load of 80 lb/ft (1.17 kN/m). The modulus of elasticity of timber (see Table A-11) is equal to 1 760 kips/in^2 (1 200 kN/cm^2).

SOLUTION By Eq. (9-25) the maximum deflection of a simply supported beam due to a uniformly distributed load is

$$y_{max} = \frac{5}{384} \frac{wL^4}{EI}$$

Here

$$w = 80 \frac{\text{lb}}{\text{ft}} = \frac{80}{12} = 6.67 \frac{\text{lb}}{\text{in}}$$

$$L = 20 \times 12 = 240 \text{ in}$$

$$E = 1\ 760\ 000 \frac{\text{lb}}{\text{in}^2}$$

$$I = \frac{1}{12} bh^3 = \frac{1}{12} \times 2 \times 12^3 = 288 \text{ in}^4$$

and

$$y_{max} = \frac{5 \times 6.67 \times 240^4}{384 \times 1\ 760\ 000 \times 288} = 0.57 \text{ in} \qquad (2.2 \text{ cm})$$

The permissible deflection is usually limited by *Uniform Building Codes* to

$$y_{all} \leqq \frac{L}{360} = \frac{20 \times 12}{360} = 0.67 \text{ in} \qquad (2.6 \text{ cm})$$

Hence the stiffness of our joists is adequate.

New Terms Appearing in Chapter 9

Bending stresses
Pure bending
Neutral plane
Neutral axis (NA)
Lever arm of internal couple
Section modulus, S
Moment of inertia, I
Allowable bending stress
Plastic or ultimate design of structures
Plastic section modulus, Z
Plastic hinge
Ultimate moment, M_u
Deflections of beams
Flexural rigidity, EI

New Structural Element Appearing in Chapter 9

A *joist-floor* consists of narrow, closely spaced beams called *joists.*

Quiz Number 8

A. TRUE OR FALSE

1. For a homogeneous beam in bending, the neutral axis is a centroidal axis of the beam cross section.
2. The bending stresses are maximum at points on the neutral axis.
3. Compressive stresses develop in all fibers of a bent beam.
4. In the working-stress method, bending stresses vary linearly over the depth of a beam cross section.
5. The larger the distance y of a fiber from the neutral plane, the smaller the bending stress in that fiber.
6. The section modulus S of a cross section has a dimension of in^4 or cm^4.
7. The section modulus S for a rectangular cross section of width b and depth h is equal to

$$S = \frac{1}{6} b \times h$$

8. As a result of bending, a maximum shear stress develops in the fibers at the neutral plane of a beam.
9. The larger the bending moment M, the larger the bending stresses in the extreme fibers of a beam.
10. As a result of bending, tensile and compressive stresses develop in a beam.
11. Two values of the section modulus must be used for sections which are not symmetrical about their neutral axis.
12. A solid rectangular section is less efficient in bending than a wide-flange section with the same area.
13. A tubular section is less efficient in bending than a solid circular section.
14. A beam cross section becomes more efficient in bending if we remove material close to the neutral axis and place it as far from the neutral axis as possible.
15. The allowable bending stress in steel beams is equal to

$$\sigma_{all} = \frac{\sigma_y}{(SF)_y}$$

16. In beams under bending and shear, the largest bending stresses develop at the section with the largest bending moment M, while the largest shear stresses develop in the section with the largest shear force V.
17. In working-stress design we assume that under service loads bending stresses vary linearly over the depth of a beam cross section from zero at the neutral axis to a maximum at the top and bottom fibers.

18. The distribution of shear stresses in a beam of rectangular cross section is parabolic.
19. The shear stress is zero in the top and bottom fibers of a beam.
20. In plastic or ultimate-strength design of beams in bending, a cross section is assumed to attain its full bending capacity when all its fibers reach the yield point.
21. The larger the moment of inertia I of a beam cross section, the larger the beam deflections.
22. The largest deflections in a simply supported beam develop in the neighborhood of its supports.
23. The complete design of a beam involves design for bending, shear, and deflection.
24. Excessive deflections are never harmful to the serviceability of beams.

B. MULTIPLE CHOICE

1. If the width b of a rectangular cross section is 12 cm and its depth h is 4 cm, the section modulus S equals
   (*a*) 144 cm^3 (*b*) 16 cm^3 (*c*) 32 cm^3 (*d*) 196 cm^3
2. If we increase the section modulus S of a beam, its moment capacity
   (*a*) increases (*b*) decreases (*c*) remains the same
3. The bending stresses in the cross section of a beam
   (*a*) increase with the distance y from the neutral axis
   (*b*) decrease with the distance y from the neutral axis
   (*c*) remain constant all over the cross section
4. The flexural rigidity of a beam is equal to
   (*a*) AE (*b*) AF (*c*) ES (*d*) EI
5. In the working-stress method the bending stresses along the depth of the cross section of a beam
   (*a*) change abruptly
   (*b*) change linearly
   (*c*) change parabolically
   (*d*) remain constant all over the cross section
6. The shear stresses on the cross section of a beam in bending
   (*a*) change abruptly
   (*b*) change linearly
   (*c*) change parabolically
   (*d*) remain constant all over the cross section
7. If the bending moment at a section of a beam is M and its section modulus is S, the maximum bending stress is
   (*a*) $\sigma = M \times S$ (*b*) $\sigma = \frac{M}{S}$
   (*c*) $\sigma = \frac{S}{M}$ (*d*) $\sigma = \frac{M}{EI}$
8. If the bending moment in a beam is positive, the largest tensile bending stress develops
   (*a*) in the top fibers

(*b*) in the bottom fibers
(*c*) at the neutral axis

9. If a shear force V acts at a section of a beam of rectangular cross section $b \times h$, the largest shear stress due to V is equal to

   (*a*) $\frac{3}{2} \times \frac{V}{bh}$ (*b*) $\frac{V}{bh}$ (*c*) $2\,Vbh$ (*d*) $\frac{2}{3} \times \frac{V}{bh}$

10. The moment of inertia of a rectangular section of width b and depth h is

    (*a*) $\frac{1}{12} b \times h^3$ (*b*) $\frac{1}{6} b \times h^2$

    (*c*) $\frac{1}{12} b \times h^2$ (*d*) $\frac{b \times h}{2}$

C. SHORT ANSWERS

1. The largest deflection in a cantilever beam loaded by a uniformly distributed load develops at the ______________.
2. A simply supported beam is loaded with a uniformly distributed load over its entire length. Its largest deflection develops at ____________.
3. If the section modulus of a beam $S = 100\ \text{in}^3$ and the allowable bending stress $\sigma_{all} = 1\,000\ \text{lb/in}^2$, the allowable maximum moment is

   $$M_{all} = \_\_\_\_\_\_\_\ \text{kip}\cdot\text{in}$$

4. If a bending moment $M = 100\ \text{kip}\cdot\text{in}$ acts on a beam section of section modulus $S = 100\ \text{in}^3$, the largest bending stresses developing at the top and bottom fibers of the cross section are

   $$\sigma_{max} = \_\_\_\_\_\_\_\ \frac{\text{lb}}{\text{in}^2}$$

5. What is the required section modulus S_{req} in a beam, if $M_{max} = 100\ \text{kip}\cdot\text{in}$ and $\sigma_{all} = 1\,000\ \text{lb/in}^2$?

   $$S_{req} = \_\_\_\_\_\_\_\ \text{in}^3$$

6. The flexural rigidity of a beam is equal to _______.
7. If we wish to design a beam in bending, the bending moment governing the design must be computed at the ______________ section of the beam.
8. The moment of inertia I of a rectangular section with $b = 4$ in and $h = 6$ in is

   $$I = \_\_\_\_\_\_\_\ \text{in}^4$$

9. The moment of inertia I of a rectangular section with $b = 6$ in and $h = 4$ in is

   $$I = \_\_\_\_\_\_\_\ \text{in}^4$$

10. If a rectangular beam has dimensions $b = 10$ cm and $h = 30$ cm and the allowable shear stress of the beam material is $\tau_{all} = 6.0\ \text{kN/cm}^2$, what is the allowable value of the maximum shear force V_{max} in the beam?

    $$V = \_\_\_\_\_\_\_\ \text{kN}$$

EXERCISES

Work out all problems by the working-stress method.

9-1 A W 12 × 79 (310 × 1 150) steel beam spans 28 ft (8.5 m) and is simply supported at both ends. It is loaded by a total uniformly distributed load w = 2 000 lb/ft (20 kN/m) over its entire length. Compute the maximum bending stress in this beam and compare it with an allowable stress in bending σ_{all} = 24 kips/in^2 (16.5 kN/cm^2).

9-2 A cantilever beam is 8 ft (2.4 m) long. It is loaded by its own weight and a vertical concentrated force F = 400 lb (1.78 kN) located at its free end. The beam is made of timber of unit weight 50 lb/ft^3 (7.86 kN/m^3). Its cross section is triangular with a base b = 6 in (15 cm) and a depth h = 12 in (30 cm). Compute the maximum tensile and compressive stresses in the beam.

9-3 Four 2- × 12-in planks are glued one on top of the other to build up a cross section with a width b = 12 in and a depth h = 8 in. Compute the allowable moment M_{all} which can be applied to this section if the allowable bending stress is 2 000 lb/in^2.

9-4 Assemble the planks described in Exercise 9-3 to build up the cross section shown in Fig. 9-27 and compute its allowable moment M_{all} for the allowable stress used in Exercise 9-3.

Hint: Compute first the moment of inertia of the section by subtracting from the moment of inertia of a solid rectangle 12 × 16 in that of a solid rectangle 8 × 12 in; then compute the section modulus of the section, $S = I/y$ (y = 8 in); and finally, the allowable moment $M_{all} = \sigma_{all} \times S$.

Compare the efficiency of the section in this problem with that in Exercise 9-3.

9-5 A simply supported beam with the cross section of Exercise 9-4 (Fig. 9-27) spans 14 ft (4.27 m) and is loaded with a uniformly distributed load w = 1 200 lb/ft (17.5 kN/m) over its entire length (including the weight of the beam). Compute the maximum bending stress in the beam and its maximum deflection, if E = 1 760 000 lb/in^2 (1 214 000 N/cm^2). Compare the computed deflection with an allowable deflection equal to 1/360 of the span ($y_{all} = L/360$).

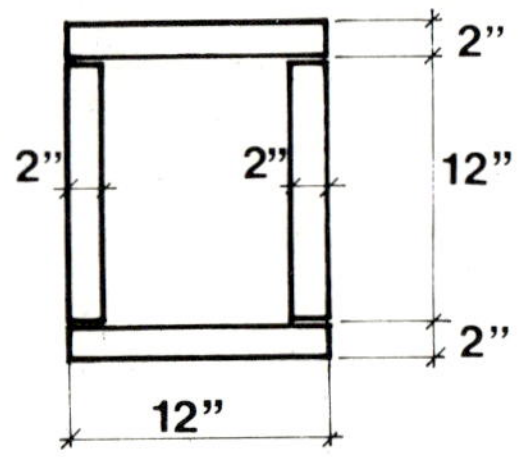

FIGURE 9-27

9-6 The cantilever beams of a balcony are spaced 4 ft (1.20 m) on center and cantilever 8 ft (2.4 m). The total load on the floor supported by these beams is 250 lb/ft^2 (12 kN/m^2), including the weight of the floor and beams (Fig. 9-28). Select the lightest wide-flange steel shape for the beams, if the allowable bending stress σ_{all} = 24 000 lb/in^2 (16.5 kN/cm^2).

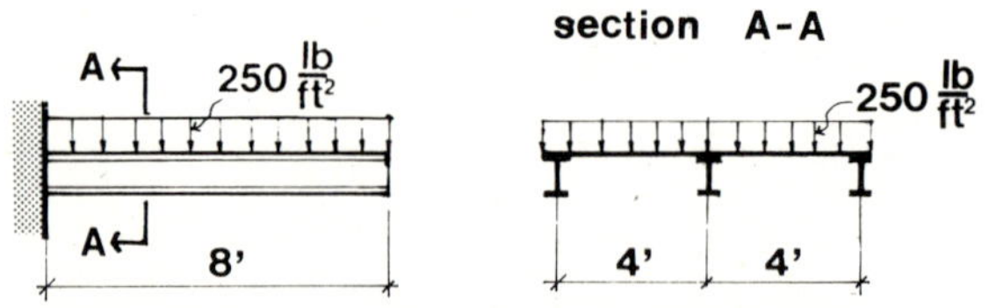

FIGURE 9-28

9-7 A simply supported beam of Douglas fir spans 15 ft (4.57 m). It is loaded by two concentrated forces F = 10 000 lb (44.5 kN) acting at the third points of its span (5 ft from the supports). Design the beam with a ratio $b/h = \frac{1}{2}$ and for an allowable bending stress of 1 250 lb/in^2 (862 N/cm^2). Round up the dimensions to the nearest full inch.

Typical truss joint with four truss members bolted together (*Courtesy of Clark Engineering, Minneapolis*).

Bolted and Welded Connections

10

10.1 INTRODUCTION

Structures are composed of individual elements. Frames consist of beams and columns; trusses consist of the bars in the upper and lower chords, the diagonals, and the verticals; the beams of a floor system are connected to the girders, which in turn are framed into the columns. All these members must be so connected to each other that they can carry and transmit safely the forces and moments due to the loads. In steel elements, such connections are achieved primarily by means of bolts and welds. Timber elements are connected by so-called connectors, nails, or glue. In concrete the elements are usually connected by welds and bolts, or built monolithically.

10.2 BEARING-TYPE BOLTED CONNECTIONS

Bolts consist of a cylindrical body (shank) with an attached head (Fig. 10-1). A portion of the body is threaded to receive a nut. Washers may or may not be placed under the nuts.

A joint in which one bolt connects two plates is shown in Fig. 10-2 and is called a *lap joint*. In a *bearing-type* bolted connection an axial force F tends to shear the bolt at section m-m. The bolt resists this tendency by developing shear stresses over the shearing area. Minor bending stresses also develop in this area as a result of the small

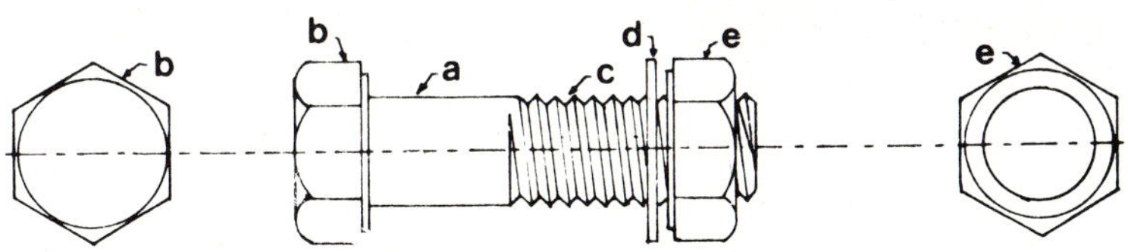

FIGURE 10-1 Bolt; *a*, shank, *b*, head; *c*, thread; *d*, washer; *e*, nut.

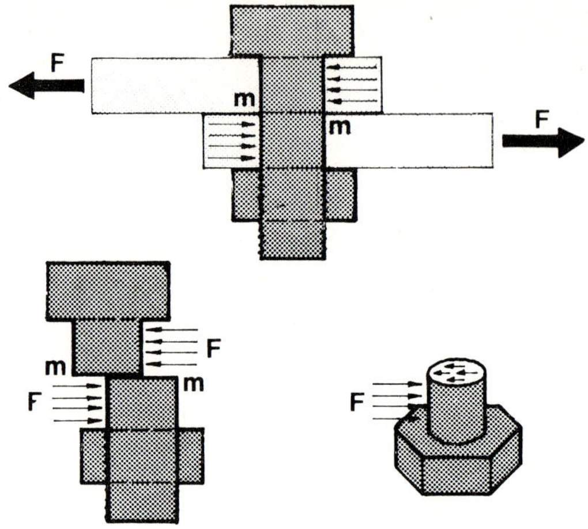

FIGURE 10-2 Bearing-type bolt connecting two plates in a lap joint. Bolt is in single shear.

bending moment of the couple of the forces F, which are not aligned, but these stresses do not govern the design of the bolt.

As proved by tests, we can assume that the shear stress at the section *m-m* is uniformly distributed over the entire area of the section, so that it can be computed by dividing the total shearing force F at the section *m-m* by the area A of the bolt cross section

$$\tau = \frac{F}{A} \tag{10-1}$$

Since the cross-sectional area of a bolt of diameter d is

$$A = \frac{\pi d^2}{4}$$

the shear stress in the bolt is

$$\tau = \frac{F}{\pi d^2/4}$$

If the two plates are connected by several bolts, we assume approximately that the total axial force F in the plate is equally subdivided among the bolts. The force F_1 in an individual bolt of a group of n bolts is thus

$$F_1 = \frac{F}{n}$$

and the shear stress in each bolt is

$$\tau = \frac{F_1}{\pi d^2/4} = \frac{F}{n(\pi d^2/4)} \tag{10-2}$$

Equation (10-2) is used to establish the number of bolts necessary for the safe transfer of the force F from plate to plate through the connection. The shear stress in a

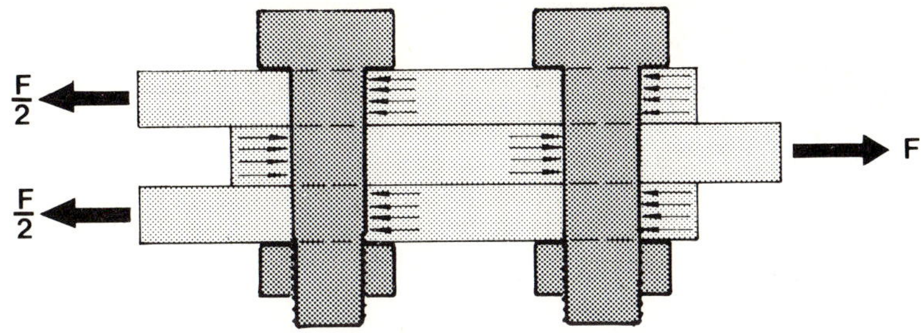

FIGURE 10-3 Bolts in double shear.

bolt must be smaller than or, at most, equal to the allowable shear stress of the bolt material

$$\tau = \frac{F}{n\,(\pi d^2/4)} \leqq \tau_{\text{all}} \tag{10-3}$$

from which we obtain the required number of bolts

$$n \geqq \frac{F}{\tau_{\text{all}}\,(\pi d^2/4)} \tag{10-4}$$

Equation (10-4) serves to establish the number of bearing-type bolts when shear governs. Allowable stresses for various types of bolts must be selected from design handbooks and specifications.

When a bolt resists shear on a single shear area, as that shown in Fig. 10-2, it is called a *bolt in single shear*.

Each bolt in the connection of Fig. 10-3 has two resisting shear areas: such bolts are called *bolts in double shear*. If the axial force to be transmitted through the joint is F, the shear stress in each shear resisting area of a bolt in double shear is

$$\tau = \frac{F}{2n\,(\pi d^2/4)}$$

and

$$n = \frac{F}{\tau_{\text{all}}\;2(\pi d^2/4)} \tag{10-5}$$

where n is the number of bolts in the joint.

When transmitting a force through the joint, bearing-type bolts bear with part of their surface against the surface of the hole in the plate. Thus *compressive* stresses develop in the bolt and in the plate. The actual distribution of these stresses over the cylindrical bolt surface is nonlinear and complicated. For practical design purposes, we assume that the *bearing area* is rectangular and equal to (Fig. 10-4)

$$A_c = d \times t \tag{10-6}$$

where d = diameter of bolt
t = thickness of plate

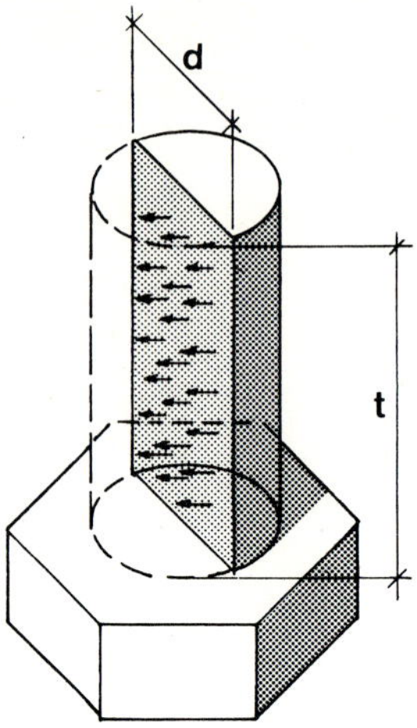

FIGURE 10-4

Thus the normal compressive stress in bearing is

$$\sigma_c = \frac{F}{d \times t}$$

If there are n_c such bearing areas $d \times t$ in a joint, the compressive stress in bearing is given by

$$\sigma_c = \frac{F}{n_c \times d \times t} \tag{10-7}$$

Equation (10-7) is used to establish the number of bolts when *bearing stress* governs. The bearing stress computed by Eq. (10-7) must be smaller than or, at most, equal to the allowable stress of the material in bearing

$$\sigma_c = \frac{F}{n_c \times d \times t} \leqq \sigma_{c,\text{all}} \tag{10-8}$$

From Eq. 10-8 the number of bolts when bearing governs is

$$n_c \geqq \frac{F}{\sigma_{c,\text{all}} \times d \times t} \tag{10-9}$$

The allowable stress of the material in bearing is usually taken as

$$\sigma_{c,\text{all}} = 1.35\ \sigma_y \text{ to } 1.5\ \sigma_y \tag{10-10}$$

where σ_y is the yield stress in compression of the plate material.

A final check must be made to ascertain whether the member to be connected can safely transfer the required force F, even if it is weakened by the bolt holes. The area resisting the force F is equal to the total area of the member less the area of holes in the weakened section (Fig. 10-5). We have

$$A_t = t(b - n_1 d_1) \tag{10-11}$$

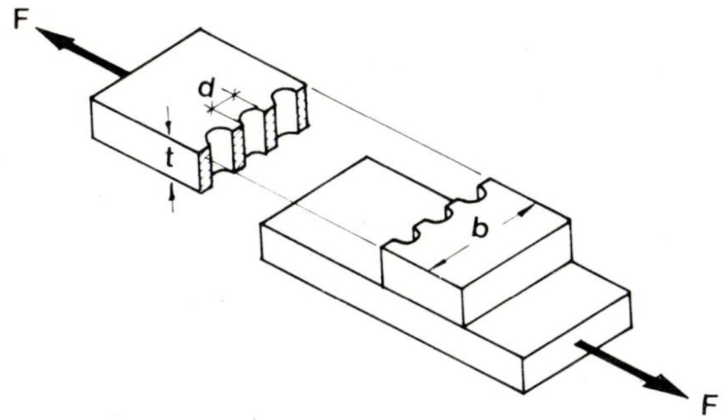

FIGURE 10-5 Area of the plate weakened by holes.

where t = thickness of plate
b = width of plate
n_1 = number of holes in the section
d_1 = diameter d of bolt plus an increment due to drilling or punching operations

In this book we assume

$$d_1 = d + \tfrac{1}{8} \text{ in}$$

From Eq. (10-11), the normal stress due to the force F in the cross section of the area A_t is

$$\sigma_t = \frac{F}{A_t} = \frac{F}{t(b - n_1 d_1)} \tag{10-12}$$

This stress must be smaller than or, at most, equal to the allowable tensile or compressive stress of the plate material

$$\frac{F}{t(b - n_1 d_1)} \leqq \sigma_{t,c,\text{all}} \tag{10-13}$$

10.3 FRICTION-TYPE BOLTED CONNECTIONS

Friction-type connections are assembled with special washers and tightened firmly so that enough friction develops between the plates, to prevent the bolts to bear against the plates. Thus friction-type bolted connections are not designed for bearing. Shear stress and tension in the bolt due to tightening are the governing stresses in the bolt design, while the stresses in the plates weakened by the holes must be checked as for a bearing-type bolted connection. Allowable stresses for friction-type connections must be selected from design handbooks and specifications (e.g., "Steel Construction Manual" of the American Institute of Steel Construction). The allowable stresses are subject to occasional changes as soon as the proper conclusions from the most recent research are made. The designer should be sure to use the most recent specifications.

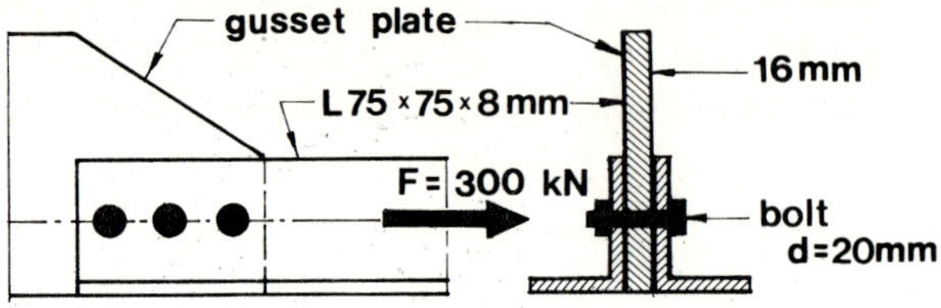

FIGURE 10-6

Illustrative Problem 10-1

In Illustrative Prob. 8-3 we designed the cross section of a tie-rod transferring a tensile force $T = 300$ kN (67.4 kips). As cross section, two equal-leg angles L 75 X 75 X 8 (the dimensions are in millimeters) were selected with a total area $A = 2 \times 11.0 = 22.0$ cm^2. Establish the number of bearing-type bolts needed to connect the two angles to a *gusset plate* 16 mm thick, as shown in Fig. 10-6. Use bolts with a diameter $d = 20$ mm and the following stresses:

(*a*) *Shear*

$$\tau_{\text{all}} = 13.8 \frac{\text{kN}}{\text{cm}^2} \quad \left(20 \frac{\text{kips}}{\text{in}^2}\right)$$

(*b*) *Bearing*

$$\sigma_{c,\,\text{all}} = 33.75 \frac{\text{kN}}{\text{cm}^2} \quad \left(40 \frac{\text{kips}}{\text{in}^2}\right)$$

(*c*) *Tension in the angles*

$$\sigma_{t,\,\text{all}} = 15 \frac{\text{kN}}{\text{cm}^2} \quad \left(22 \frac{\text{kips}}{\text{in}^2}\right)$$

Notice that the bolts are in double shear.

SOLUTION

(*a*) *Design for Shear* The number of bolts in double shear is computed by Eq. (10-5)

$$n \geqq \frac{F}{\tau_{\text{all}} \times 2(\pi d^2/4)} = \frac{300}{13.8 \times 2(\pi 2^2/4)} = 3.46 \qquad \text{(i.e., 4 bolts)}$$

(*b*) *Design for Bearing* The number of bearing-type bolts needed in bearing is found from Eq. (10-9):

$$n_c \geqq \frac{F}{\sigma_{c,\,\text{all}} \times d \times t} = \frac{300}{33.75 \times 2.0 \times 1.6} = 2.8 \text{ bolts} \qquad \text{(i.e., 3 bolts)}$$

where $d = 20$ mm $= 2.0$ cm and $t = 16$ mm $= 1.6$ cm are the diameter of the bolt and the thickness of the gusset plate, respectively. We could have checked bearing against the leg of one angle of thickness $t_1 = 8$ mm = 0.8 cm. However, the force to be transmitted in bearing through one angle is $F_1 = F/2 = 300/2 = 150$ kN. Thus, for the number of bolts governed by bearing, we again obtain

$$n_c \geqq \frac{F_1}{\sigma_{c,\text{all}} \times d \times t} = \frac{150}{33.75 \times 2.0 \times 0.8} = 2.8 \text{ bolts}$$

as before. Thus use four bolts with $d = 20$ mm.

(c) *Tensile Stress* Now check the tensile stress in the angles when their cross section is weakened by one bolt hole of diameter $d_1 = 20 \text{ mm} + \frac{1}{8} \text{ in} = 20 \text{ mm} + 3 \text{ mm} = 23 \text{ mm} = 2.3$ cm. Since the gross area of one angle is 11.0 cm^2 (see Table A-9) and the thickness of its leg is 8.0 mm = 0.8 cm, the net resisting area of one angle is

$$A_t = 11.0 - 2.3 \times 0.8 = 11 - 1.84 = 9.16 \text{ cm}^2$$

By Eq. (10-12) we have

$$\sigma_t = \frac{F_1}{A_t} = \frac{150}{9.16} = 16.38 \frac{\text{kN}}{\text{cm}^2} > \sigma_{t,\text{all}} = 15 \frac{\text{kN}}{\text{cm}^2}$$

Since the tensile stress in the angles is larger than the allowable tensile stress of their material, we have two choices:

1. Select larger angles made of the same material with

$$\sigma_{t,\text{all}} = 15 \frac{\text{kN}}{\text{cm}^2}$$

2. Select steel with a higher allowable stress, which should be at least

$$\sigma_{t,\text{all}} = 16.38 \frac{\text{kN}}{\text{cm}^2}$$

Illustrative Problem 10-2

The *butt* joint in Fig. 10-7 consists of two 1.0-in.-diameter (25-mm) bearing-type bolts on each side of the joint. The main plates to be connected are $10 \times \frac{3}{4}$ in

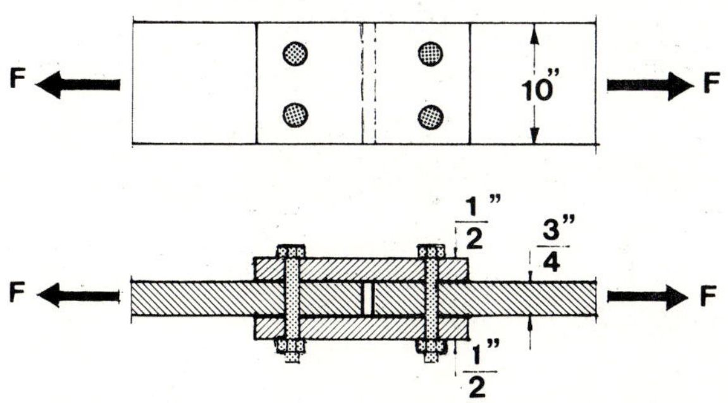

FIGURE 10-7 A bolted butt joint in Illustrative Prob. 10-2.

(254 × 19 mm), and the thickness of the cover plates is $\frac{1}{2}$ in (12.5 mm). Find the safe load which can be carried by the joint if the allowable stresses are as follows.

(*a*) *Shear*

$$\tau_{all} = 15\,000 \frac{\text{lb}}{\text{in}^2} \qquad \left(10.3 \frac{\text{kN}}{\text{cm}^2}\right)$$

(*b*) *Bearing*

$$\sigma_{c,all} = 45\,000 \frac{\text{lb}}{\text{in}^2} \qquad \left(31.0 \frac{\text{kN}}{\text{cm}^2}\right)$$

(*c*) *Tension*

$$\sigma_{t,all} = 20\,000 \frac{\text{lb}}{\text{in}^2} \qquad \left(13.80 \frac{\text{kN}}{\text{cm}^2}\right)$$

Assume

$$d_1 = d + \tfrac{1}{8} \text{ in} = 1.125 \text{ in}$$

Remember that the required strength of the joint must be developed on either side of the joint, that is, by two bolts.

SOLUTION

(*a*) *Shear* From Eq. (10-5) we can compute the unknown force F. In our case we have $n = 2$, and the bolts are in double shear. Thus

$$F = 2n \times \tau_{all} \times \frac{\pi d^2}{4} = 2 \times 2 \times 15\,000 \times \frac{\pi \times 1^2}{4} = 47\,124 \text{ lb} \qquad (209 \text{ kN})$$

(*b*) *Bearing against the Main Plates to be Connected* Use Eq. (10-9), in which $d = 1.0$ in, $t = \frac{3}{4}$ in, $n_c = 2$, and $\sigma_{c,all} = 45\,000$ lb/in^2. We have

$$F = n_c \times \sigma_{c,all} \times d \times t = 2 \times 45\,000 \times 1.0 \times 0.75 = 67\,000 \text{ lb} \qquad (300 \text{ kN})$$

It is easy to check that bearing in the cover plates does not govern the value of the allowable force F. Each cover plate transmits a force $F/2$, but their thickness is equal to $\frac{1}{2}$ in. By Eq. (10-9) we obtain

$$\frac{F}{2} = n_c \times \sigma_{c,all} \times d \times t$$

$$= 2 \times 45\,000 \times 1.0 \times 0.5$$

$$= 45\,000 \text{ lb}$$

so that

$$F = 2 \times 45\,000 = 90\,000 \text{ lb}$$

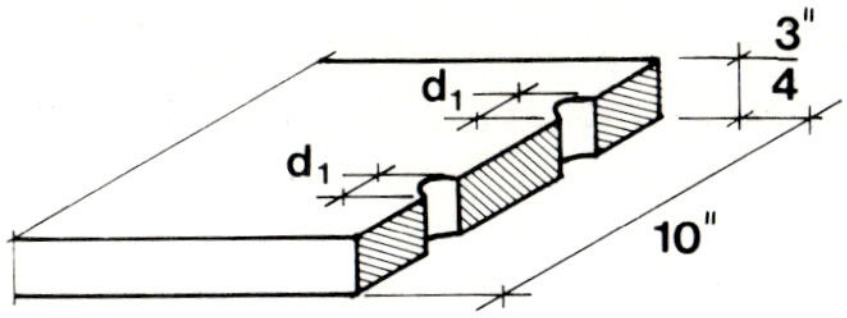

FIGURE 10-8

This value of F cannot govern, since it would cause bearing stresses above allowable values in the main plates.

(c) *Tension* For tension in the main plate $10 \times \frac{3}{4}$ in, weakened by two holes $d_1 \times t$ (Fig. 10-8), from Eq. (10-13) we have

$$F = \sigma_{t,\,\mathrm{all}} \times t(b - n_1 d_1) = 20\,000 \times 0.75\ (10 - 2 \times 1.125)$$

$$= 15\,000\ (10 - 2.25) - 15\,000 \times 7.75 = 116\,250 \text{ lb} \qquad (517 \text{ kN})$$

As a safe load on the joint we select the smallest of the three above values of F, that is, $F = 47\,124$ lb (209 kN), thus assuring that all stresses are within allowable limits.

Illustrative Problem 10-3

Compute the diameter of a bolt to connect a hanger 8 mm thick to two strips 5 mm thick (Fig. 10-9). The hanger must transfer a force $F = 25$ kN. The allowable stress is as follows.

(a) *Shear*

$$\tau_{\mathrm{all}} = 12\,\frac{\text{kN}}{\text{cm}^2}$$

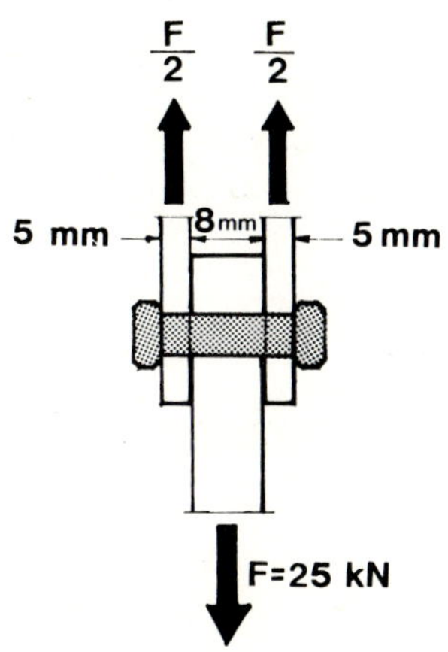

FIGURE 10-9

Detail of a short hanger connection on the Tower Bridge in London (*Photo by L. Cerny*).

(*b*) *Bearing*

$$\sigma_{c,\,\text{all}} = 30\,\frac{\text{kN}}{\text{cm}^2}$$

The bolt is considered to be of the bearing type and to act in double shear.

SOLUTION

(*a*) *Shear* Use Eq. (10-5) in the form

$$\frac{F}{2n\,(\pi d^2/4)} = \tau_{\text{all}} \qquad (n = 1.0)$$

from which the required diameter d of the bolt becomes

$$d^2 = \frac{4F}{2 \times 1 \times \pi \times \tau_{\text{all}}} = \frac{4 \times 25}{2\pi \times 12} = 1.32\ \text{cm}^2$$

and

$$d = \sqrt{1.32} = 1.15 \text{ cm} = 11.5 \text{ mm}$$

(*b*) *Bearing* Use Eq. (10-8) in the form

$$\frac{F}{n_c \times d \times t} = \sigma_{c,\text{all}}$$

where $t = 8$ mm = 0.8 cm and $n_c = 1.0$. Thus

$$d = \frac{F}{t \times \sigma_{c,\text{all}}} = \frac{25}{0.8 \times 30} = 1.04 \text{ cm} = 10.4 \text{ mm}$$

Use the larger of the two computed values of d and round it up to the next full millimeter, so that the bolt diameter becomes $d = 12$ mm.

10.4 WELDED JOINTS

Almost all metals may be connected by heating the two parts which are to be jointed and by depositing molten material between them. This process of joining two pieces together by fusion is called *welding.* Figure 10-10 shows what is known as a *side-fillet lap weld* with so-called corner returns (to prevent stress concentrations in the corners of the top plate). In the *fillet lap weld* the fused metal is deposited in the angle between the two plates. Figure 10-11 shows a side-fillet lap weld combined with

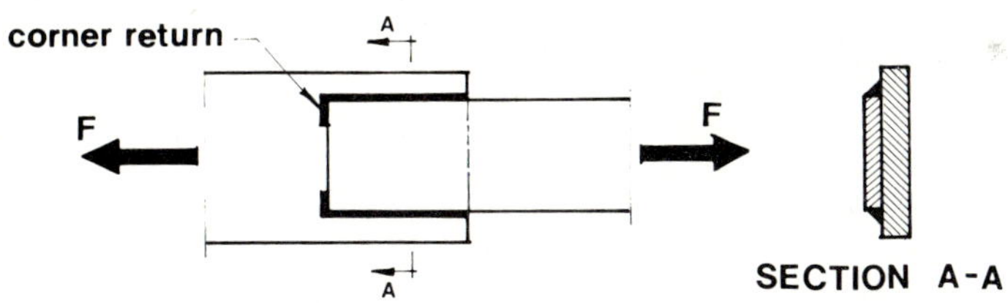

FIGURE 10-10 Lap joint with side-fillet lap welds and corner returns.

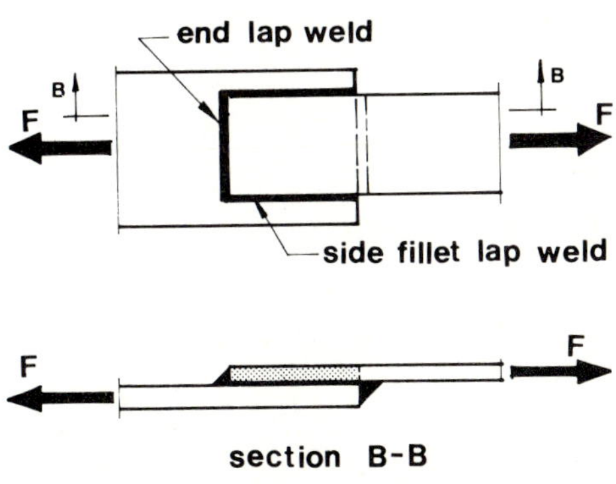

FIGURE 10-11 Lap joint with side- and end-fillet lap welds.

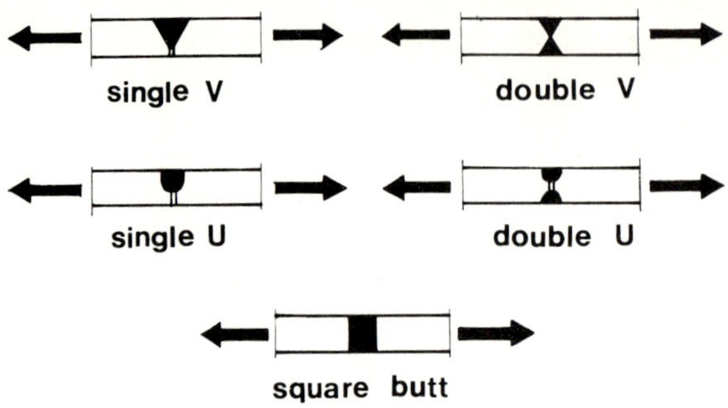

FIGURE 10-12 Types of butt welds.

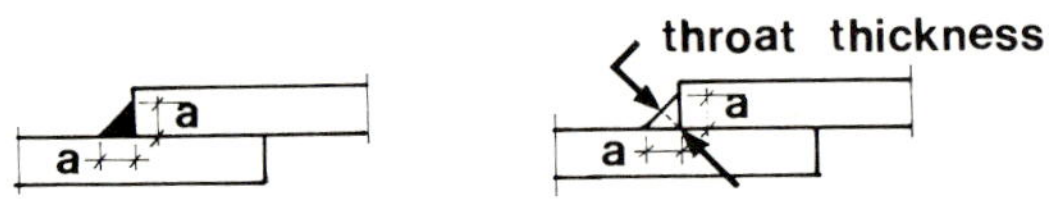

FIGURE 10-13 Throat thickness of a fillet lap weld.

the *end lap welds* at the ends of each plate. In this manner a larger total length of weld is achieved, thus decreasing the length of the required overlap of the two plates and thus saving plate material. Figure 10-12 shows some types of so-called butt welds.

When a tensile or compressive axial force F acts on the fillet lap-welded joint of Fig. 10-10 or 10-11, shear stresses develop in the welds. Fillet welds are usually considered to resist loads on the *throat area*, which is the smallest cross section of the weld. The throat area is the product of the *throat thickness* (Fig. 10-13) and the length of the weld. The throat thickness is taken to be

$$t = 0.707a$$

where a is called the *leg of a weld* or *weld size*. If the allowable shear stress of a weld material is τ_{all}, the allowable load F_1 for a weld 1.0 in long with a size a can be computed by

$$F_1 = \tau_{all} \times t \times 1.0 = \tau_{all} \times 0.707 \times a \tag{10-14}$$

Equation (10-14) has been used to compute the values of the allowable load F_1 for τ_{all} = 15 000 lb/in^2 (22 kN/cm^2) and different weld sizes a in Table 10-1.

TABLE 10-1 Allowable Loads for Fillet Welds
For τ_{all} = 15 000 lb/in²

a, in	F_1, lb/in	a, mm	F_1, kN/cm
3/16	1 980	4	6.2
1/4	2 650	6	9.3
5/16	3 310	8	12.4
3/8	3 960	10	15.5
7/16	4 640	12	18.7
1/2	5 300	14	21.8
9/16	5 960	16	24.9
5/8	6 630	18	28.0
11/16	7 290	20	31.1
3/4	7 950	22	34.2

A butt joint of two plates of equal size, when made of *matching weld material*, has the same strength as the plates.

Welding is a delicate process, to be executed by *certified welders*. The designer must follow the most recent specifications on weld sizes, throat areas, allowable stresses, materials, welding procedures (types of electrodes, electric current data), and so on.

Illustrative Problem 10-4

Design the length of the fillet weld, when both side- and end-fillet welds are to be used, to connect a plate 3 × 0.5 in to the gusset plate shown in Fig. 10-14. The axial force on the plate is $F = 80\ 000$ lb (356 kN). The size of the weld is $a = 0.5$ in, and the allowable shear stress of the weld material is $\tau_{all} = 15\ 000$ lb/in^2.

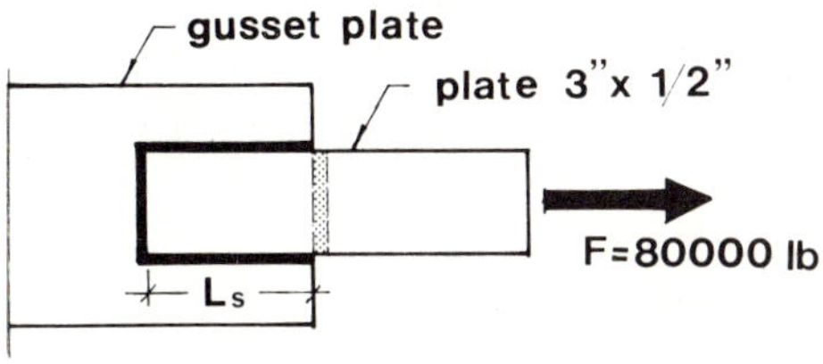

FIGURE 10-14

SOLUTION According to Table 10-1, the allowable load of a 0.5-in weld which is 1.0 in long is $F_1 = 5\ 300$ lb. Then the necessary length of the weld L is computed from the equation

$$F = F_1 \times L$$

or

$$80\ 000 = 5\ 300\,L$$

from which

$$L = \frac{80\ 000}{5\ 300} = 15.1 \text{ in}$$

The two end-fillet welds provide the length 2 × 3 = 6 in. Thus the length of the two side-fillet welds is

$$L_s = \frac{15.1 - 6}{2} = \frac{9.1}{2} = 4.55 \text{ in}$$

Provide a length of side fillet welds $L_s = 5$ in (12.7 cm).

New Terms Appearing in Chapter 10

Bolted joints
Lap joints
Bolt in single shear
Bolt in double shear

Bearing of bolts
Bearing-type bolts
Friction-type bolts
Gusset plate
Butt joint
Welded joints
Side-fillet weld
End-fillet weld
Lap weld
Butt weld
Throat of a weld
Throat thickness
Leg (size) of a weld

EXERCISES

10-1 The truss tension member in Fig. 10-15 consists of an equal-leg angle $L\ 5 \times 5 \times 7/16$ $(130 \times 130 \times 20)$ made of A-36 steel. The member is connected to the gusset plate, which is $\frac{1}{2}$ in thick (13 mm) by five $\frac{7}{8}$-in bolts (22 mm). Compute the allowable force F which can be safely applied to the member, if the allowable stresses are

(*a*) $\tau_{all} = 15\ 000\ \dfrac{\text{lb}}{\text{in}^2} \quad \left(10\ \dfrac{\text{kN}}{\text{cm}^2}\right)$

(*b*) $\sigma_{c,all} = 48\ 600\ \dfrac{\text{lb}}{\text{in}^2} \quad \left(33.5\ \dfrac{\text{kN}}{\text{cm}^2}\right)$

(*c*) $\sigma_{t,all} = 22\ 000\ \dfrac{\text{lb}}{\text{in}^2} \quad \left(15\ \dfrac{\text{kN}}{\text{cm}^2}\right)$

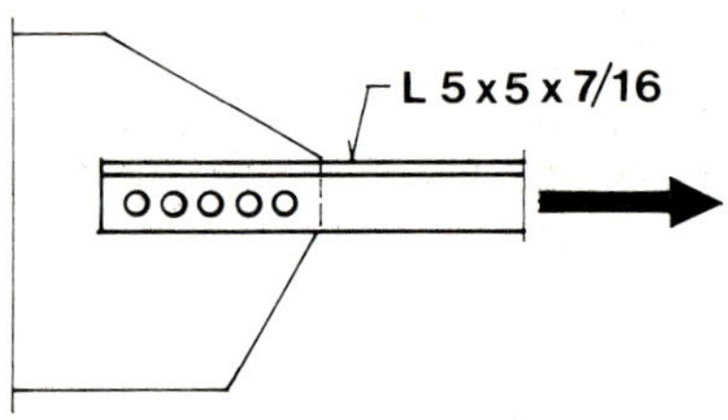

FIGURE 10-15

10-2 A butt joint of two plates $5 \times \frac{1}{2}$ in is obtained by means of two cover plates $5 \times \frac{3}{8}$ in and two bolts on each side of the joint (Fig. 10-16). The joint has to transfer a force of $F = 30\ 000$ lb (133.4 kN). Compute the necessary diameter of the bearing-type bolts if the allowable stresses are the same as in Exercise 10-1. (Notice that the bolts are in double shear.)

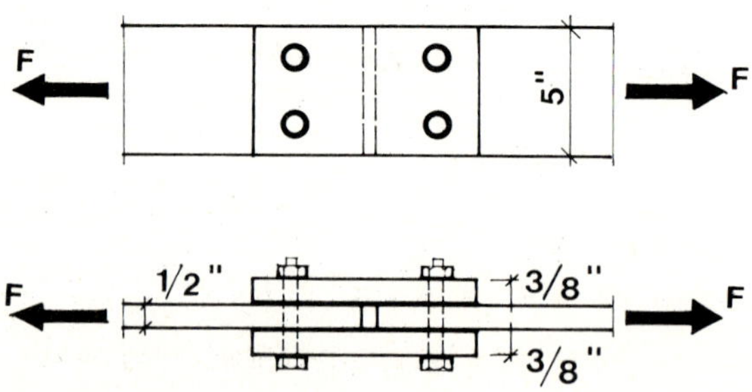

FIGURE 10-16

10-3 Compute the necessary number of 20-mm-diameter bolts needed for each side of the butt joint in Fig. 10-17, acted on by a force $F = 360$ kN. The allowable stresses are the same as in Exercise 10-1. The bolts are of the bearing type and in double shear.

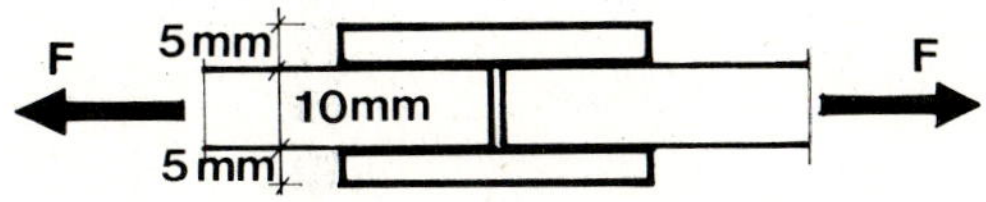

FIGURE 10-17

10-4 An American Standard Channel C 9 × 20 (200 × 290) is to transfer a force of 125 000 lb (556 kN) into a gusset plate by the combination of two end-fillet welds and two side-fillet welds (Fig. 10-18). The weld size is $\frac{3}{8}$ in (10 mm). What is the required length L of the side welds? Use Table 11-1 and round up L to the nearest whole inch.

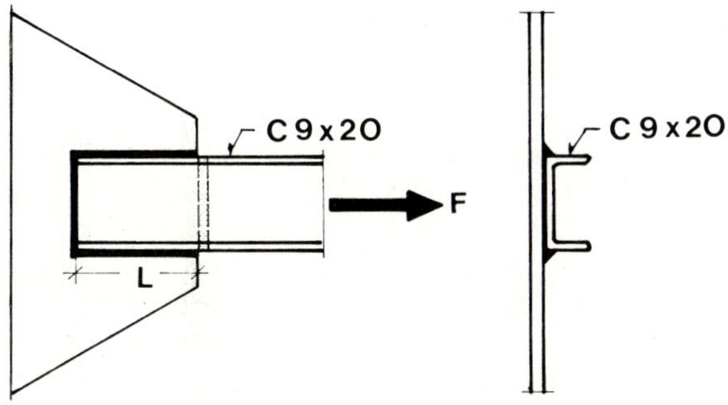

FIGURE 10-18

10-5 Two 5 × $\frac{1}{4}$-in plates are lap jointed by top and bottom end-fillet welds. What allowable load will the $\frac{3}{16}$-in weld carry according to Table 10-1? Check the strength of the plate in tension if it is made of A-36 steel with a $\sigma_{t,\text{all}} = 22\ 000\ \text{lb/in}^2$ (15 kN/cm^2).

One of four corner-columns supporting Northwestern Life Insurance Building in Minneapolis (*Photo by L. Cerny*).

Design of Axially Loaded Columns

11

11.1 INTRODUCTION

Slender compressed structural elements, as columns or truss members, can fail under the action of compressive axial forces much smaller than tensile axial forces. When a slender bar is subjected to axial tension, it tends to lengthen and straighten; the same bar under axial compression tends, instead, to bend out laterally. You may break a long, thin glass rod by applying to it a considerable tensile force; however, if you apply a compressive force to the same rod (Fig. 11-1), you may bend it out of its straight shape without too much effort, and a further small increase of the compressive force may break it. The slenderer the bar, the smaller the compressive axial force needed to bend it and break it. The bending of a compressed member out of its originally straight shape is called *buckling*. When this happens, we say that the bar has lost its *stability* by buckling.

If a straight bar buckles (bends laterally) under a compressive axial force F, bending moments develop in it (Fig. 11-2). The bending moment at a section x is equal to the product of the compressive force F times the lever arm y, which is the ordinate of

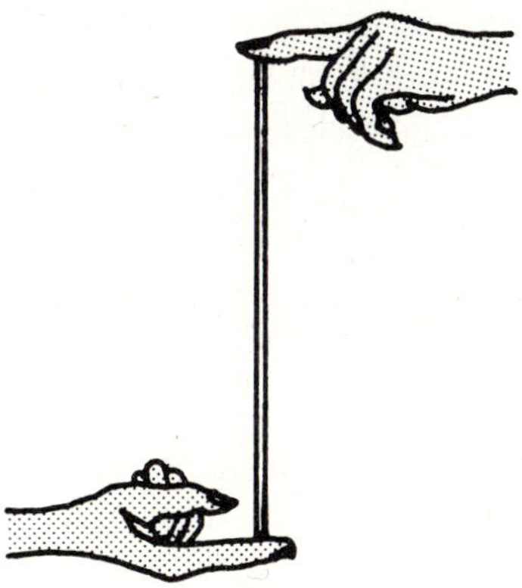

FIGURE 11-1

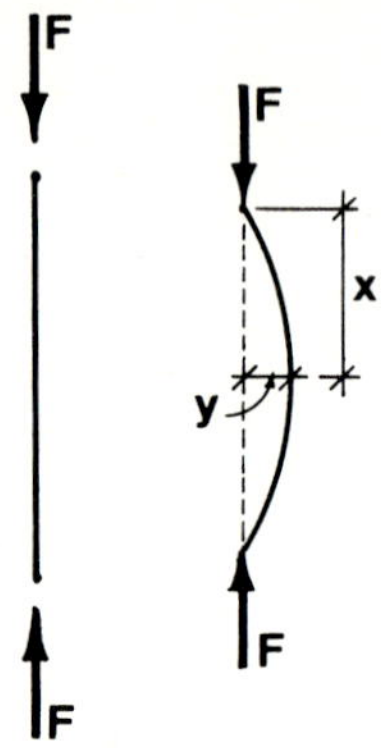

FIGURE 11-2 Buckling of a slender bar under compressive axial forces.

the deflection curve at the abscissa x. Thus, in addition to the axial force, bending moments act on the bar, which may bend it out still further, thus increasing the ordinates y and the moments. If this happens, a "chain reaction" develops and the process eventually results in the failure of the bar.

The phenomenon of loss of stability in slender compression members was not clearly understood in the past and caused the failure of many structures. When the loss of stability in compressed bars became understood through the work of the Swiss physicist Leonhard Euler in 1744, it became apparent that a compressive axial force F cannot buckle the bar unless it reaches a so-called critical value. Forces below this value cannot buckle the bar; forces above it do. Thus there is a specific critical value of the compressive force for each given bar at which the bar just starts buckling.

The value of the *critical force* is also called, after its discoverer, *Euler's load*, and its expression is called *Euler's formula.* After his discovery of the mathematical expression for the critical load, Euler suggested a design procedure for slender structural elements under compressive axial forces, that would make them safe against buckling failure (a failure due to loss of stability by buckling).

The basic expression for the critical force derived by Euler applies to straight slender columns supported at the ends by pins or hinges (Fig. 11-3*a*). (We have assumed this support condition to exist, for example, at the ends of truss members.) When the force F acting on a pin-connected column reaches its critical value, the column buckles and deflects as shown by the dashed line in Fig. 11-3*a*. According to Euler, the critical value of the force for such a column is expressed by the following equation, whose derivation is beyond the scope of this book:

$$F_{cr} = \frac{\pi^2 EI}{L^2} \tag{11-1}$$

where F_{cr} = Euler's load, lb or N

E = modulus of elasticity of the column material, lb/in^2 or N/cm^2

I = smallest moment of inertia of the column cross section about a centroidal axis, in^4 or cm^4

L = length of the column between the hinges, in or cm

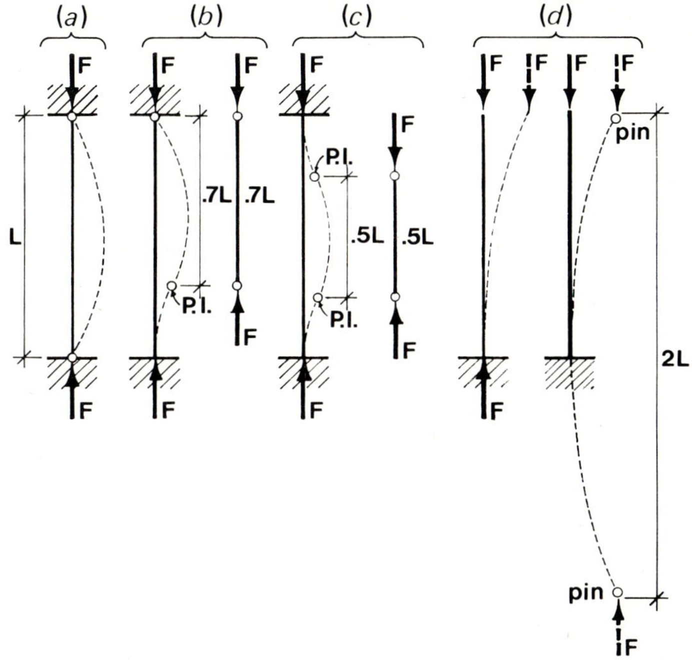

FIGURE 11-3 Effective lengths of columns with different end support conditions.

Illustrative Problem 11-1

A steel bar has a square cross section with 2-in (5-cm) sides and is 100 in (250 cm) long. It is pin connected at both ends and subjected to an axial load F. At what value of F will the bar fail?

(1) in tension

(2) in compression

Assume the modulus of elasticity of steel to be $E = 30\ 000\ 000\ \text{lb/in}^2$ (20 700 kN/cm²) and its yield stress $\sigma_y = 36\ 000\ \text{lb/in}^2$ (25 kN/cm²).

SOLUTION First, the stress in the bar under tension is not influenced by its slenderness, that is, by the cross-sectional properties and its length. The ultimate value of the tensile axial force for the bar is computed from Eq. (9-12),

$$F_u = A \times \sigma_y$$

where A is the area of the bar cross section,

$$A = 2 \times 2 = 4\ \text{in}^2$$

Thus

$$F_u = 4 \times 36\ 000 = 144\ 000\ \text{lb} \qquad (640\ \text{kN})$$

Second, if this slender bar is compressed, Euler's formula [Eq. (11-1)] must be used to calculate the critical value of force. Thus

$$F_{cr} = \frac{\pi^2 EI}{L^2}$$

where $\pi = 3.1416$

$$I = \tfrac{1}{12} 2^4 = 1.333 \text{ in}^4 \dagger$$

and

$$L = 100 \text{ in}$$

Thus

$$F_{cr} = \frac{\pi^2 \times 30\,000\,000 \times 1.33}{100^2} = 39\,480 \text{ lb} \qquad (176 \text{ kN})$$

This value is much lower than the ultimate value of F_u = 144 000 lb calculated for the bar in tension.

If we increase the length of the bar to L = 150 in (375 cm), the critical force drops to

$$F_{cr} = \frac{\pi^2 \times 30\,000\,000 \times 1.33}{150^2} = 17\,550 \text{ lb} \qquad (78 \text{ kN})$$

You can grasp the consequences of incorrectly assuming that, because steel has the same *strength* in tension as in compression, the ultimate value of the compressive force on the bar could be 144 000 lb instead of 17 550 lb.

11.2 COLUMNS WITH DIFFERENT SUPPORT CONDITIONS

Euler's formula [Eq. (11-1)] can be applied to *other than hinged* end-support conditions (Fig. 11-3*b*, *c*, *d*), provided we investigate the shape of the deflection curve assumed by the buckled column. Depending on the shape of the actual buckling curve, an adjustment must be made to the length L in Eq. (11-1). The shapes of the buckling curve for various end-support conditions of the column are shown in Fig. 11-3 by the dashed curves.

It can be proved, and it is physically obvious, that a column built in at one end and pin connected (hinged) at the other (Fig. 11-3*b*) is more rigid and therefore stronger against buckling than the same column when pinned at both ends (Fig. 11-3*a*). The "effective length" L_e of this column, to be substituted in Eq. (11-1) for L, is $0.7L$,

† See the formula for the calculation of the moment of inertia I of a rectangular section in Table A-10.

where L is the total length of the column from support to support. Thus

$$L_e = KL = 0.7L$$

where K is the *effective length factor* (in this case $K = 0.7$).

The length $KL = 0.7L$ is the distance between the top hinge support and the inflection point (PI) of the deflection curve in Fig. 11-3*b*.

Bending moments in pins are always zero—pins do not transmit them. We have previously observed from the deflection-bending moment relationship that bending moments are zero in points of inflection. Therefore, points of inflection on the deflection curves in Fig. 11-3 can be thought of as imaginary pins.

Substituting $KL = 0.7L$ for L in Eq. (11-1), we obtain the critical force for the pinned-fixed column,

$$F_{cr} = \frac{\pi^2 EI}{(0.7L)^2} = \frac{\pi^2 EI}{0.49L^2} \simeq 2 \frac{\pi^2 EI}{L^2} \tag{11-2}$$

Equation (11-2) indicates that the pinned-fixed column shown in Fig. 11-3*b* is *twice* as strong against buckling as the pinned column in Fig. 11-3*a*. In other words, a pinned-fixed column with length L has the same strength as a pinned column of length equal to $0.7L$.

A column fixed at both ends (Fig. 11-3*c*) has two points of inflection on its buckling deflection curve; they are a distance $0.5L$ apart, so that the effective length of this column is

$$L_e = KL = 0.5L$$

and

$$F_{cr} = \frac{\pi^2 EI}{(0.5L)^2} = 4\left(\frac{\pi^2 EI}{L^2}\right) \tag{11-3}$$

A column fixed at both ends is *four* times stronger than a pinned column of the same length. A pinned column of the same buckling strength as a fixed-fixed column of length L has a length equal to $0.5L$.

Finally, a column with one end fixed and the other end *free* (a cantilever column) has

$$L_e = KL = 2L$$

and

$$F_{cr} = \frac{\pi^2 EI}{(2L)^2} = \frac{1}{4}\left(\frac{\pi^2 EI}{L^2}\right) \tag{11-4}$$

Thus a cantilever column of length L (Fig. 11-3*d*) is *four times* weaker than a pinned column of the same length. (This is easily understood if one realizes that the cantilever column behaves like the upper half of a *pinned* column of length $2L$ such as that in Fig. 11-3*d*, in which the two inflection points are $2L$ apart.)

11.3 STRESSES IN COMPRESSED COLUMNS; RANGE OF VALIDITY OF EULER'S FORMULA

A critical stress σ_{cr} may be obtained by dividing the critical force F_{cr} by the area of the column's cross section

$$\sigma_{cr} = \frac{F_{cr}}{A} \tag{11-5}$$

where

$$F_{cr} = \frac{\pi^2 EI}{(KL)^2} \tag{11-6}$$

Substituting F_{cr} from Eq. (11-6) in Eq. (11-5), we have

$$\sigma_{cr} = \frac{\pi^2 EI}{(KL)^2 A} = \frac{\pi^2 E}{(KL)^2 A/I} \tag{11-7}$$

The quantity I/A is usually denoted r^2 (it has the dimensions of a length squared, as you can and should check), and its square root r is called the *radius of gyration* of the column's cross section. Thus

$$r^2 = \frac{I}{A}$$

and

$$r = \sqrt{\frac{I}{A}} \qquad \text{(in or cm)} \tag{11-8}$$

Since a column, if free to buckle in any direction, buckles about an axis with the *smallest* moment of inertia I_{min}, the radius of gyration r corresponding to I_{min} is the smallest radius of gyration r_{min} of the column's cross section. Thus Eq. (11-7) becomes

$$\sigma_{cr} = \frac{\pi^2 E}{(KL/r_{min})^2} = \frac{\pi^2 E}{\lambda^2} \tag{11-9}$$

where

$$\lambda = \frac{KL}{r_{min}} \tag{11-10}$$

is the ratio of the effective length KL to the smallest radius of gyration r_{min} of the column's cross section. The ratio λ is called the *effective slenderness ratio.* Notice that

λ

λ is small for columns with a small KL (short columns) or with a large r_{min}, that is, columns with a large I_{min} and a small area A. Columns with a large λ have large lengths and/or small r_{min}, that is, small I and large A. This means that λ *is large for slender* columns and *small for sturdy* columns.

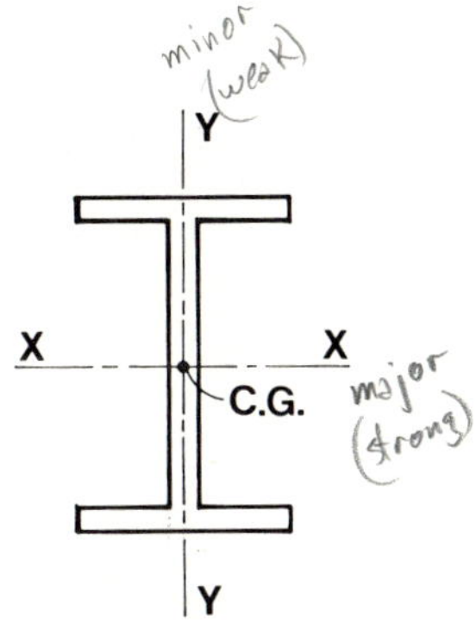

FIGURE 11-4

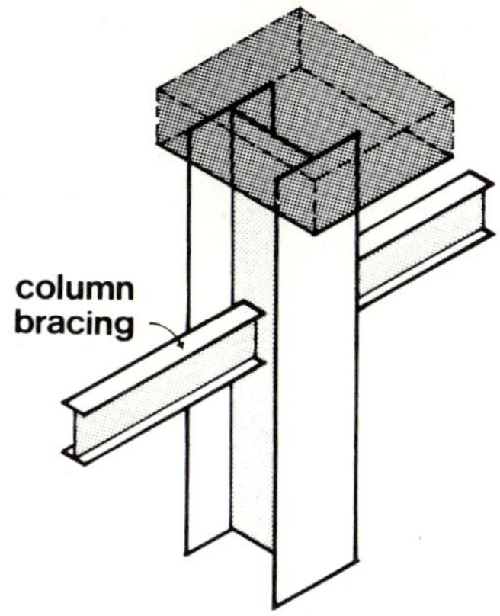

FIGURE 11-5

To facilitate the use of Eq. (11-9), the areas A, the moments of inertia I about the most significant centroidal axes, and the corresponding radii of gyration r for the most commonly used steel sections are given in Tables A-4 through A-9. For example, from Table A-4 a wide-flange section W 12 × 120 has an area $A = 35.3$ in^2. This section has two axes of symmetry, X-X and Y-Y (Fig. 11-4). The X-X axis of this section is called its *major* or *strong axis*, since the moment of inertia I_x about this axis is the largest among all the moments of inertia about any centroidal axis of the section. From Table A-4 we find that for this section, $I_x = 1\,070$ in^4 and the corresponding radius of gyration is $r_x = 5.51$ in. The Y-Y axis of this section is called its *minor* or *weak axis*, since the moment of inertia I_y about this axis is the smallest among all the moments of inertia about any centroidal axis of the section. From Table A-4, we find that $I_y = 345$ in^4 and the corresponding radius of gyration is $r_y = r_{\min} = 3.13$ in. A column built from this section would buckle about the weak, Y-Y, axis, unless buckling about this axis is prevented by some sort of bracing (Fig. 11-5).

According to Eq. (11-8) and Table A-10, the radii of gyration of a rectangular section $b \times h$ about the X-X and Y-Y axes are given by (Fig. 11-6*a*)

$$r_x = \sqrt{\frac{I_x}{A}} = \sqrt{\frac{bh^3}{12bh}} = \sqrt{\frac{1}{12} \times h^2} = 0.29h$$

and

$$r_y = \sqrt{\frac{I_y}{A}} = \sqrt{\frac{hb^3}{12bh}} = \sqrt{\frac{1}{12} \times b^2} = 0.29b$$

where

$$I_x = \frac{1}{12}bh^3 \qquad I_y = \frac{1}{12}hb^3 \qquad \text{and} \qquad A = bh$$

Hence for $b < h$,

$$r_y = r_{\min}$$

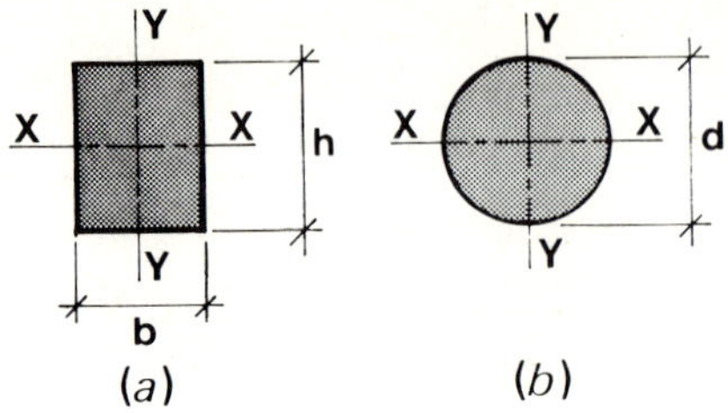

FIGURE 11-6

For a circular section of diameter d (Fig. 11-6b), we have

$$r_x = \sqrt{\frac{I_x}{A}} = \sqrt{\frac{\pi d^4/64}{\pi d^2/4}} = \sqrt{\frac{1}{16} d^2} = 0.25d$$

where the X-X axis is *any* diametral axis.

Let us now return to Eq. (11-9) and investigate its meaning. We notice that the larger the effective slenderness ratio $\lambda = KL/r$, the smaller the stress σ_{cr} needed to buckle the column. If we denote σ_{cr} in Eq. (11-9) as y, λ as x and $\pi^2 E$ as B (a constant), we will write Eq. (11-9) as

$$y = \frac{B}{x^2} \tag{11-11}$$

Which in the x, y plane is the equation of a curve called a *hyperbola.* We call the curve y, defined by Eq. (11-11), *Euler's hyperbola.* The graph of Euler's hyperbola can be plotted by selecting different values of $\lambda = KL/r$ and computing σ_{cr} from Eq. (11-9), using for E the modulus of elasticity of the column's material. In Table 11-1, several points of Euler's hyperbola have been computed for a column made of A-36 steel with $E = 30\ 000\ 000\ \text{lb/in}^2$. For example, for $\lambda = 100$,

$$\sigma_{cr} = \frac{\pi^2 E}{\lambda^2} = \frac{3.1416^2 \times 30\ 000\ 000}{100^2} = 29\ 609\ \frac{\text{lb}}{\text{in}^2}$$

TABLE 11-1

$\lambda = \frac{KL}{r}$	λ^2	$\sigma_{cr} = \frac{\pi^2 E}{\lambda^2}, \frac{\text{lb}}{\text{in}^2}$
10	100	2 960 880
25	625	473 740
50	2 500	118 435
75	5 625	52 638
100	10 000	29 609
150	22 500	13 160
200	40 000	7 400

If we plot the data of Table 11-1 in a cartesian coordinate system, with the values of λ as abscissas on the horizontal axis and the corresponding values of σ_{cr} on the vertical axis as ordinates, we obtain the graphical representation of Euler's hyperbola shown in Fig. 11-7. We see from this figure that for small values of the effective slenderness ratio λ, that is, for short, sturdy columns, the values of σ_{cr} are very high. Theoretically, for $\lambda = 0$ (infinitely rigid short columns), σ_{cr} would be infinitely large ($\sigma_{cr} = \infty$). However, we know from previous chapters that the largest stress which can develop in any member at failure is σ_y, the yield stress of the column material (point A in Fig. 11-7, where for A-36 steel $\sigma_{cr} = \sigma_y = 36\,000$ lb/in^2). Moreover, we know that E, the modulus of elasticity of the column material, is constant only within the limits of proportionality (see the stress-strain diagram in Fig. 8-10, earlier). For steel columns, this limit is usually taken as $\sigma_y/2$ (point B in Fig. 11-7, where $\sigma_y/2 = 18\,000$ lb/in^2). Hence σ_{cr} as computed by Euler's formula is valid only up to point B.

We conclude that Euler's hyperbola is valid for values of λ to the right of point B. Between points A and B, Euler's hyperbola does not have physical meaning (since it leads to stresses unrealistically high) and is replaced by an experimentally determined curve.

In the design of columns, the actual stress σ in a column must be a fraction of the critical stress σ_{cr} and must always be equal to or less than the allowable stress σ_{all}

$$\sigma \leqq \frac{\sigma_{cr}}{\text{SF}} = \sigma_{all} \tag{11-12}$$

where SF is the factor of safety against failure of the column in buckling.

The allowable stress σ_{all} in columns depends on σ_{cr}, which in turn depends on the effective slenderness ratio $\lambda = KL/r$. Table A-15 shows the allowable stresses for

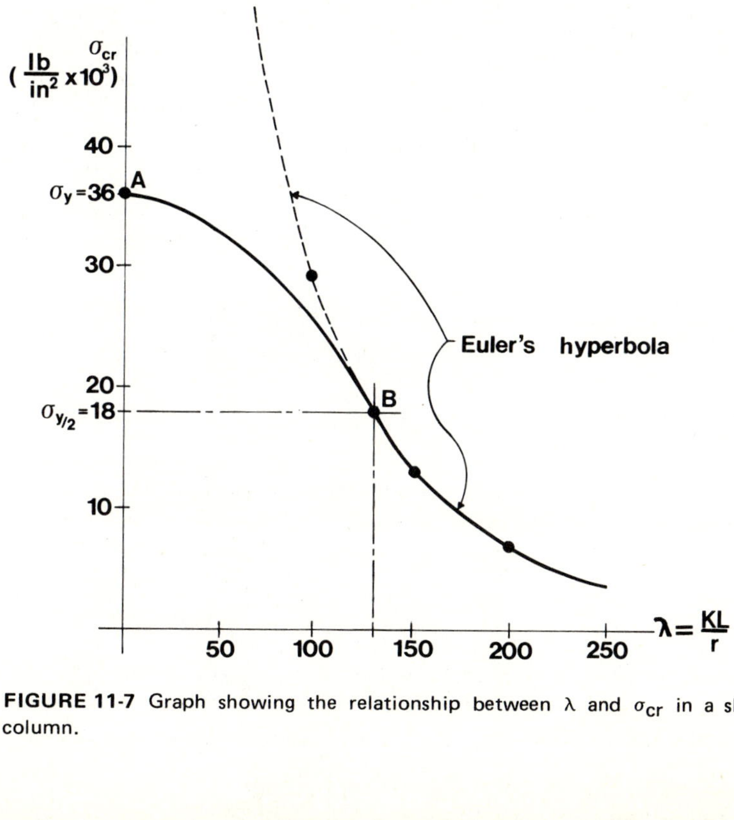

FIGURE 11-7 Graph showing the relationship between λ and σ_{cr} in a slender axially loaded column.

different values of $\lambda = KL/r$ which may be used to design columns made of A-36 steel. For example, if $KL/r = 20$ (short, sturdy column), $\sigma_{all} = 20.60$ kips/in^2. If, however, $KL/r = 120$, the allowable stress drops to $\sigma_{all} = 10.28$ kips/in^2. The largest effective slenderness ratio for steel columns allowed by specifications is $KL/r = 200$.

Illustrative Problem 11-2

Compute the allowable compressive axial force that can be safely applied to a W 12 X 120 column made of A-36 steel, if the column is 20 ft long (610 cm), pinned at the top and fixed at the bottom (Fig. 11-8).

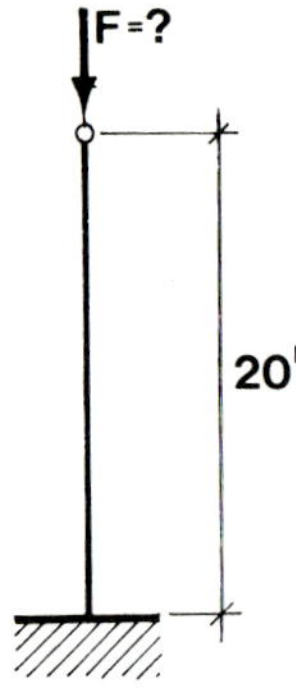

FIGURE 11-8

Sloped columns in a steel building (*Photo by L. Cerny*).

SOLUTION The significant properties of the **W** 12 X 120 section are found in Table A-4:

$$A = 35.3 \text{ in}^2 \qquad r_y = r_{min} = 3.13 \text{ in}$$

The effective length factor $K = 0.7$, so that

$$\lambda = \frac{KL}{r_{min}} = \frac{0.7 \times 20 \times 12}{3.13} = 53.67$$

From Table A-15 we find that, for $\lambda = 53.67 \simeq 54$, the allowable compressive axial stress on the column is $\sigma_{all} = 17.99$ kips/in^2. The allowable force on the column is

$$F_{all} = A \times \sigma_{all} = 35.3 \times 17.99 = 635 \text{ kips} \qquad (2\ 830 \text{ kN})$$

Illustrative Problem 11-3

Compute the allowable compressive force for the column described in Illustrative Prob. 11-2, assuming that the column is braced against buckling about its minor axis, but that buckling about its major axis is not prevented.

SOLUTION From Table A-4 we have

$$r_{max} = r_x = 5.51 \text{ in}$$

and

$$\lambda = \frac{KL}{r_x} = \frac{0.7 \times 20 \times 12}{5.51} = 30.5$$

From Table A-15 we find that, for $\lambda = 30.5$

$$\sigma_{all} = 19.9 \frac{\text{kips}}{\text{in}^2}$$

and

$$F_{all} = A \times \sigma_{all} = 35.3 \times 19.9 = 702 \text{ kips} \qquad (3\ 120 \text{ kN})$$

The use of lateral bracing increased the capacity of the column from 635 kips in Illustrative Prob. 11-2 to 702 kips.

Illustrative Problem 11-4

The column in Fig. 11-9 is a **W** 14 X 202 section made of A-36 steel. It is pin connected at its ends and loaded by an axial force $F = 1\ 000$ kips (4 450 kN). Since it is a very long column, it is braced at midheight by girts against buckling about the minor axis (Fig. 11-10). The bracing is light and does not prevent buckling about the major axis over the full length of the column. Check the adequacy of the column.

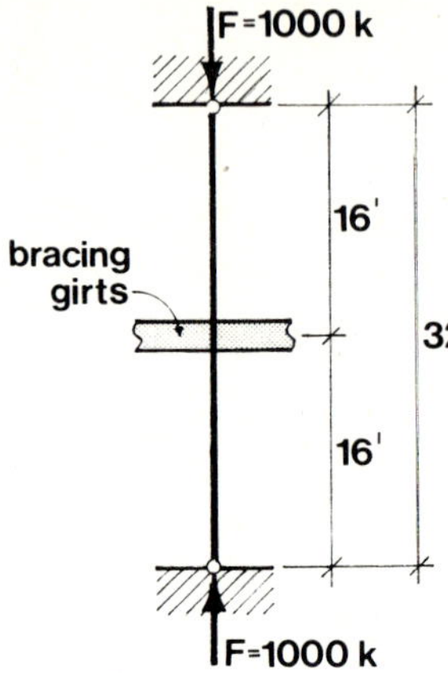

FIGURE 11-9

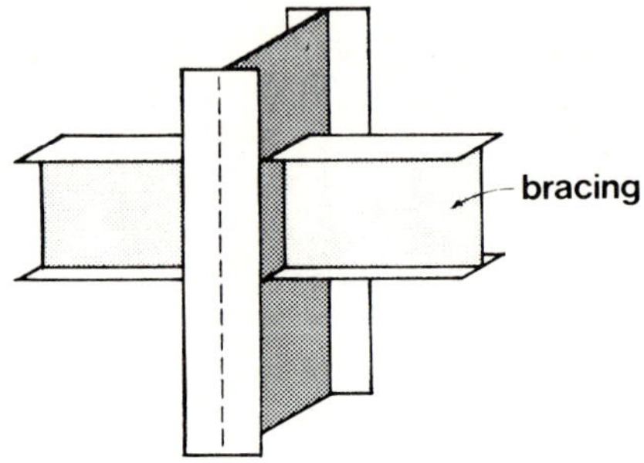

FIGURE 11-10

SOLUTION From Table A-4 we obtain

$$A = 59.4 \text{ in}^2 \qquad r_{max} = r_x = 6.54 \text{ in} \qquad r_{min} = r_y = 4.06 \text{ in}$$

The column may buckle in either of two modes:

1. Buckling about the major axis is not prevented by the bracing. Thus the effective length is the entire length of the column ($L = 32$ ft), as shown in Fig. 11-11, and

$$\lambda = \frac{KL}{r_x} = \frac{1.0 \times 32 \times 12}{6.54} = \frac{384}{6.54} = 58.72$$

The corresponding allowable axial stress from Table A-15 is

$$\sigma_{all} = 17.55 \ \frac{\text{kips}}{\text{in}^2}$$

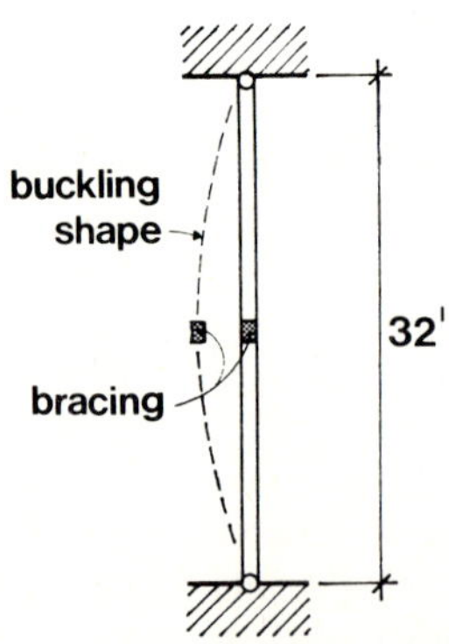

FIGURE 11-11 Buckling of the column about the major axis.

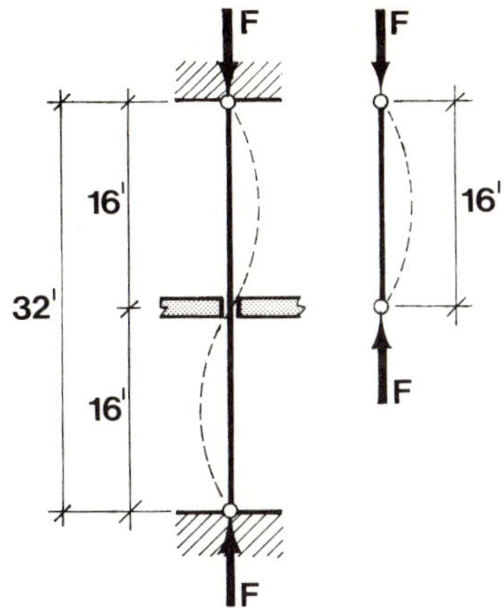

FIGURE 11-12 Buckling of the column about the minor axis.

and the capacity of the column is

$$F_{all} = A \times \sigma_{all} = 59.4 \times 17.55 = 1\,042 \text{ kips} \qquad (4\,640 \text{ kN})$$

which is larger than the applied force of 1 000 kips.

2. Because of the bracing at midheight, buckling of the column about its minor axis develops in the mode shown in Fig. 11-12: the 32-ft column behaves as two pinned columns, each 16 ft long.† Thus

$$\lambda = \frac{KL}{r_y} = \frac{1.0 \times 16 \times 12}{4.06} = 47.3$$

The corresponding allowable axial stress from Table A-15 is

$$\sigma_{all} = 18.58 \frac{\text{kips}}{\text{in}^2}$$

and

$$F_{all} = A \times \sigma_{all} = 59.4 \times 18.58 = 1\,104 \text{ kips} \qquad (4\,911 \text{ kN})$$

This capacity is also larger than the required $F = 1\,000$ kips, and the column section is adequate.

Note: If bracing were not provided, the effective length of the column with respect to buckling about the minor axis would be 32 ft. Hence λ would be

$$\lambda = \frac{KL}{r_y} = \frac{1.0 \times 32 \times 12}{4.06} = 94.6$$

the corresponding allowable stress (Table A-15) would be only

$$\sigma_{all} = 13.65 \frac{\text{kips}}{\text{in}^2}$$

†If the bracing were not at midheight, the middle "pin" would not be at the same location as the bracing and the problem could not be so easily solved.

and

$$F_{all} = A \times \sigma_{all} = 59.4 \times 13.65 = 811 \text{ kips}$$

would be less than 1 000 kips. Without lateral bracing, the column section would be inadequate.

Illustrative Problem 11-5

An unbraced 16-ft (488-cm) column of *A*-36 steel is pinned at both ends and must support an axial load of 180 kips (800 kN). Select the *lightest* wide-flange section among the **W** 14 sections in Table A-4 adequate for this column.

SOLUTION The design without the use of the manuals, handbooks, or charts is generally done by a *trial-and-error* process. First, we select a trial section and check its adequacy. For this column let us try **W** 14 × 30, which has

$$A = 8.83 \text{ in}^2 \qquad r_{min} = r_y = 1.49 \text{ in}$$

As the column is unbraced, buckling occurs about the minor axis, so that

$$\lambda = \frac{KL}{r_y} = \frac{1.0 \times 16 \times 12}{1.49} = 129$$

and, from Table A-15

$$\sigma_{all} = 8.97 \frac{\text{kips}}{\text{in}^2}$$

The allowable axial force is

$$F_{all} = A \times \sigma_{all} = 8.83 \times 8.97 = 79 \text{ kips} < 180 \text{ kips}$$

This section is insufficient.

Let us try a heavier section, **W** 14 × 48, which has

$$A = 14.1 \text{ in}^2 \qquad \text{and} \qquad r_y = 1.91 \text{ in}$$

Its effective slenderness ratio is

$$\lambda = \frac{KL}{r_y} = \frac{1.0 \times 16 \times 12}{1.91} = 100.5$$

for which, from Table A-15

$$\sigma_{all} = 12.92 \frac{\text{kips}}{\text{in}^2}$$

and

$$F_{all} = A \times \sigma_{all} = 14.1 \times 12.92 = 182 \text{ kips} > 180 \text{ kips}$$

The **W** 14 × 48 column has sufficient capacity to carry safely the required axial force of 180 kips.

11.4 TIMBER COLUMNS

Timber columns are always considered to be pin supported ($K = 1.0$). The allowable axial stress for timber columns is given by the formula†

$$\sigma_{all} = \frac{0.30E}{(L/d)^2} \tag{11-13}$$

where E = modulus of elasticity of timber, lb/in²
L = actual length of the column, in
d = the least dimension of the column section, in

Equation (11-13) is based on Euler's formula, with the inclusion of appropriate safety factor. Timber codes usually limit the ratio L/d to 50 or less:

$$\frac{L}{d} \leqslant 50 \tag{11-14}$$

The allowable stress σ_{all} computed from Eq. (11-13) should not be higher than the basic allowable stress for timber in compression parallel to the grain as given, for example, in Table A-12 for Douglas fir (1.0 – 1.5 kips/in²).

Illustrative Problem 11-6

A 10-in × 10-in column of Douglas fir is 18 ft long (548 cm). The basic allowable stress in compression parallel to the grain for the timber is 1 200 lb/in² (830 N/cm²) and E = 1 760 kips/in² (1 200 kN/cm²). Compute the allowable axial force on the column.

SOLUTION Notice that a column of *nominal* size 10 × 10 in has an actual or *dressed* size due to the surface finishing of only 9.5 × 9.5 in. Thus $d = 9.5$ in, $L = 18 \times 12 = 216$ in, and

$$\frac{L}{d} = \frac{216}{9.5} = 22.74 < 50$$

The allowable axial stress, by Eq. (11-13), is

$$\sigma_{all} = \frac{0.30 \times 1\,760\,000}{(22.74)^2} = 1\,021\ \frac{\text{lb}}{\text{in}^2} < 1\,200\ \frac{\text{lb}}{\text{in}^2}$$

The allowable force on the column is

$$F_{all} = A \times \sigma_{all} = 9.5 \times 9.5 \times 1\,021 = 92\,145\text{ lb} = 92.145\text{ kips} \qquad (410\text{ kN})$$

Illustrative Problem 11-7

A square timber column of Douglas fir has a length of 20 ft (610 cm). Find its nominal dimensions if it is loaded by an axial force F = 60 kips (267 kN).

†"Timber Construction Manual," Wiley, New York, 1974.

Assume $E = 1\ 760$ kips/in^2 (1 200 kN/cm^2) and the basic allowable stress in compression parallel to grain to be 1 050 lb/in^2 (730 N/cm^2).

SOLUTION This is a design problem which is usually started by selecting a trial section from a handbook.

First, ignore buckling and compute the required area using the basic allowable stress $\sigma_{all} = 1\ 050$ lb/in^2:

$$A = \frac{F}{\sigma_{all}} = \frac{60\ 000}{1\ 050} = 57.14 \text{ in}^2$$

The required side for a square column is

$$a = \sqrt{57.14} = 7.56 \text{ in}$$

Let us try a column with a nominal size 8 × 8 in and a dressed size 7.5 × 7.5 in. Then

$$\frac{L}{d} = \frac{20 \times 12}{7.5} = \frac{240}{7.5} = 32 < 50$$

For $E = 1\ 760\ 000$ lb/in^2, Eq. (11-13) gives

$$\sigma_{all} = \frac{0.3E}{(L/d)^2} = \frac{0.3 \times 1\ 760\ 000}{32^2} = 515 \frac{\text{lb}}{\text{in}^2}$$

The allowable stress in buckling for the column is smaller than the basic σ_{all} = 1 050 lb/in^2 and the required area is greater than 57.14 in^2. Assuming tentatively the required area to be proportional to the allowable stresses, we choose

$$A_{req} = 57.14 \times \frac{1\ 050}{515} = 116.5 \text{ in}^2$$

or

$$a = \sqrt{116.5} = 10.8 \text{ in}$$

A square column 10 × 10 in nominal size, with a dressed size 9.5 × 9.5 in, gives

$$\frac{L}{d} = \frac{240}{9.5} = 25.26$$

and

$$\sigma_{all} = \frac{0.3 \times 1\ 760\ 000}{(25.26)^2} = 827.30 \frac{\text{lb}}{\text{in}^2}$$

The axial capacity of the column is

$$F_{all} = (9.5)^2 \times 827.30 = 74\ 663 \text{ lb} > 60\ 000 \text{ lb (required)}$$

New Terms Appearing in Chapter 11

Slenderness effects in columns
Buckling
Loss of stability (by buckling)
Critical value of a compressive force
Euler's formula (for the critical value of a compressive force)
Effective length of a column, KL
Effective length factor, K
Radius of gyration, r
Effective slenderness ratio, $\lambda = KL/r$
Strong axis (of a section), major axis
Weak axis (of a section), minor axis
Bracing (against buckling)
Girt
Euler's hyperbola
Nominal size of a timber section
Dressed size of a timber section

Structural Elements Appearing in Chapter 11

Columns. These are (1) supported by a pin at each end (hinge or pin supported), (2), supported by a pin at one end and built in (fixed) at the other end, (3) fixed at both ends, or (4) fixed at one end and free at the other end (cantilever).

Bracing. This consists of light structural elements and is used to decrease the effective length of columns, usually in the direction of their minor axis.

Girt. This is a light structural element used to brace columns against buckling (usually about the minor axis of their section). In addition, girts are often used to support walls.

Quiz Number 9

This quiz contains questions for Chaps. 10 and 11.

A. TRUE OR FALSE

1. The slenderness of a column is measured by its smallest radius of gyration.
2. The radius of gyration r about a centroidal axis of a section is the product of the moment of inertia of the section about that axis and the area of the section.
3. A bolt is in single shear when only one cross-sectional plane resists shear.
4. The throat area of a fillet weld is the smallest cross section of the weld.
5. A short, sturdy column fails in yielding rather than as a result of loss of stability in buckling.
6. The larger the modulus of elasticity E of a column material, the smaller the critical compressive force on the column.
7. The throat area of a weld is the product of the throat thickness and the length of the weld.

8. The size of a fillet weld is governed by its throat area.
9. If a bolt resists shear in two shear planes, it is called a *bolt in single shear*.
10. If a side-fillet weld is close to the corner of a connecting member, it should end with a corner return.
11. In connecting two tension members by bolts we must check that the cross section of the members weakened by the holes can still carry the design force.
12. A pin-supported column is more resistant to buckling than is a cantilever column of the same length, cross section, and material.
13. The effective length of a column fixed at both ends is equal to one-half of the total length of the column.
14. Euler's formula for the critical value of a compressive force is needed to design short and sturdy columns.
15. The number of bearing-type bolts in a connection is governed by shear and bearing.
16. There are two basic types of bolts: friction bolts and bearing bolts.
17. The number of friction-type bolts is not governed by bearing.
18. The graphical representation of the relationship between the effective slenderness ratio λ and the critical stress σ_{cr} in a cartesian coordinate system is a parabolic curve.
19. In Chap. 10 we discussed the design of two kinds of joints: the lap joint and the butt joint.
20. Bracing is usually employed to prevent buckling of a column about the strong axis of its section.

B. MULTIPLE CHOICE

1. A structural member is connected to an adjacent structure by a set of single-shear bolts of diameter d. What is the number of bolts n required by shear, if the tensile force on the connection is F and the allowable stress of the bolts in shear is τ_{all}?

   (*a*) $n = F \times \dfrac{\pi d^2}{4} \times \tau_{all}$

   (*b*) $n = F \times d \times t \times \tau_{all}$

   (*c*) $n = \dfrac{F}{\tau_{all} \times d}$

   (*d*) $n = \dfrac{F}{\tau_{all} \times \pi d^2/4}$

2. What is the required number of bolts n_c in bearing for the connection described in question 1, if the thickness of the member is t and the allowable stress in bearing is $\sigma_{c,all}$?

   (*a*) $n_c = F \times E$

(*b*) $n_c = \dfrac{F}{\sigma_{c,\,\text{all}} \times d \times t}$

(*c*) $n_c = F \times d \times t \times \tau_{\text{all}}$

(*d*) $n_c = F \times d \times t \times \sigma_{c,\,\text{all}}$

3. Identify the correct expression for Euler's formula giving F_{cr} for a pin-connected column of length L, if the modulus of elasticity of the column material is E and the smallest moment of inertia of the column cross section is I.

(*a*) $F_{\text{cr}} = \dfrac{\sigma_{\text{all}}}{EI}$

(*b*) $F_{\text{cr}} = \sigma_{\text{all}} \times A$

(*c*) $F_{\text{cr}} = \dfrac{\pi^2 EI}{L^2}$

(*d*) $F_{\text{cr}} = \dfrac{\pi EI}{4L}$

4. The correct definition of the effective slenderness ratio λ is

(*a*) $\lambda = \dfrac{L}{A}$

(*b*) $\lambda = \dfrac{KL}{r}$

(*c*) $\lambda = F_{\text{cr}} \times E$

(*d*) $\lambda = \dfrac{L}{I}$

5. The correct definition of the radius of gyration is

(*a*) $r = \sqrt{\dfrac{I}{A}}$ (*b*) $r = \dfrac{L}{I}$

(*c*) $r = \dfrac{I}{A}$ (*d*) $r = A \times I$

6. Euler's hyperbola is the graphical representation in a cartesian coordinate system of the relationship between
   (*a*) the stress and the strain of the column material
   (*b*) column areas for different E
   (*c*) the length of a column and its smallest radius of gyration
   (*d*) the effective slenderness ratio λ and the critical stress σ_{cr}
7. The critical stress σ_{cr} of a short sturdy steel column equals
   (*a*) σ_y (*b*) infinity
   (*c*) zero (*d*) $\sigma_y/2$
8. Euler's hyperbola (steel column) is valid for
   (*a*) all values of λ from $\lambda = 0$ to $\lambda = \infty$

(*b*) small values of λ (close to $\lambda = 0$)
(*c*) critical stresses σ_{cr} equal to or less than $\sigma_y/2$
(*d*) critical stresses $\sigma_{cr} = \sigma_y$

9. If the effective slenderness ratio λ of a column increases, the allowable axial stress σ_{all} in the column
(*a*) decreases (*b*) increases (*c*) remains unchanged
10. The effective slenderness ratio for a cantilever column of length L and smallest radius of gyration r_y is

(*a*) $\lambda = \dfrac{L}{r_y}$

(*b*) $\lambda = \dfrac{2L}{r_y}$

(*c*) $\lambda = \dfrac{0.7L}{r_y}$

(*d*) $\lambda = \dfrac{0.5L}{r_y}$

C. SHORT ANSWERS

1. A compressive axial force which causes the failure of a column in buckling is called its ______________ load.
2. The discoverer of the formula for the critical load of a compressive column was ________.
3. A column which is not braced against buckling along its length buckles about its __________________ axis.
4. We have discussed the slenderness properties of
(*a*) cantilever columns
(*b*) pin-supported columns
(*c*) fixed columns
(*d*) columns pinned at one end and fixed at the other
Arrange these columns in the order of their growing buckling capacity, if all of them have the same actual length L, cross section, and material.
5. A column fixed at both ends is ________ times stronger than an identical column pinned at both ends.
6. A cantilever column is ________ times weaker than an identical column pinned at both ends.
7. The larger the effective slenderness ratio of a column, the ____________ the allowable axial stress in the column.
8. The effective length factor K for the following columns is
(*a*) pin-supported column; $K =$ ________
(*b*) cantilever column; $K =$ ________
(*c*) column fixed at both ends; $K =$ ________
(*d*) column pinned at one end and fixed at the other; $K =$ ________

9. The critical stress for a steel column with $\lambda = 0$ is σ_{cr} = ________ .
10. The allowable stress in buckling for a timber column of length L, smallest cross-sectional dimension d and modulus of elasticity E is

$$\sigma_{all} = \underline{\hspace{3cm}}$$

11. The critical stress in buckling for a steel column with $\lambda = KL/r$ and modulus of elasticity E is σ_{cr} = ________ .

EXERCISES

11-1 An unbraced column of A-36 steel is made of a **W** 36 × 300 section. It is fixed at the bottom, is pinned at the top, and is 70 ft (2 134 cm) long. What is the allowable compressive axial force on this column?

11-2 Find the capacity of the column described in Exercise 11-1, if it is braced in the weak direction so that its effective length is KL = 30 ft = 360 in in the weak direction, while it remains unchanged in the strong direction (KL = 0.7 × 70 × 12 = 588 in).

11-3 Find the capacities of an A-36 steel column, made of a **W** 12 × 120 section and with an actual length of 24 ft (731 cm), under the following end conditions:
(*a*) pinned at both ends
(*b*) fixed at both ends
(*c*) pinned at one end and fixed at the other
(*d*) cantilevered

11-4 An A-36 column, pin supported at both ends and 18 ft long, is loaded by an axial compressive force F = 260 kips. Select a proper size for this column among the **W** 10 sections.

11-5 A timber column of Douglas fir has a 12 × 12 in nominal size and an unbraced length of 20 ft. Compute its allowable axial load, if the basic allowable stress in compression parallel to grain is 1 500 lb/in^2 and E = 1 760 000 lb/in^2.

11-6 A 10- × 12-in nominal-sized column of Douglas fir has an unsupported length of 16 ft. Compute its allowable axial load, assuming a basic allowable stress in compression parallel to grain of 1 500 lb/in^2 and E = 1 760 000 lb/in^2. Remember that buckling about the weak axis governs.

11-7 Design a square column of Douglas fir 20 ft long for an axial force of 150 kips. Assume E = 1 700 000 lb/in^2 and the basic allowable stress in compression σ_{all} = 1 300 lb/in^2.

Construction of reinforced concrete columns for the Seattle domed stadium (*Courtesy of Bethlehem Steel Corporation*).

Additional Topics in Structural Analysis

12

12.1 REINFORCED AND PRESTRESSED CONCRETE

In the preceding chapters the basic principles of statics and strength of materials were presented and then used to analyze or design simple structural elements made of metal or wood. The design of concrete structures requires additional considerations because concrete, while possessing considerable strength in compression, is weak in tension: its tensile strength is less than one tenth of its compressive strength.

In order to increase the carrying capacity of concrete members, *steel reinforcing bars* are placed in them in those regions where tensile stresses develop. Reinforced concrete structures are commonly used all over the world, and theories for their analysis and design have been developed.

The first requirement in the design of reinforced concrete structures is to identify the regions of tensile stresses. For example, the bottom fibers of the beam shown in Fig. 12-1*a* are in tension since its bending moments are positive. Therefore, the main steel reinforcing bars must be placed close to the bottom of the beam. In a cantilever beam, such as that shown in Fig. 12-1*b*, tensile stresses due to the negative moments develop in the top fibers, and the main reinforcing bars must be placed close to the top of the beam. Concrete beams without reinforcement would fail as a result of tension at very low load levels.

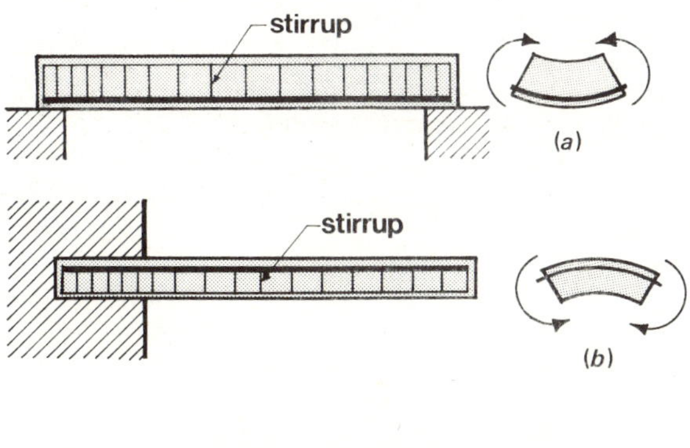

FIGURE 12-1 (*a*) Main reinforcement for positive bending moments is at the bottom of the beam; (*b*) main reinforcement for negative bending moments is at the top.

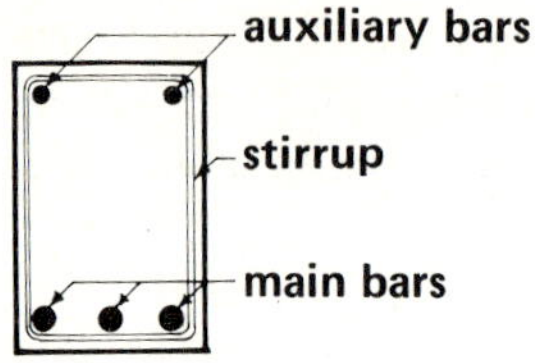

FIGURE 12-2

The largest shear stresses in a beam develop in the sections where the largest shear forces develop. The shear strength of concrete is also very low. Thus *shear reinforcement* or *stirrups* are used to transfer shear in concrete beams. Stirrups sew the beam layers transversely, so that shear cracks cannot develop (Fig. 12-2).

Prestressed concrete is a special kind of reinforced concrete. The principle used in a prestressed concrete beam is based on the superposition of two kinds of stresses (Fig. 12-3):

1. *Flexural (bending) stresses.* These are caused by the *applied loads.* If concrete were equally strong in tension and compression, their maximum values would occur in the top and bottom beam fibers (their sign depends on the sense of the bending moments M). For a positive bending moment acting at a section, these stresses would be

$$\sigma_c = -\frac{M}{S} \quad \text{and} \quad \sigma_T = +\frac{M}{S} \tag{12-1}$$

(see Chap. 9). The distribution of these flexural stresses along the beam depth is shown in Fig. 12-3*a*.

2. *Stresses due to prestressing.* Prestressing is used to introduce in the beam compressive stresses to counterbalance the undesirable tensile stresses that would be

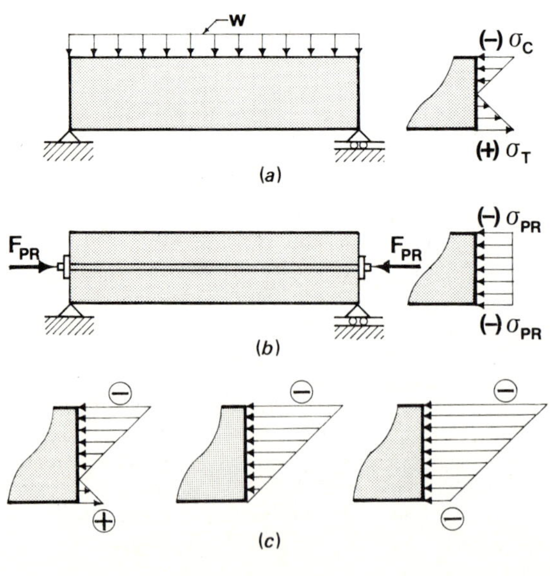

FIGURE 12-3 Principle of prestressed concrete: (*a*) flexural stresses due to applied loads; (*b*) stresses due to axial prestressing; (*c*) three cases of combined stresses due to applied loads plus prestressing.

developed by the applied loads. It may be achieved by threading a cable (tendon) through a duct embedded in the concrete beam and then stretching it by applying a large tensile force to it. The stretched (elongated) cable is then anchored at the anchor plates at the beam ends, thus producing a compressive force on the anchor plates and consequently compressive stresses in the beam. If the cable is located along the beam's centroidal axis, uniformly distributed compressive stresses develop in its cross sections. These compressive prestressing stresses are equal to

$$\sigma_{PR} = -\frac{F_{PR}}{A} \tag{12-2}$$

where F_{PR} = prestressing force
A = area of cross section

The distribution of the prestressing stresses is shown in Fig. 12-3*b*.

By superposition of flexural and prestressing stresses, the unfavorable tensile stresses due to the applied load can be reduced, canceled, or changed to compressive stresses, as shown in Fig. 12-3*c*. For given loads, the resulting value of stresses depends on the magnitude of the prestressing force.

A prestressing force along the centroidal axis of the beam is not as efficient as one applied close to the bottom fibers (Fig. 12-4*a*). The force F_{PR} counteracting the undesirable tensile stresses due to, say, gravity loads produces two kinds of stresses:

1. When F_{PR} acts along the centroidal axis of the beam (see Fig. 12-3*b*), it produces the uniformly distributed compressive stress in all cross sections along the beam, as given by Eq. (12-2),

$$\sigma_{PR} = -\frac{F_{PR}}{A}$$

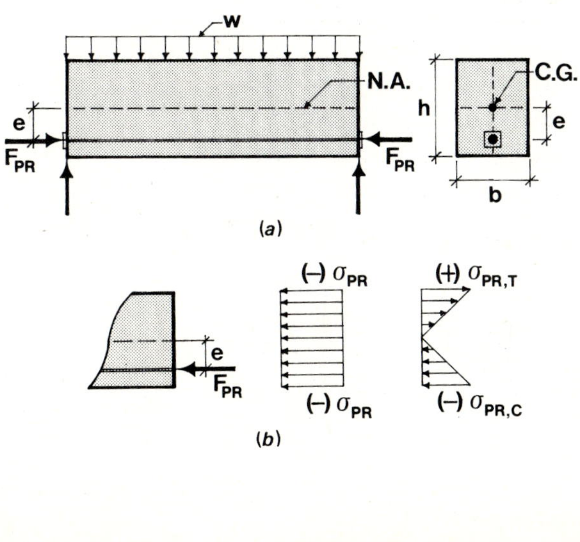

FIGURE 12-4 Stresses in the beam due to eccentric prestressing. (*a*) prestressing force F_{PR} is applied at eccentricity *e*. (*b*) Combination of two kinds of stresses due to eccentric prestressing.

2. When the force F_{PR} acts at a section, at a distance e from the beam axis, called the *eccentricity*, it also develops at that section a moment

$$M_{PR} = F_{PR} \times e \tag{12-3}$$

and hence produces additional bending stresses in its extreme fibers (Fig. 12-4*b*) given by Eq. (12-1),

$$\sigma_{PR,T} = +\frac{M_{PR}}{S} \qquad \sigma_{PR,C} = -\frac{M_{PR}}{S} \tag{12-4}$$

The final stresses in the extreme fibers of the cross section are obtained by the superposition of the bending stresses due to the applied load w and the stresses due to prestressing, given by Eqs. (12-1), (12-2), and (12-4). For the top fibers, we have

$$\sigma = \sigma_C + \sigma_{PR} + \sigma_{PR,T} \tag{12-5}$$

For the bottom fibers, we have

$$\sigma = \sigma_T + \sigma_{PR} + \sigma_{PR,C}$$

Thus, by prestressing with an eccentricity below the beam axis, the tensile stresses at the bottom fibers due to the applied loads are counteracted by the compressive stresses σ_{PR} and $\sigma_{PR,C}$, while the top fibers are protected against excessive compression by the tensile stress component $\sigma_{PR,T}$. Eccentric prestressing of flexural members is much more effective than prestressing along the beam axis.

The most favorable curve for the prestressing cables in a simply supported beam loaded by a uniformly distributed load is a parabola (Fig. 12-5). The moment diagram due to the applied uniformly distributed load has a parabolic shape (see Chap. 7, p. 162). When the eccentricity e of the prestressing force varies with a parabolic law, the diagram of the M_{PR} moments [Eq. (12-3)] is also a parabola but of sign opposite to that of the load, and the stresses due to the load are equally reduced at all beam sections. Thus the parabolic path of the prestressing cable is the most efficient for this type of loading. There is no moment at the supports of a simply supported beam due to the applied loads; if the eccentricity of the prestressing force at the supports is made zero ($e = 0$), the only stresses at the support cross sections are the uniform compressive stresses σ_{PR} given by Eq. (12-2).

Prestressed concrete members have larger capacity than do those of ordinary

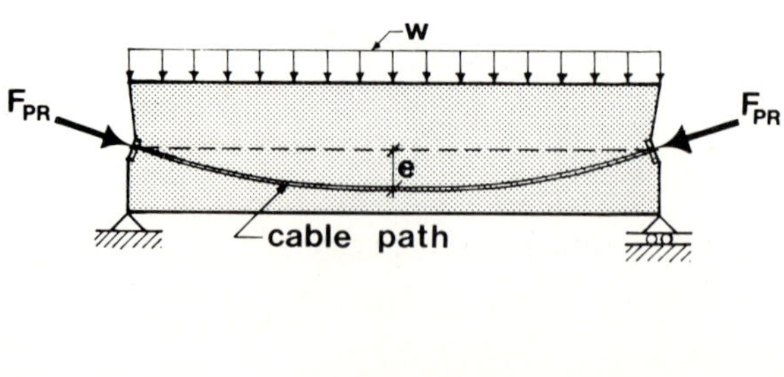

FIGURE 12-5 Parabolic prestressing cable.

reinforced concrete. As a result, structures with larger spans and slender members can be built by using the prestressing.

Illustrative Problem 12-1

A simply supported concrete beam of rectangular cross section 18 × 36 in (46 × 91 cm) spans 40 ft (12.2 m) and is loaded by a total uniformly distributed load of 2.0 kips/ft (32.7 kN/m), which includes the beam weight. Compute the value of an *axially* applied compressive force F_{PR} such that no tensile stress develops at the midspan section.

SOLUTION For stresses due to applied load, the bending moment at midspan is

$$M_{max} = \frac{1}{8} wL^2 = \frac{1}{8} \times 2.0 \times 40^2 = 400 \text{ kip}\cdot\text{ft}$$

The section modulus of the section (assumed to react equally in tension and compression) is

$$S = \frac{1}{6} bh^2 = \frac{1}{6} \times 18 \times 36^2 = 3\ 888 \text{ in}^3$$

The bending stresses due to the applied load are then

$$\sigma_C = -\frac{M}{S} = -\frac{400 \times 12 \times 1\ 000}{3\ 888} = -1\ 235 \frac{\text{lb}}{\text{in}^2} \qquad \text{(compression in top fibers)}$$

$$\sigma_T = +1\ 235 \frac{\text{lb}}{\text{in}^2} \qquad \text{(tension in bottom fibers)}$$

To cancel the tensile stress $\sigma_T = +1\ 235$ lb/in², the prestressing force F_{PR} acting along the beam axis must develop a compressive stress

$$\sigma_{PR} = -1\ 235 \frac{\text{lb}}{\text{in}^2}$$

Hence, from Eq. (12-2), the required compressive prestressing force

$$F_{PR} = \sigma_{PR} \times A = -1\ 235 \times 18 \times 36 = -800\ 280 \text{ lb}$$

$$= -800.28 \text{ kips} \qquad (-3\ 560 \text{ kN})$$

With the prestressing force acting along the centroidal beam axis, prestressing stress is distributed uniformly over the cross section.

The resulting stress in the bottom fibers of the midspan section is then

$$\sigma = \sigma_T + \sigma_{PR} = +1\ 235 - 1\ 235 = 0$$

and the stress in the top fibers is

$$\sigma = \sigma_C + \sigma_{PR} = -1\ 235 - 1\ 235 = -2\ 470 \frac{\text{lb}}{\text{in}^2} \qquad \left(-1\ 700 \frac{\text{N}}{\text{cm}^2}\right)$$

The diagram of the resulting stresses in the midspan section is shown in Fig. 12-6.

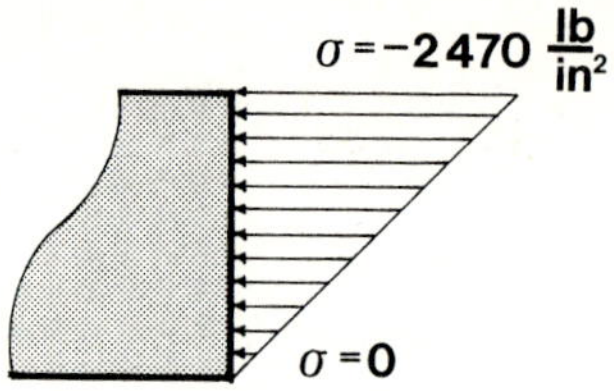

FIGURE 12-6

Illustrative Problem 12-2

Compute the required prestressing force for the beam described in Illustrative Prob. 12-1, if the cable at midspan is 2 in above the bottom face of the beam.

SOLUTION The eccentricity of the prestressing force is

$$e = 18 - 2 = 16 \text{ in}$$

and

$$M_{PR} = 16F_{PR}$$

Thus the tensile stress due to bending moment M_{PR}

$$\sigma_{PR,T} = +\frac{M_{PR}}{S} = +\frac{16F_{PR}}{3\,888} \qquad \text{(top fibers)}$$

and the compressive stress due to bending moment M_{PR}

$$\sigma_{PR,C} = -\frac{M_{PR}}{S} = -\frac{16F_{PR}}{3\,888} \qquad \text{(bottom fibers)}$$

The uniformly distributed compressive stress due to the prestress is

$$\sigma_{PR} = -\frac{F_{PR}}{A} = -\frac{F_{PR}}{18 \times 36} = -\frac{F_{PR}}{648}$$

The bending stresses due to the applied loads are those computed in Illustrative Prob. 12-1, that is,

$$\sigma_C = -1\,235 \frac{\text{lb}}{\text{in}^2} \qquad \text{(top fibers)}$$

$$\sigma_T = +1\,235 \frac{\text{lb}}{\text{in}^2} \qquad \text{(bottom fibers)}$$

The condition of zero resulting stress at the bottom fibers due to the combined effect of load and prestress is [Eq. (12-5)]

$$\sigma = \sigma_T + \sigma_{PR} + \sigma_{PR,C} = 0$$

or

$$+1\,235 - \frac{F_{PR}}{648} - \frac{16F_{PR}}{3\,888} = 0$$

and after simplification

$$1\,235 - 0.00566F_{PR} = 0$$

from which

$$F_{PR} = \frac{1\,235}{0.00566} = 218\,260 \text{ lb} = 218.26 \text{ kips} \qquad (971 \text{ kN})$$

Notice the reduction in the value of F_{PR} compared to that in the example of Illustrative Prob. 12-1 (F_{PR} = 800.28 kips). By selecting a proper path of the cables, considerable savings in prestressing steel can be achieved.

The resulting top-fiber stresses are now

$$\sigma = \sigma_C + \sigma_{PR} + \sigma_{PR,T}$$

$$\sigma = -1\,235 - \frac{218\,260}{648} + \frac{16(218\,260)}{3\,888}$$

$$= -\,1\,235 - 337 + 898 = -674\,\frac{\text{lb}}{\text{in}^2} \qquad \left(-465\,\frac{\text{N}}{\text{cm}^2}\right)$$

There is a considerable reduction in the value of the top fiber stresses compared to that in Illustrative Prob. 12-1, where it was equal to 2 470 lb/in^2. Thus, in addition to saving steel, concrete of much lower quality can be used for the beam, which may also be made slenderer.

The values of F_{PR} computed in Illustrative Probs. 12-1 and 12-2 are purely theoretical. In practice, higher values of F_{PR} must be used because concrete tends to shrink in time as a result of compressive stresses, and as the concrete shrinks, the tension in the prestressing tendons decreases. Initial higher F_{PR} values are used to compensate for concrete shrinkage.

12.2 ANALYSIS OF STATICALLY INDETERMINATE STRUCTURES

Reactions and internal forces in statically *indeterminate* structures cannot be evaluated solely by the methods of statics discussed in the previous chapters of this book.

We have used so far three conditions of equilibrium for a coplanar force system acting on a two-dimensional structure or a free body cut out of it. These conditions are (Sec. 6.1)

$$\Sigma V = 0 \qquad \Sigma H = 0 \qquad \Sigma M = 0 \tag{12-6}$$

Equations (12-6) are sufficient for computation of the three unknown reactions or internal forces in statically *determinate* structures.

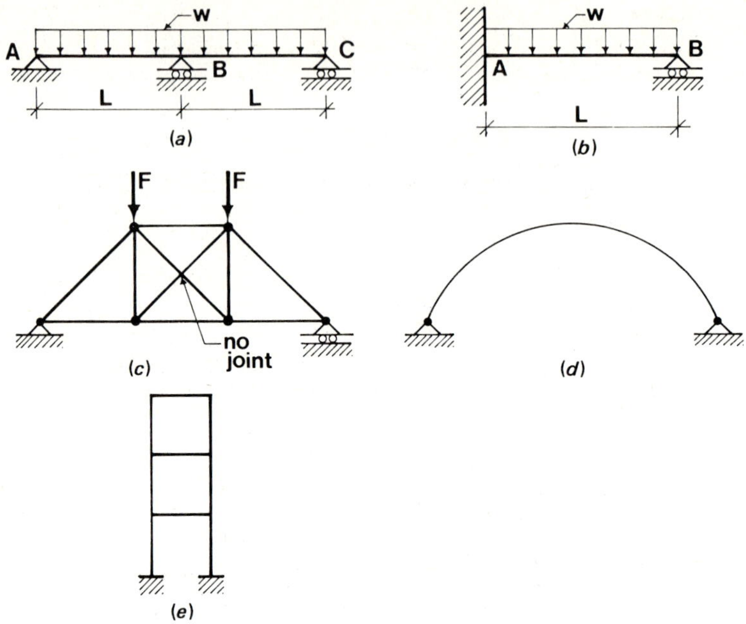

FIGURE 12-7 Statically indeterminate structures: (*a*) two-span continuous beam; (*b*) fixed-hinged beam; (*c*) statically indeterminate truss; (*d*) two-hinged arch; (*e*) statically indeterminate multistory rigid frame.

Statically indeterminate structures, instead, have more external reactions or internal forces than do comparable statically determinate structures, and we need additional equations to solve the problem completely.

Some types of statically indeterminate structures are shown in Fig. 12-7. The beam shown in Fig. 12-7*a* is called a *continuous beam.* Such beams span continuously over one or more interior supports and may be fixed or simply supported at their ends. The interior support *B* of the beam in Fig. 12-7*a* is *redundant*: if it were removed, the beam would become a simply supported, statically determinate beam with a span $2L$. If the support *B* is a *roller support*, only one (vertical) component of reaction can develop in it, and we say that this beam is *once* statically indeterminate. Besides the equations of statics [Eqs. (12-6)], we need one additional equation to determine its reactions.

To analyze this beam, we may first remove the redundant support *B* and solve the simply supported beam *AC* of span $2L$ (Fig. 12-8*a*) loaded by the uniformly distributed load *w*. The downward deflection at midspan of this beam, due to the uniformly distributed load, is, by Eq. (9-25),

$$y_{B,w} = \frac{5w(2L)^4}{384EI} \quad \text{(downward)} \tag{12-7}$$

In the actual structure shown in Fig. 12-7*a*, however, there is a rigid support at *B*, and therefore no deflection can occur at *B*. To achieve a zero deflection at *B* we

Double-deck continuous steel bridge over the Mississippi River at the University of Minnesota in Minneapolis (*Photo by L. Cerny*).

must apply an *upward* reaction R_B to the simply supported beam AC in Fig. 12-8 of a magnitude that will cancel the deflection $y_{B,w}$. The deflection at midspan of the simply supported beam AC due to an upward concentrated force R_B is (see Table A-14, case 1)

$$y_{B,R} = \frac{R_B(2L)^3}{48EI} \quad \text{(upward)} \tag{12-8}$$

Since the deflection at B in the original statically indeterminate beam in Fig. 12-7*a* must be zero, the absolute values of the deflections in Eqs. (12-7) and (12-8) must be equal:

$$y_{B,w} = y_{B,R} \tag{12-9}$$

or

$$\frac{5w(2L)^4}{384EI} = \frac{R_B(2L)^3}{48EI}$$

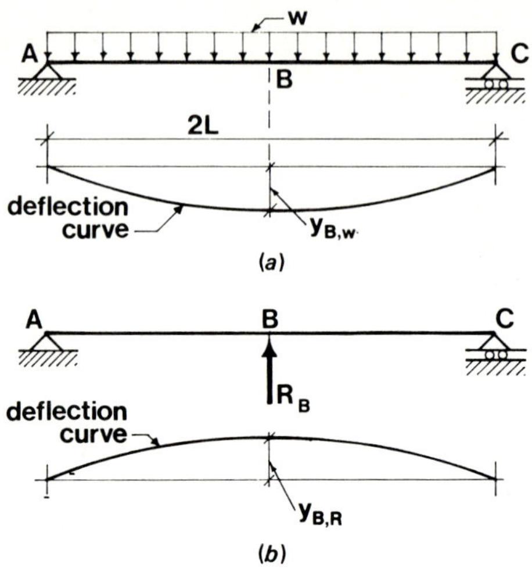

FIGURE 12-8 Solution of the two-span continuous beam: (*a*) statically determinate beam obtained by the removal of support at *B*; (*b*) reaction R_B replaces the effect of support at *B*.

from which the value of the unknown reaction R_B is

$$R_B = \tfrac{5}{8}\, w(2L) = \tfrac{10}{8}\, wL$$

where L is the length of the equal spans of the statically indeterminate beam in Fig. 12-7*a*. Notice that the value of R_B does not depend on EI.

The resulting reactions and the bending-moment diagram for the original statically indeterminate continuous beam are obtained by superposition of the two cases of Fig. 12-8*a*, *b*. The reactions and bending moments for the two cases are shown in Fig. 12-9. The positive bending moment at B due to the uniformly distributed load of case *a* is (Fig. 12-9*a*)

$$M_{B,w} = +\tfrac{1}{8}\, w(2L)^2 = +\tfrac{1}{2}\, wL^2$$

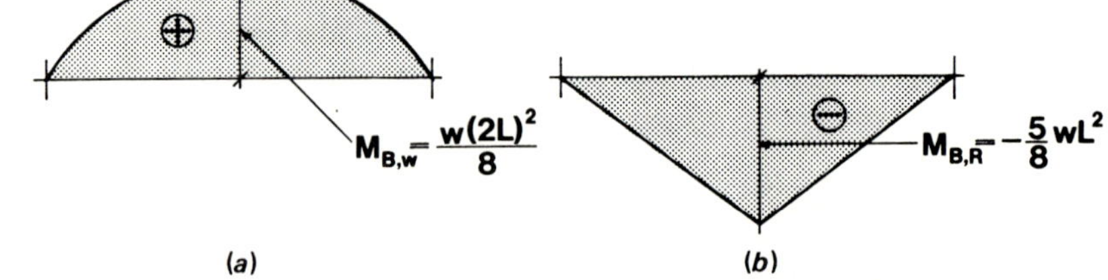

FIGURE 12-9 (*a*) Bending-moment diagram on the statically determinate beam due to uniformly distributed load; (*b*) bending-moment diagram on the statically determinate beam due to reaction R_B at *B*.

The negative bending moment at B due to the concentrated force $R_B = \frac{10}{8} wL$ in case b is (Fig. 12-9b)

$$M_{B,R} = -R_{A,R} \times L = -\tfrac{5}{8} wL \times L = -\tfrac{5}{8} wL^2$$

The bending moment over the support B in the statically indeterminate beam is

$$M_B = M_{B,w} + M_{B,R} = +\tfrac{1}{2} wL^2 - \tfrac{5}{8} wL^2$$
$$= \tfrac{4-5}{8} wL^2 = -\tfrac{1}{8} wL^2$$

The resulting reactions of the continuous beam are (Fig. 12-9)

$$R_A = R_{A,w} + R_{A,R} = +wL - \tfrac{5}{8} wL$$
$$= +\tfrac{3}{8} wL$$
$$R_B = \tfrac{10}{8} wL$$
$$R_c = R_A = +\tfrac{3}{8} wL \qquad \text{(because of symmetry)}$$

To check the correctness of the reaction values, we may check whether their sum is equal to the total uniformly distributed load, or $2wL$:

$$R_A + R_B + R_c = 2wL$$
$$\tfrac{3}{8} wL + \tfrac{10}{8} wL + \tfrac{3}{8} wL = \tfrac{16}{8} wL = 2wL$$

Thus equilibrium of the vertical forces is satisfied.

The resulting bending moment diagram for the continuous beam is shown in Fig. 12-10a. By plotting the vertical coordinates of this bending-moment diagram on a horizontal line, we obtain the bending-moment diagram of Fig. 12-10b.

Similar methods are used to analyze other statically indeterminate structures.

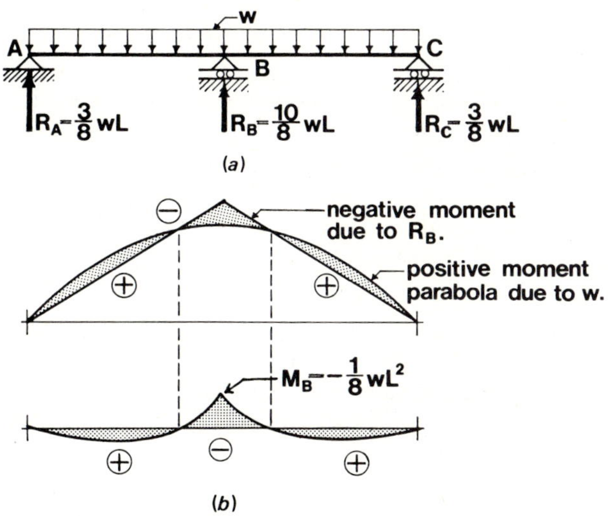

FIGURE 12-10 (*a*) Resulting bending-moment diagram is obtained by superposition of cases (*a*) and (*b*) in Fig. 12-9; (*b*) the same diagram plotted on the horizontal reference line.

The beam in Fig. 12-7*b* is once statically indeterminate, since the removal of the roller support at *B* reduces it to a statically determinate cantilever.

The truss of Fig. 12-7*c* is also once statically indeterminate since, after the removal of one of the two crossing (diagonal) bars in the center panel, the truss becomes statically determinate.

The *two-hinge arch* shown in Fig. 12-7*d* is once statically indeterminate, since by changing one hinge into a roller support it becomes statically determinate. Multistory *rigid frames*, like that in Fig. 12-7*e*, are structures with high statical indeterminancy. Their analysis is complex and is often performed with the help of an electronic computer.

12.3 EXPERIMENTAL STRESS-ANALYSIS OF STRUCTURES

The structural systems used in modern urban, transportation, nuclear, hydraulic, underground, and other structures have become more and more complicated, and such are also their analysis and design, which is often obtained with the help of the most efficient computers and computer programs. But when the structure to be analyzed is of exceptional importance or when the analytical assumptions (idealizations) made in the calculations are uncertain, or when the validity of a new theory is to be verified, the analysis is sometimes checked experimentally. Such an analysis consists of measuring the stresses, the strains, the deformations (deflections), or other quantities of interest on the real structure (prototype) or on a *model* of the real structure made in smaller scale out of a *model material.* Materials used in actual structures are seldom used to build models. Typical model materials are fine-aggregate concrete, aluminum, steel, plastics, gypsum, rubber, gelatin, and others, which are chosen according to their required mechanical properties. If we want to use a model of a structure which develops stresses within the elastic limit, the model material must also exhibit elastic properties, such as those represented by Hooke's law. The modulus of elasticity of the model material, though, should be relatively low, so that large forces are not needed to produce large enough deformations, stresses, or strains measurable with the required accuracy. Model materials should also be easy to machine, shape, cast, or glue, so that the model may be easily manufactured.

The model of a structure, once built, must be supported as the prototype and loaded by loads distributed like those on the prototype.

Different experimental techniques have been developed to obtain the desired mechanical quantities by model analysis. *Strain gauges*, used to measure strains at different points of models, are commonly of the *electrical resistance* type. A unidirectional strain gauge of this type is shown in Fig. 12-11. It consists of a thin resistance wire bent into a number of parallel segments and glued to a paper base or cast

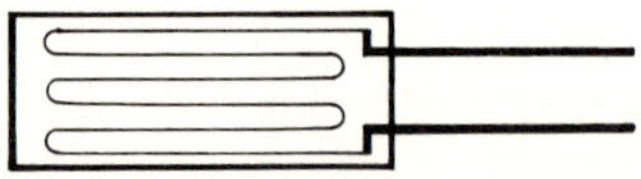

FIGURE 12-11 Electrical resistance strain gauge.

in plastic foil. Thicker wires are attached to the ends of the resistance element, which are soldered to the wires from the measuring or recording instrument. The strain gauge itself is glued to the surface of the model or the prototype at a point where, and in the direction in which, the strain is to be measured.

As the model or prototype deforms under load, the wire in the strain gauge deforms with it. When a tensile strain develops at the measured point, the wire of the strain gauge elongates, its cross section *decreases*, and, as a consequence, its electrical (ohmic) resistance *increases.* If the strain is compressive, the wire shortens, its cross section increases, and its electrical resistance decreases. If R is the resistance of the strain gauge wire before deformation and ΔR is the change of this resistance caused by the loading of the model or the prototype, the $\Delta R/R$ ratio may be shown to be proportional to the strain:

$$\frac{\Delta R}{R} = C\epsilon \qquad (12\text{-}10)$$

where C is the constant of proportionality between the unit change of electrical resistance $\Delta R/R$ of the strain gauge wire and the strain ϵ at the point in the given direction.

The unit change of electrical resistance $\Delta R/R$ can be measured with great accuracy by an instrument called a *Wheatstone bridge.* Instruments are available that are capable of measuring and recording simultaneously the output of hundreds of strain gauges, thus monitoring the changes of strain at various locations of a model or of a real structure.

Once the unidirectional strain ϵ at a point is known, the corresponding stress σ can be computed by Hooke's law,

$$\sigma = E\epsilon$$

where E is the modulus of elasticity of the model or the prototype material, whenever the stress field is unidirectional, such as in the bars of a truss or in fibers of a beam.

In another experimental method of stress analysis called *photoelasticity*, models are made of *optically sensitive* materials which, when stressed and placed in a field of

FIGURE 12-12 Isochromatic lines of a disk loaded by two forces F across its diameter.

FIGURE 12-13 Isochromatic lines of a simply supported beam loaded by two equal concentrated forces.

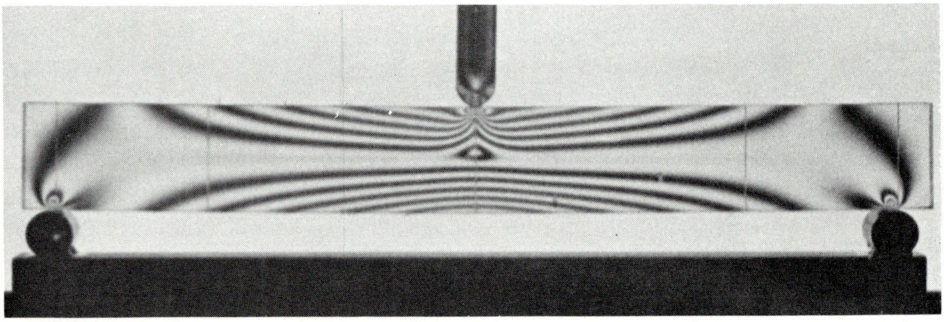

FIGURE 12-14 Isochromatic lines of a simply supported beam loaded by one concentrated force at midspan.

FIGURE 12-15 Isochromatic lines in the model of a shearwall of a multistory building, supported on two columns at the level of main floor.

polarized light,† show colored fringes related to the direction and magnitude of the stresses caused by the load. Some of the so-called isochromatic lines or fringes thus obtained are shown in Figs. 12-12 through 12-15.

Figure 12-12 shows the isochromatic lines in a disk loaded by two concentrated, equal and opposite forces acting across the disk diameter. Figure 12-13 shows the photoelastic model of a simply supported beam loaded by two equal concentrated forces, while in Fig. 12-14 the same beam is loaded by one concentrated force acting at midspan. Finally, Fig. 12-15 shows the isochromatic lines in a transverse bearing wall (a so-called shearwall) of a multistory building, supported on two columns at the level of the main floor. The openings in the wall model represent the corridors passing through the walls in the longitudinal direction of the building.

†While white natural light vibrates in all directions perpendicular to the light ray, polarized light vibrates in only one direction. Polarized light is obtained by means of *Polaroid filters*, similar to the antiglare lenses used in Polaroid sunglasses.

Bibliography

1. Beer, F. P., and E. R. Johnston: "Statics," 3d ed., McGraw-Hill Book Company, New York, 1977.
2. Jensen, A., and H. H. Chenoweth: "Applied Engineering Mechanics," 3d ed., McGraw-Hill Book Company, New York, 1972.
3. Bassin, M. G., S. M. Brodsky, and H. Wolkoff: "Statics and Strength of Materials," 2d ed., McGraw-Hill Book Company, New York, 1969.
4. Jensen, A., and H. H. Chenoweth: "Statics and Strength of Materials," 2d ed., McGraw-Hill Book Company, New York, 1967.

Answers to Exercises and Quizzes

Answers to Exercises (all exercises are completely solved in the *Solution Manual*)

CHAPTER 2

2-1 Dead load = 136 lb/ft^2; live load = 40 lb/ft^2; total load = 176 lb/ft^2.

2-2 The pressure under the footing p = 1 705 lb/ft^2.

2-3 The pressure on the contact area p = 2 870 lb/in^2.

2-4 The weight of the beam = 120.4 lb/ft.

2-5 (*a*) F = 126 000 lb; (*b*) F = 31 500 lb; (*c*) F = 63 000 lb.

2-6 The load on the joist = 67 lb/ft.

2-7 The total wind force on the narrow side = 1 152 kips. The total wind force on the wide side = 2 303 kips.

CHAPTER 3

3-1 F_2 = 31 kips; F_3 = 81.5 kips (the arrows of both reactions are heading toward the point of their intersection with the force F_1).

3-2 The total vertical force = 23.6 kips (downward).

3-3 The vertical component = 13.7 kips (downward). The horizontal component = 8.2 kips (to the right).

3-4 The force in member AB = 267 lb (tension). The force in member BC = 480 lb (compression).

3-5 (*a*) R = 3.2 kips (downward, almost vertical; makes an angle of 7° to the right from the vertical wall); (*b*) R = 47 kips (downward; makes an angle of 41° to the left from the vertical).

3-6 (*a*) R = 14.2 kips; downward; in the plan (Fig. 3-39*a*) the resultant makes an angle of 28° to the left from the vertical; (*b*) E = 14.2 kips [the same magnitude and direction as the resultant R in part (*a*), but of opposite sense].

3-7 R = 12.4 kips (acting downward, exactly in the longitudinal axis of the column).

CHAPTER 4

4-1 $M_c = 0$; $M_B = 0$; the force system is in equilibrium.

4-2 $M_B = -18\ 300$ lb·ft (counterclockwise); $M_C = -22\ 200$ lb·ft (counterclockwise).

4-3 $Q = 10$ kips.

4-4 $T = 33.3$ lb.

4-5 p = 544.5 lb/in^2; the force on the piston $P = 3\ 849$ lb; area of the piston $A = 7.07$ in^2; the arms of T and P are 2 in and 5.2 in, respectively.

4-6 $T = 288$ lb; the arms of T and W are 6 ft and 3.46 ft, respectively.

4-7 $R = 160$ kips, downward, to the right. The resultant is located about 20 ft to the right from F_2 and makes approximately 18° with the vertical.

4-8 $R = 70$ kips, downward, vertical, located 4.3 ft to the right from F_2.

CHAPTER 5

5-1 $\bar{y} = 8.74$ in (above the bottom, on the vertical axis of symmetry).

5-2 (*a*) $\bar{y} = 16.1$ in (above the bottom; on the vertical axis of symmetry); (*b*) $\bar{y} = 3.29$ m (above the bottom; on the vertical axis of symmetry); (*c*) $\bar{x} = 9.29$ cm (from the left side), $\bar{y} = 19.29$ cm (from the bottom).

5-3 SF = 1.812.

CHAPTER 6

6-1 $R_A = 33$ kN; $R_B = 36$ kN (both upward).

6-2 $R_H = 1$ kN (at support A, to the right); $R_A = 4.17$ kN (at support A, upward); $R_B = 7.83$ kN (at support B, upward).

6-3 $R_H = 50$ kN (to the left); $R_A = 160$ kN (upward); $M_A = 1\ 080$ kN·m (counterclockwise).

6-4 $R_H = 0$; $R_A = 16$ kN (upward); $M_A = 64$ kN·m (counterclockwise).

6-5 $R_H = 20$ kips (to the left); $R_A = 26.25$ kips (upward); $R_B = 33.75$ kips (upward).

6-6 $R_A = 150$ kips (upward); $R_B = 50$ kips (upward).

6-7 $R_A = 200$ kips (up); $R_B = 200$ kips (up); $R'_B = 200$ kips (down); $R_C = 250$ kips (up); $R_D = 50$ kips (down); $R_E = R'_E = R_F = 0$.

CHAPTER 7

7-1 $N_1 = -2.0$ kN; $N_2 = +2.0$ kN; $N_3 = -2.25$ kN; $N_4 = N_5 = 0$; $N_6 = -2.25$ kN; $N_7 = 0$; $N_8 = -2.0$ kN.

7-2 $N_1 = -20$ kN; $N_2 = N_4 = 0$; $N_3 = -20$ kN; $N_5 = -10$ kN.

7-3 Reactions $R_A = R_B = 10$ kips (upward); $N_1 = N_{17} = -10$ kips; $N_2 = N_{16} = 0$; $N_3 = N_{15} = +8.5$ kips; $N_4 = N_{14} = -6$ kips; $N_5 = N_{13} = -6$ kips; $N_6 = N_{12} = +6$ kips; $N_7 = N_{11} = +2.83$ kips; $N_8 = N_{10} = -8$ kips; $N_9 = -4$ kips.

7-4 $N_1 = N_{12} = +17.5$ kips; $N_2 = N_{13} = -18.8$ kips; $N_3 = N_{11} = +2.0$ kips; $N_4 = N_8 = +17.5$ kips; $N_5 = N_9 = -8.1$ kips; $N_6 = N_{10} = -10.8$ kips.; $N_7 = +6$ kips.

7-5 (*a*) $R_A = 5$ kips; $R_B = 15$ kips (both upward); $V'_A = +5$ kips; $V_c = +5$ kips; $V_B = -15$ kips; critical section is 7.5 ft from support B; $M_A = M_B = 0$; $M_C = 50$ kip·ft; $M_{max} = 56.28$ kip·ft; (*b*) At the fixed end $R_H = 0$; $R_A = 40$ kN; $M_A = -84$ kN·m; $V'_A = +40$ kN; $V_B = +40$ kN; $V_C = +20$ kN; $M_C = 0$; $M_B = -36$ kN·m; (*c*) $R_A = 1.0$ kip; $R_B = 5.0$ kips (both upward); $V'_A = +1.0$ kip; $V_C = 1.0$ kip; $V'_C = -3$ kips; $V_B = -3.0$ kips; $V'_B = +2.0$ kips; $V_D = +2.0$ kips; $M_A = M_D = 0$; $M_C = +4$ kip·ft; $M_B = -8$ kip·ft; (*d*) $R_A = R_B = 20$ kN (upward); $V'_C = V_A = -20$ kN; $V'_A = V_B = 0$; $V'_B = +20$ kN; $V_D = +20$ kN; $M_C = M_D = 0$; $M_A = M_B = -12$ kN·m; (*e*) $R_A = R_B = 5$ kips; $V_C = 0$;

$V_A = -2$ kips; $V'_A = +3$ kips; $V_B = -3$ kips; $V'_B = +2$ kips; $V_D = 0$; $M_C = M_D = 0$; $M_A = M_B = -2$ kip·ft; $M_{max} = +2.5$ kip·ft (at midspan).

CHAPTER 8

8-1 W 12 × 22 (W 310 × 320) or W 14 × 22 (W 350 × 320).

8-2 $F = 47\ 124$ lb.

8-3 Required thickness $t = 2.3$ cm; use $t = 2.5$ cm; $\sigma = 9.26$ kN/cm^2.

8-4 Five strips 80 × 10 mm.

8-5 Use four **L** 4 × 4 × $\frac{5}{8}$ ($A = 4 \times 4.61 = 18.44$ in^2).

8-6 $F = 27.5$ kips; $\Delta L = 0.08$ in.

8-7 We need nine posts. Divide the platform into nine equal squares (8 × 8 ft). Place the posts in the centers of the squares.

CHAPTER 9

9-1 $M_{max} = 2\ 352\ 000$ lb·in, $S = 107$ in^3; $\sigma_{max} = 21\ 981$ lb/in^2 $< \sigma_{all}$.

9-2 $M_{max} = 43\ 200$ lb·in, $\sigma_{max} = +1\ 200$ lb/in^2 (top fibers); $\sigma_{max} = -600$ lb/in^2 (bottom fibers).

9-3 $M_{all} = 256\ 000$ lb·in.

9-4 $S = 368$ in^3; $M_{all} = 736\ 000$ lb·in; the beam is 2.875 times more efficient than the one described in Exercise 9-3.

9-5 $\sigma_{max} = 959$ lb/in^2; $y = 0.2$ in; $y_{all} = 0.47$ in.

9-6 W 8 × 20 (W 210 × 290); $S_x = 17$ in^3.

9-7 $b = 9$ in; $h = 18$ in.

CHAPTER 10

10-1 $F_s = 45\ 099$ lb; $F_c = 93\ 023$ lb; $F_t = 82\ 355$ lb; $F_{all} = F_s = 45\ 099$ lb.

10-2 Required diameter for bolts in shear = 0.798 in. Required diameter for bolts in bearing = 0.62 in. Use $\frac{7}{8}$-in bolts. Allowable tensile force on joint $T = 33\ 000$ lb $> 30\ 000$ lb.

10-3 $n_s = 5.73$ bolts (shear); $n_c = 5.373$ bolts (bearing). Use six bolts each side of the joint.

10-4 Total weld length = 31.56 in. Required length of side welds = 6.78 in. Use 7.0-in-long side welds.

10-5 End welds can safely carry 19.8 kips; the plates can safely carry 27.5 kips.

CHAPTER 11

11-1 $F_{all} = 560$ kips (buckling about minor axis).

11-2 $F_{all} = 1\ 211.5$ kips (buckling about minor axis).

11-3 (*a*) 493 kips; (*b*) 660 kips; (*c*) 600 kips; (*d*) 155 kips.

11-4 W 10 × 60 ($F_{all} = 264$ kips).

11-5 $F_{all} = 160.3$ kips (use the dimensions of the dressed size, 11.5 × 11.5 in).

11-6 $F_{all} = 141.2$ kips.

11-7 Use 12 × 12 in ($F_{all} = 155$ kips).

The Key to Quizzes

QUIZ NUMBER 1

A. True-False (T-F)—1 point for each correct "true-false" answer (maximum 23 points).

1. F	2. F	3. T	4. F	5. F	6. F
7. T	8. T	9. F	10. T	11. F	12. T
13. F	14. T	15. T	16. F	17. T	18. F
19. T	20. F	21. T	22. F	23. T	

B. Multiple choice—2 points for each correct "multiple choice" answer (maximum 18 points).

1. *d*	2. *b*	3. *a*	4. *d*	5. *b*	6. *c*
7. *b*	8. *a*	9. *b*			

C. Short answers—2 points for each correct "short" answer (maximum 30 points).

1. static loads 2. dynamic loads 3. statics
4. reactions 5. cross-sectional dimensions and deformations
6. deformations (or deflections) 7. internal 8. pressure (or stress)
9. 40 10. 2 11. 3 12. pitch 13. shape
14. wind tunnel 15. dead

Maximum possible points earned = 71

Grading: Under 60 points = F (failure)
60 to 62 = D
63 to 65 = C
66 to 68 = B
69 to 71 = A

Advice: If your grade falls below B, study the chapter once more.

QUIZ NUMBER 2

A. True-False (maximum 38 points).

1. F	2. T	3. F	4. T	5. F	6. F
7. T	8. F	9. T	10. F	11. T	12. F
13. T	14. F	15. T	16. F	17. F	18. T
19. F					

B. Multiple choice (maximum 20 points).

1. *b*	2. *c*	3. *a*	4. *d*	5. *c*	6. *b*
7. *c*	8. *d*	9. *a*	10. *a*		

C. Short answers (maximum 28 points).

1. collinear 2. resultant 3. equilibrant 4. force components
5. tensile 6. tip to tail 7. free body
8. The horizontal force F_3 = 18 kips acting to the left, and the vertical upward force F_4 = 26 kips.
9. The horizontal force F_3 = 18 kips acting to the right.
10. Connect point A with point C. Arrow goes from A to C.
11. (*a*) Connect $B \equiv G$ with A. Arrow goes from G to A.
    (*b*) Connect D with A. Arrow goes from D to A.
    (*c*) Connect D with A. Arrow goes from D to A.
    (*d*) Connect D with A. Arrow goes from D to A.

Maximum possible points earned = 86

Grading: Under 66 points = F
66 to 70 = D
71 to 75 = C
76 to 80 = B
81 to 86 = A

Advice: If your grade falls below B, study the chapter once more.

QUIZ NUMBER 3

A. True or False (maximum 18 points).

1. F	2. F	3. T	4. T	5. F	6. T
7. No	8. F	9. F	10. T	11. F	12. T
13. T	14. T	15. F	16. T	17. T	18. F

B. Multiple choice (maximum 18 points).

1. *b*	2. *b*	3. *a*	4. *b*	5. *a*	6. *d*
7. *a*	8. *d*	9. *c*			

C. Short answers (maximum 20 points).

1. moment 2. equilibrium
3. the sum of the moments of all forces of the system about that point. 4. 0.5 5. No.
6. Yes. 7. smaller 8. string polygon 9. pole 10. No.

Maximum possible points earned = 56

Grading: Under 43 points = F
43 to 46 = D
47 to 49 = C
50 to 52 = B
53 to 56 = A

Advice: If your grade falls below B, study the chapter once more.

QUIZ NUMBER 4

A. True or False (maximum 14 points).

1. T	2. F	3. T	4. F	5. F	6. T
7. F	8. T	9. F	10. T	11. T	12. F
13. F	14. T				

B. Multiple choice (maximum 12 points).

1. *c*	2. *b*	3. *a*	4. *b*	5. *b*	6. *c*

C. Short answers (maximum 14 points).

1. M_s/M_o 2. centroidal axis 3. right triangle
4. greater 5. 12.48 kips/ft of wall 6. $\overline{x} = 1$ ft; $\overline{y} = 3$ ft
7. Yes.

Maximum possible points earned = 40

Advice: If your score drops below 35 points, study the chapter once more.

QUIZ NUMBER 5

A. True or False (maximum 21 points).

1. T 2. T 3. F 4. F 5. F 6. F
7. T 8. T 9. F 10. F 11. T 12. F
13. T 14. F 15. T 16. T 17. F 18. T
19. T 20. T 21. F

B. Multiple choice (maximum 20 points).

1. *c* 2. *c* 3. *b* 4. *d* 5. *c* 6. *b*
7. *a* 8. *d* 9. *a* 10. *d*

C. Short answers (maximum 20 points).

1. rotational; translational 2. rotates and translates
3. three 4. 10 kips 5. pin 6. roller (sliding)
7. fixed 8. precast 9. statically determinate
10. masonry (walls); doors; windows

Maximum possible points earned = 61
Grading: Under 45 points = F
45 to 49 = D
50 to 54 = C
55 to 57 = B
58 to 61 = A
Advice: If your score falls below B, study the chapter once more.

QUIZ NUMBER 6

A. True or False (maximum 20 points).

1. F 2. T 3. F 4. F 5. T 6. T
7. T 8. T 9. F 10. F 11. T 12. F
13. T 14. F 15. F 16. T 17. F 18. T
19. F 20. T

B. Multiple choice (maximum 20 points).

1. *a* 2. *a* 3. *d* 4. *a* 5. *c* 6. *b*
7. *b* 8. *c* 9. *b* 10. *a*

C. Short answers (maximum 20 points).

1. negative 2. negative 3. negative 4. positive
5. rectangle, positive 6. negative 7. a right triangle, negative
8. at the fixed end 9. $N_5 = 0$ 10. $N_4 = 0$

Maximum possible points earned = 60
Grading: Under 45 points = F
45 to 49 = D
50 to 54 = C
55 to 57 = B
58 to 60 = A
Advice: If your score falls below B, study the chapter once more.

QUIZ NUMBER 7

A. True or False (maximum 20 points).

1. F	2. F	3. F	4. T	5. F	6. F
7. T	8. T	9. T	10. F	11. T	12. F
13. T	14. T	15. F	16. F	17. T	18. F
19. T	20. T				

B. Multiple choice (maximum 20 points).

1. *a*	2. *c*	3. *e*	4. *d*	5. *b*	6. *c*
7. *a*	8. *a*	9. *b*	10. *c*		

C. Short answers (maximum 20 points).

1. brittle	2. three	3. 20 000 lb/in^2	4. (see Fig. 8-12)
5. three	6. 10 in^2	7. $\epsilon = \sigma/E$	8. exchange E for timber and steel
9. 0.001	10. 30 000 lb/in^2		

Maximum possible points earned = 60

Grading: Under 45 points = F
45 to 49 = D
50 to 54 = C
55 to 57 = B
58 to 60 = A

Advice: If your score falls below B, study the chapter once more.

QUIZ NUMBER 8

A. True or False (maximum 24 points).

1. T	2. F	3. F	4. T	5. F	6. F
7. F	8. T	9. T	10. T	11. T	12. T
13. F	14. T	15. T	16. T	17. T	18. T
19. T	20. T	21. F	22. F	23. T	24. F

B. Multiple choice (maximum 20 points).

1. *c*	2. *a*	3. *a*	4. *d*	5. *b*	6. *c*
7. *b*	8. *b*	9. *a*	10. *a*		

C. Short answers (maximum 20 points).

1. free end	2. midspan	3. 100	4. 1 000
5. 100	6. *EI*	7. critical	8. 72
9. 32	10. 1200 kN		

Maximum possible points earned = 64

Grading: Under 49 points = F
49 to 52 = D
53 to 56 = C
57 to 60 = B
61 to 64 = A

Advice: If your score falls below B, study the chapter once more.

QUIZ NUMBER 9

A. True or False (maximum 20 points).

1. F	2. F	3. T	4. T	5. T	6. F
7. T	8. T	9. F	10. T	11. T	12. T
13. T	14. F	15. T	16. T	17. T	18. F
19. T	20. F				

B. Multiple choice (maximum 20 points).

1. *d*	2. *b*	3. *c*	4. *b*	5. *a*	6. *d*
7. *a*	8. *c*	9. *a*	10. *b*		

C. Short answers (maximum 22 points).

1. critical	2. Euler	3. minor (weak)	4. *a*, *b*, *d*, *c*
5. four	6. four	7. smaller	8. *a*, 1.0; *b*, 2; *c*, 0.5; *d*, 0.7
9. σ_y	10. $0.3E/(L/d)^2$	11. $\pi^2 E/\lambda^2$	

Maximum possible points earned = 62

Grading: Under 49 points = F
49 to 51 = D
52 to 55 = C
56 to 59 = B
60 to 62 = A

Advice: If your score falls below B, study the chapter once more.

Appendix Tables

TABLE A-1(A) Unit Weights of Structural Materials
Weights per unit volume

MATERIAL	POUNDS PER CUBIC FOOT, lb/ft^3	kN/m^3
	Metals	
Aluminum, cast, hammered	165	26
Brass, cast, rolled	534	84
Copper, cast, rolled	556	87
Iron, cast, pig	450	70
Structural steel, rolled	490	77
Lead	710	112
	Timber	
Softwoods		
Fir, Douglas	32	5.0
Pine, red	30	4.7
Pine, yellow, long-leaf	44	6.9
Pine, yellow, short-leaf	38	6.0
Redwood	26	4.1
Spruce	27	4.2
Hardwoods		
Ash, white, red	40	6.3
Hickory	49	7.7
Maple, hard	43	6.8
Oak	54	8.5

TABLE A-1(A) Continued

MATERIAL	POUNDS PER CUBIC FOOT, lb/ft^3	kN/m^3
	Concrete	
Plain (nonreinforced)		
Lightweight	32–110	5.0–17.0
Slag	130	20.7
Stone (normal weight)	144	22.6
Reinforced (with steel)		
Structural lightweight	65–120	10–19
Slag	140	21.7
Stone (normal weight)	150	23.5
	Soils	
Clay, dry	65	10
Clay, damp, plastic	110	17.0
Clay and gravel, dry	100	15.7
Earth, dry	76–95	11.9–14.9
Earth, moist	78–95	12.3–15.1
Mud	110–115	17–18
Sand and gravel	90–120	14.1–18.9
	Liquids	
Water (4°C, maximum density)	62.4	9.8
Gasoline	42	6.6
Oils	50–58	7.85–9.11
	Masonry	
Cement (Portland)	90	14.1
Concrete, cinder	100	15.7
Concrete, slag	130	20.7
Concrete, stone	144	22.6
Brick masonry, soft	100	15.7
Brick masonry, common	120	18.9
Brick masonry, pressed	140	22.0
Asphalt	80	12.7
Glass	156	24.5
Pitch	70	10.8

TABLE A-1(B) Weights of Structural Materials per Unit Area of Floors, Ceilings, and Roofs

MATERIAL	lb/ft^2	N/m^2
Ceilings		
Acoustical fiber tile	1.0	48
Channel suspended system	1.0	48
Plaster (per 1 in thickness)	10.0	480 (per 1 cm)
Floors		
Hardwood (per 1 in thickness)	4.0	75 (per 1 cm)
Plywood (per 1 in thickness)	3.0	56 (per 1 cm)
Asphalt mastic (per 1 in thickness)	12.0	226 (per 1 cm)
Cement finish (per 1 in thickness)	12.0	226 (per 1 cm)
Ceramic tile, $\frac{3}{4}$ in	10.0	250 (per 1 cm)
Concrete (per 1 in thickness)		
Lightweight	6–10	110–190 (1 cm)
Stone	12.0	226 (1 cm)
Reinforced	12.5	235 (1 cm)
Flexicore, 6 in (15 cm) slab	46.0	2 200
Linoleum, $\frac{1}{4}$ in	1.0	75 (per 1 cm)
Terazzo finish, 1 in	13.0	245 (per 1 cm)
Vinyl tile, $\frac{1}{8}$ in	1.4	210 (per 1 cm)
Steel deck	2–10	95–480
Roofs		
Corrugated steel	1–5	48–250
Copper or tin	1–5	48–250
Three-ply ready roofing	1	48
Three-ply felt and gravel	5.5	260
Five-ply felt and gravel	6	290
Corrugated asbestos cement	3	145
Shingles		
Wood (1 in)	3	56 (per 1 cm)
Asphalt (ca. $\frac{1}{4}$ in)	2	96
Clay-tile	9–14	430–670
Slate, $\frac{1}{4}$ in	10	480
Sheathing		
Wood, $\frac{3}{4}$ in	3	75 (per 1 cm)
Gypsum (per 1 in)	4	75 (per 1 cm)
Insulation		
Loose	0.5	9
Poured in place	2.0	37
Rigid	2.5	47

TABLE A-2 Uniform and Concentrated Loads on Structures

CATEGORY	DESCRIPTION	UNIFORMLY DISTRIBUTED LOAD		CONCENTRATED LOAD†	
		lb/ft²	kN/m²	lb	kN
Residential	Private dwellings, apartments, hotel guest rooms	40	1.9	0	0
Schools	Classrooms	40	1.9	1 000	4.45
	Corridors	100	4.8	0	0
Offices		50	2.4	2 000	8.9
Hospitals	Rooms and wards	40	1.9	1 000	4.45
Libraries	Reading rooms	60	2.9	1 000	4.45
	Stack rooms	125	6.0	1 500	6.70
Exit facilities, public	Corridors, stairways, fire escapes	100	4.8	0	0
Stores	Retail	75	3.6	2 000	8.9
	Wholesale	100	4.8	3 000	13.3
Garages	General storage and/or repair	100	4.8	—‡	—
	Private pleasure car storage	50	2.4	2 000	8.9
Assembly areas and auditoriums, dance halls, gymnasiums, playgrounds, plazas, public terraces, etc.	Fixed seating areas	50	2.4	0	0
	Movable seating and other areas	100	4.8	0	0
	Stage areas and enclosed platforms	125	6.0	0	0

†Concentrated load shown (except as stated below under Footnote ‡), shall be placed on any space 2½ feet (0.75 m) square, wherever this load on an otherwise unloaded floor would produce stresses greater than those caused by the corresponding uniform load required.

‡Two or more concentrated loads should be investigated. Loads are to be spaced 5 ft (1.5 m) on center without any uniform live load. Each load shall be 40 percent of the gross weight of the largest vehicle to be accommodated.

TABLE A-3 Wind Pressures for Various Height Zones above Ground of Rectangular Buildings

Metric equivalents are rounded

HEIGHT ZONES		BASIC WIND PRESSURE FROM WIND MAP													
FEET	METERS	lb/ft^2, **20**	kN/m^2, **0.95**	lb/ft^2, **25**	kN/m^2, **1.20**	lb/ft^2, **30**	kN/m^2, **1.45**	lb/ft^2, **35**	kN/m^2, **1.70**	lb/ft^2, **40**	kN/m^2, **1.95**	lb/ft^2, **45**	kN/m^2, **2.20**	lb/ft^2, **50**	kN/m^2, **2.40**
<30	<9	15	0.70	20	0.95	25	1.20	25	1.30	30	1.45	35	1.70	40	1.95
30–49	9–15	20	0.95	25	1.20	30	1.50	35	1.70	40	1.95	45	2.20	50	2.50
50–99	15–30	25	1.20	30	1.45	40	1.85	45	2.10	50	2.40	55	2.70	60	3.00
100–499	30–150	30	1.45	40	1.95	45	2.20	55	2.50	60	2.90	70	3.20	75	3.60
500–1199	150–360	35	1.70	45	2.20	55	2.50	60	2.90	70	3.40	80	3.70	90	4.20
⩾1200	>360	40	1.95	50	2.40	60	2.85	70	3.35	80	3.90	90	4.30	100	4.80

TABLE A-4 W Shapes (Wide-Flange Shapes)—Properties for Designing (British Units)

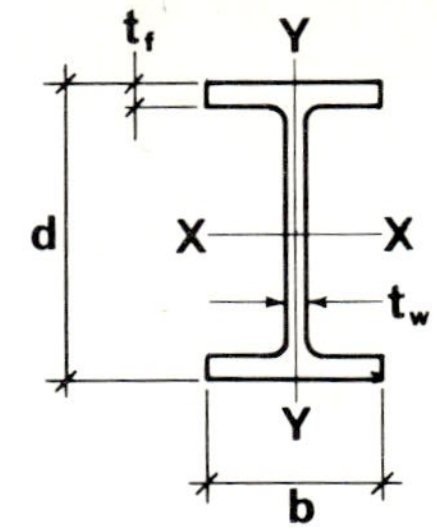

DESIGNATION	WEIGHT lb/ft	DEPTH d, in	WIDTH b, in	THICKNESS: WEB t_w, in	THICKNESS: FLANGE t_f, in	AREA A, in^2	MOMENT OF INTERTIA I: ABOUT X-X I_x, in^4	MOMENT OF INTERTIA I: ABOUT Y-Y I_y, in^4	SECTION MODULUS S: S_x, in^3	SECTION MODULUS S: S_y, in^3	RADIUS OF GYRATION r: r_x, in	RADIUS OF GYRATION r: r_y, in
W 36 X 300	300	36.72	16.655	0.945	1.680	88.3	20 300	1 300	1 110	156	15.2	3.83
W 36 X 230	230	35.88	16.471	0.761	1.260	67.7	15 000	940	837	114	14.9	3.73
W 36 X 150	150	35.84	11.972	0.625	0.940	44.2	9 030	270	504	45	14.3	2.47
W 33 X 240	240	33.50	15.865	0.830	1.400	70.6	13 600	933	813	118	13.9	3.64
W 33 X 141	141	33.31	11.535	0.605	0.960	41.6	7 460	246	448	42.7	13.4	2.43
W 33 X 118	118	32.86	11.484	0.554	0.738	34.8	5 900	187	359	32.5	13.0	2.32
W 30 X 210	210	30.38	15.105	0.775	1.315	61.9	9 890	757	651	100	12.6	3.50
W 30 X 99	99	29.64	10.458	0.522	0.670	29.1	4 000	128	270	24.5	11.7	2.10
W 27 X 145	145	26.88	13.965	0.600	0.975	42.7	5 430	443	404	63.5	11.3	3.22
W 27 X 84	84	26.69	9.963	0.463	0.636	24.8	2 830	105	212	21.1	10.7	2.06
W 24 X 145	145	24.49	14.043	0.608	1.020	42.7	4 570	471	373	67.1	10.3	3.32
W 24 X 100	100	24.00	12.000	0.468	0.775	29.5	3 000	223	250	37.2	10.1	2.75
W 24 X 68	68	23.71	8.961	0.416	0.582	20.0	1 820	70	153	15.6	9.53	1.87
W 24 X 55	55	23.55	7.000	0.396	0.503	16.2	1 340	28.9	114	8.25	9.10	1.34
W 21 X 96	96	21.14	9.038	0.575	0.935	28.3	2 100	115	198	25.5	8.61	2.02
W 21 X 55	55	20.80	8.215	0.375	0.522	16.2	1 140	48.3	110	11.8	8.40	1.73
W 18 X 105	105	18.32	11.792	0.554	0.911	30.9	1 850	249	202	42.3	7.75	2.84
W 18 X 55	55	18.12	7.532	0.390	0.630	16.2	891	45	98.4	11.9	7.42	1.67
W 16 X 88	88	16.16	11.502	0.504	0.795	25.9	1 220	202	151	35.1	6.87	2.79
W 16 X 40	40	16.00	7.000	0.307	0.503	11.8	517	28.8	64.6	8.23	6.62	1.56
W 14 X 730	730	22.44	17.889	3.069	4.910	215	14 400	4 720	1 280	527	8.18	4.69
W 14 X 500	500	19.63	17.008	2.188	3.501	147	8 250	2 880	840	339	7.49	4.43
W 14 X 398	398	18.31	16.590	1.770	2.843	117	6 010	2 170	657	262	7.17	4.31
W 14 X 320	320	16.81	16.710	1.890	2.093	94.1	4 140	1 640	493	196	6.63	4.17

W 14 X 202	202	15.63	15.750	0.930	1.503	59.4	2 540	980	325	124	6.54	4.06
W 14 X 103	103	14.25	14.575	0.495	0.813	30.3	1 170	420	146	57.6	6.21	3.72
W 14 X 48	48	13.81	8.031	0.339	0.593	14.1	485	51.3	70.2	12.8	5.86	1.91
W 14 X 30	30	13.86	6.733	0.270	0.383	8.83	290	19.5	41.9	5.8	5.74	1.49
W 14 X 22	22	13.72	5.000	0.230	0.335	6.49	198	7.0	28.9	2.8	5.53	1.04
W 12 X 120	120	13.12	12.320	0.710	0.106	35.3	1 070	345	163	56	5.51	3.13
W 12 X 79	79	12.38	12.080	0.470	0.736	23.2	663	216	107	35.8	5.34	3.05
W 12 X 50	50	12.19	8.077	0.371	0.641	14.7	395	56.4	64.7	14	5.18	1.96
W 12 X 22	22	12.31	4.030	0.260	0.424	6.47	156	4.64	25.3	2.31	4.91	0.847
W 10 X 60	60	10.25	10.075	0.415	0.683	17.7	344	116	67.1	23.1	4.41	2.57
W 10 X 39	39	9.94	7.990	0.318	0.528	11.5	210	44.9	42.2	11.2	4.27	1.98
W 10 X 25	25	10.08	5.762	0.252	0.430	7.36	133	13.7	26.5	4.76	4.26	1.37
W 10 X 15	15	10.00	4.000	0.230	0.269	4.41	68.9	2.88	13.8	1.44	3.95	0.809
W 8 X 40	40	8.25	8.077	0.365	0.558	11.8	146	49	35.5	12.1	3.53	2.04
W 8 X 20	20	8.14	5.268	0.248	0.378	5.89	69.4	9.22	17.0	3.5	3.43	1.25
W 8 X 10	10	7.90	3.940	0.170	0.204	2.96	30.8	2.08	7.8	1.06	3.23	0.839
W 6 X 20	20	6.20	6.018	0.258	0.367	5.88	41.5	13.3	13.4	4.43	2.66	1.51
W 6 X 12	12	6.00	4.000	0.230	0.279	3.54	21.7	2.98	7.25	1.49	2.48	0.918
W 5 X 16	16	5.00	5.000	0.240	0.360	4.70	21.3	7.51	8.53	3.00	2.13	1.26
W 4 X 13	13	4.16	4.060	0.280	0.345	3.82	11.3	3.76	5.45	1.85	1.72	0.991

TABLE A-5 W Shapes (Wide-Flange Shapes)—Properties for Designing (Metric Units)

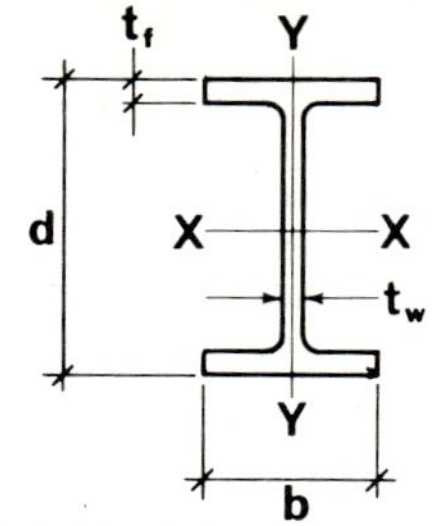

DESIGNATION	WEIGHT (PER METER), N/m	DEPTH d, mm	WIDTH b, mm	THICKNESS: WEB t_w, mm	THICKNESS: FLANGE t_f, mm	AREA A, cm^2	MOMENT OF INERTIA I: ABOUT X-X I_x, cm^4	MOMENT OF INERTIA I: ABOUT Y-Y I_y, cm^4	SECTION MODULUS S: ABOUT X-X S_x, cm^3	SECTION MODULUS S: ABOUT Y-Y S_y, cm^3	RADIUS OF GYRATION r: ABOUT X-X r_x, cm	RADIUS OF GYRATION r: ABOUT Y-Y r_y, cm
W 36 X 300† (**W** 930 X 4 380)‡	4 380	930	423	24	43	570	844 886	54 106	18 193	2 557	38.6	9.7
W 36 X 230 (**W** 910 X 3 360)	3 360	911	418	19	32	437	624 300	39 123	13 718	1 868	37.8	9.5
W 36 X 150 (**W** 910 X 2 180)	2 190	910	304	16	24	285	375 828	11 237	8 261	738	36.3	6.3
W 33 X 240 (**W** 850 X 3 500)	3 500	851	402	21	37	456	566 032	38 032	13 325	1 934	35.3	9.2
W 33 X 141 (**W** 840 X 2 060)	2 060	846	293	15	24	268	310 485	10 239	7 343	700	34.0	6.2
W 33 X 118 (**W** 840 X 1 720)	1 720	835	292	14	19	225	245 558	7 783	5 884	533	33.0	5.9
W 30 X 210 (**W** 770 X 3 060)	3 060	770	384	20	33	340	411 622	31 506	10 670	1 639	32.0	8.9
W 30 X 99 (**W** 750 X 1440)	1 440	753	266	13	17	188	166 480	5 327	4 425	402	29.7	5.3
W 27 X 145 (**W** 680 X 2 120)	2 120	683	355	15	25	276	225 997	18 438	6 622	1 041	28.7	8.2
W 27 X 84 (**W** 680 X 1 230)	1 230	678	253	12	16	160	117 785	4 370	3 475	346	27.2	5.2

†**W** means wide-flange section. First number means total rounded depth of the section in inches. Second number means rounded weight of the section in pounds per lineal foot.

‡Designation in parentheses expresses the depth of a section in millimeters and the weight per unit length of the section in Newtons per meter.

W 24 X 145 (**W** 620 X 2 120)	2 120	622	357	15	26	256	190 203	19 603	6 113	1 100	26.2	8.4
W 24 X 100 (**W** 610 X 1 460)	1 460	610	305	12	20	190	124 860	9 281	4 098	610	25.7	7.0
W 24 X 68 (**W** 600 X 990)	990	602	228	11	15	129	75 748	2 913	2 508	256	24.2	4.7
W 24 X **55** (**W** 600 X 800)	800	600	178	10	13	105	55 771	1 203	1 868	135	23.1	3.4
W 21 X 96 (**W** 540 X 1 400)	1 400	537	230	15	24	183	87 402	4 786	3 245	418	21.9	5.1
W 21 X 55 (**W** 530 X 800)	800	528	208	10	13	105	47 447	2 010	1 303	193	21.3	4.4
W 18 X 105 (**W** 460 X 1 530)	1 530	465	300	14	23	199	76 997	10 363	3 311	693	19.7	7.2
W 18 X 55 (**W** 460 X 800)	800	460	191	10	16	105	37 083	1 873	1 613	195	18.8	4.2
W 16 X 88 (**W** 410 X 1 280)	1 280	410	292	13	20	167	50 776	8 407	2 475	575	17.4	7.1
W 16 X 40 (**W** 400 X 580)	580	406	178	8	13	76	21 517	1 199	1 059	135	17.0	4.0
W 14 X 730 (**W** 570 X 10 650)	10 650	570	454	78	125	1 387	599 328	196 446	20 979	8 637	20.8	12.0
W 14 X 500 (**W** 500 X 7 300)	7 300	500	432	56	89	948	343 365	119 865	13 767	5 556	19.0	11.2
W 14 X 398 (**W** 460 X 5 800)	5 800	465	421	45	72	755	250 136	90 315	10 768	4 294	18.2	10.9
W 14 X 320 (**W** 430 X 4 670)	4 670	427	424	48	53	607	172 307	68 257	8 080	3 212	16.8	10.6
W 14 X 202 (**W** 400 X 2 950)	2 950	397	400	24	38	383	105 715	40 787	5 327	2 032	16.6	10.3
W 14 X 103 (**W** 360 X 1 500)	1 500	458	370	13	21	195	48 695	17 480	2 688	944	15.8	9.4
W 14 X 48 (**W** 350 X 700)	700	350	204	9	15	91	20 186	2 135	1 150	210	14.9	4.9
W 14 X 30 (**W** 350 X 440)	440	350	171	7	10	57	12 070	812	687	95	14.6	3.8
W 14 X 22 (**W** 350 X 320)	320	348	127	6	9	42	8 241	291	474	46	14.0	2.6
W 12 X 120 (**W** 330 X 1 750)	1 750	332	313	18	28	228	44 533	14 359	2 671	918	14.0	8.0
W 12 X 79 (**W** 310 X 1 150)	1 150	315	307	12	19	150	27 594	8 990	1 754	587	13.5	7.7
W 12 X 50 (**W** 300 X 730)	730	310	205	9	16	95	16 440	2 347	1 060	229	13.2	5.0

TABLE A-5 Continued

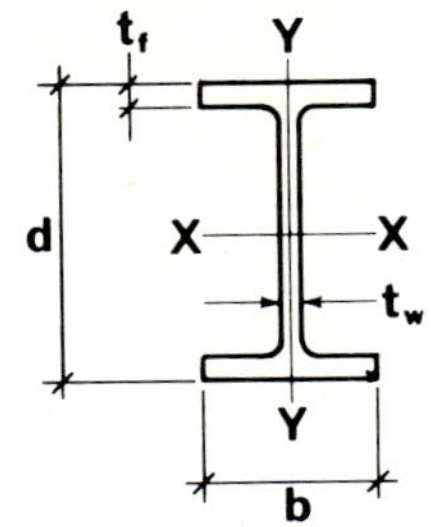

DESIGNATION	WEIGHT (PER METER), N/m	DEPTH d, mm	WIDTH b, mm	THICKNESS: WEB t_w, mm	THICKNESS: FLANGE t_f, mm	AREA A, cm^2	MOMENT OF INERTIA I: ABOUT X-X I_x, cm^4	MOMENT OF INERTIA I: ABOUT Y-Y I_y, cm^4	SECTION MODULUS S: ABOUT X-X S_x, cm^3	SECTION MODULUS S: ABOUT Y-Y S_y, cm^3	RADIUS OF GYRATION r: ABOUT X-X r_x, cm	RADIUS OF GYRATION r: ABOUT Y-Y r_y, cm
W 12 X 22 (**W** 310 X 320)	320	312	102	7	11	42	6 493	193	415	38	12.5	2.5
W 10 X 60 (**W** 260 X 880)	880	260	273	11	17	114	14 317	4 828	1 100	379	11.2	6.5
W 10 X 39 (**W** 250 X 570)	570	252	203	8	13	74	8 740	1 869	692	184	10.8	5.0
W 10 X 25 (**W** 250 X 360)	360	256	146	6	11	47	5 535	570	434	78	10.8	3.5
W 10 X 15 (**W** 250 X 220)	220	254	102	6	7	28	2 868	120	226	24	10.0	2.0
W 8 X 40 (**W** 210 X 580)	580	210	205	9	14	76	6 077	2 040	582	198	9.0	5.2
W 8 X 20 (**W** 210 X 290)	290	207	134	6	10	38	2 888	384	279	57	8.7	3.2
W 8 X 10 (**W** 200 X 150)	150	200	100	4	5	19	782	53	128	17	8.2	2.1
W 6 X 20 (**W** 160 X 290)	290	157	153	7	9	38	1 727	554	220	73	6.8	3.8
W 6 X 12 (**W** 150 X 180)	180	152	102	6	7	23	903	124	119	24	6.3	2.3
W 5 X 16 (**W** 130 X 230)	230	127	127	6	9	30	887	313	140	49	5.4	3.2
W 4 X 13 (**W** 100 X 190)	190	106	103	7	9	25	470	156	89	30	4.4	2.5

TABLE A-6 American Standard Channels—Properties for Designing (British Units)

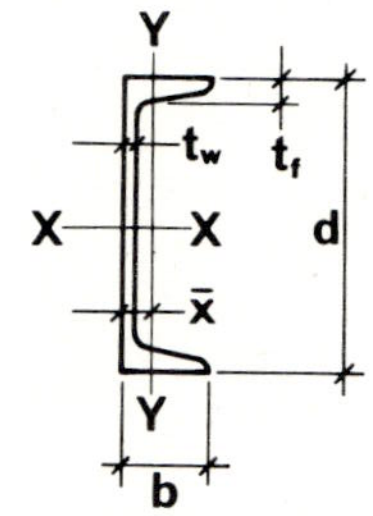

DESIGNATION	WEIGHT lb/ft	DEPTH d, in	WIDTH b, in	THICKNESS OF WEB t_w, in	AVERAGE THICKNESS OF FLANGE t_f, in	AREA A, in^2	I_x, in^4	I_y, in^4	S_x, in^3	S_y, in^3	r_x, in	r_y, in	$\bar{x}$, in
C 15 X 50	50	15	3.716	0.716	0.650	14.7	404	11	53.8	3.78	5.24	0.867	0.799
C 15 X 40	40	15	3.520	0.520	0.650	11.8	349	9.23	46.5	3.36	5.44	0.886	0.778
C 12 X 30	30	12	3.170	0.510	0.501	8.82	162	5.14	27	2.06	4.29	0.763	0.674
C 10 X 25	25	10	2.886	0.526	0.436	7.35	91.2	3.36	18.2	1.48	3.52	0.676	0.617
C 9 X 20	20	9	2.648	0.448	0.413	5.88	60.9	2.42	13.5	1.17	3.22	0.642	0.583
C 8 X 18.75	18.75	8	2.527	0.487	0.390	5.51	44	1.98	11	1.01	2.82	0.599	0.565
C 7 X 14.75	14.75	7	2.299	0.419	0.366	4.33	27.2	1.38	7.78	0.779	2.51	0.564	0.532
C 6 X 13	13	6	2.157	0.437	0.343	3.83	17.4	1.05	5.80	0.642	2.13	0.525	0.514
C 5 X 9	9	5	1.885	0.325	0.320	2.64	8.9	0.632	3.56	0.449	1.83	0.489	0.478
C 4 X 7.25	7.25	4	1.721	0.321	0.296	2.13	4.59	0.432	2.29	0.343	1.47	0.450	0.459
C 3 X 6	6	3	1.596	0.356	0.273	1.76	2.07	0.305	1.38	0.268	1.08	0.416	0.455

TABLE A-7 Angles (Equal Legs)—Properties for Designing (British Units)

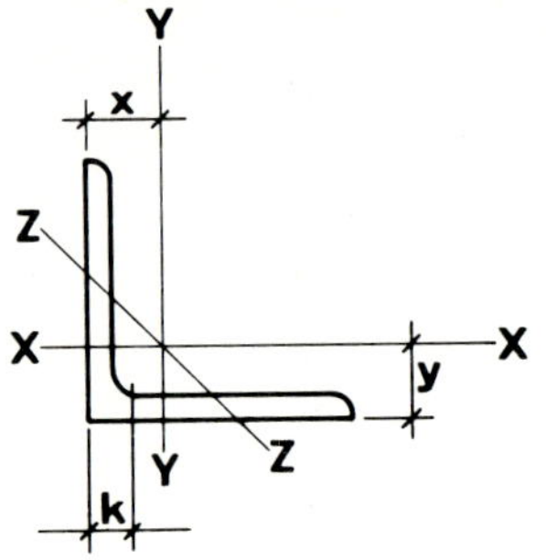

DESIGNATION	WEIGHT PER FOOT, lb/ft	AREA A, in^2	$I_x = I_y$, in^4	$S_x = S_y$, in^3	$r_x = r_y$, in	$x = y$, in
L 8 X 8 X 1	51.0	15.0	89.0	15.8	2.44	2.37
L 8 X 8 X 3/4	38.9	11.4	69.7	12.2	2.47	2.28
L 8 X 8 X 9/16	29.6	8.68	54.1	9.34	2.50	2.21
L 6 X 6 X 3/4	28.7	8.44	28.2	6.66	1.83	1.78
L 6 X 6 X 7/16	17.2	5.06	17.7	4.08	1.87	1.66
L 5 X 5 X 3/4	23.6	6.94	15.7	4.53	1.51	1.52
L 5 X 5 X 7/16	14.3	4.18	10.0	2.79	1.55	1.41
L 4 X 4 X 5/8	15.7	4.61	6.66	2.40	1.20	1.23
L 4 X 4 X 7/16	11.3	3.31	4.97	1.75	1.23	1.16
L 4 X 4 X 5/16	8.2	2.40	3.71	1.29	1.24	1.12
L 3 1/2 X 3 1/2 X 7/16	9.8	2.87	3.26	1.32	1.07	1.04
L 3 1/2 X 3 1/2 X 5/16	7.2	2.09	2.45	0.976	1.08	0.990
L 3 X 3 X 7/16	8.3	2.43	1.99	0.954	0.905	0.910
L 3 X 3 X 5/16	6.1	1.78	1.51	0.707	0.922	0.865
L 2 1/2 X 2 1/2 X 5/16	5.0	1.46	0.849	0.482	0.761	0.740
L 2 1/2 X 2 1/2 X 3/16	3.07	0.902	0.547	0.303	0.778	0.694
L 2 X 2 X 3/8	4.7	1.36	0.479	0.351	0.594	0.636
L 2 X 2 X 3/16	2.44	0.715	0.272	0.190	0.617	0.569
L 1 3/4 X 1 3/4 X 3/16	2.12	0.621	0.179	0.144	0.537	0.506
L 1 1/2 X 1 1/2 X 3/16	1.80	0.527	0.110	0.104	0.457	0.444

TABLE A-8 American Standard Channels—Properties for Designing (Metric Units)

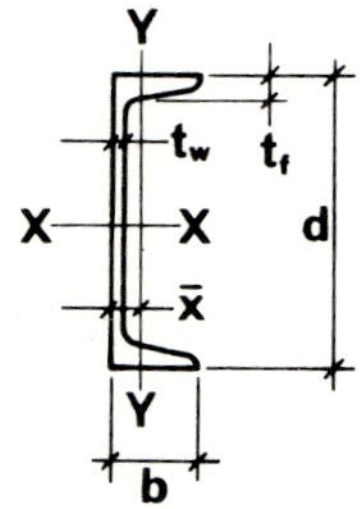

DESIGNATION	WEIGHT (PER METER), N/m	DEPTH d, mm	WIDTH b, mm	THICKNESS OF WEB t_w, mm	AVERAGE THICKNESS OF FLANGE t_f, mm	AREA A, cm^2	I_x, cm^4	I_y, cm^4	S_x, cm^3	S_y, cm^3	r_x, cm	r_y, cm	$\bar{x}$, mm
C 15 X 15 (**C** 380 X 730)	730	380	94	18	17	95	16 814	458	882	62	13.3	2.2	20
C 15 X 40 (**C** 380 X 580)	580	380	89	13	17	76	14 525	384	762	55	13.8	2.3	20
C 12 X 30 (**C** 300 X 440)	440	304	80	13	13	57	6 742	214	443	34	10.9	1.7	17
C 10 X 25 (**C** 250 X 360)	360	254	73	13	11	47	3 795	140	298	24	8.9	1.7	16
C 9 X 20 (**C** 230 X 290)	290	229	67	11	10	38	2 535	101	221	19	8.2	1.6	15
C 8 X 18.75 (**C** 200 X 270)	270	203	64	12	10	36	1 831	82	180	17	7.2	1.5	14
C 7 X 14.75 (**C** 180 X 220)	220	178	58	11	9	38	1 132	57	128	13	6.4	1.4	14
C 6 X 13 (**C** 150 X 190)	190	152	55	11	9	25	724	44	95	11	5.4	1.3	13
C 5 X 9 (**C** 130 X 130)	130	127	48	8	8	17	370	26	58	7	4.6	1.2	12
C 4 X 7.25 (**C** 100 X 110)	110	102	44	8	8	14	191	18	38	6	3.7	1.1	12
C 3 X 6 (**C** 80 X 90)	90	76	40	7	9	11	86	13	22	4	2.7	1.1	12

TABLE A-9 Angles (Equal Legs)—Properties for Designing (Metric Units)

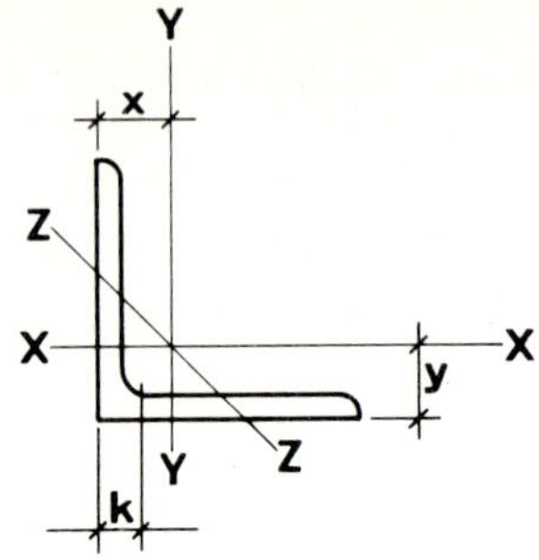

DESIGNATION	WEIGHT (PER METER), N/m	AREA A, cm^2	$I_x = I_y$, cm^4	$S_x = S_y$, cm^3	$r_x = r_y$, cm	$x = y$, mm
L 8 X 8 X 1 (**L** 200 X 200 X 25)	740	97	3 700	230	6.2	60
L 8 X 8 X 3/4 (**L** 200 X 200 X 20)	570	74	2 900	200	6.2	58
L 8 X 8 X 9/16 (**L** 200 X 200 X 14)	430	56	2 250	150	6.4	56
L 6 X 6 X 3/4 (**L** 150 X 150 X 20)	420	54	1 170	110	4.6	45
L 6 X 6 X 7/16 (**L** 150 X 150 X 11)	250	32	740	67	4.7	42
L 5 X 5 X 3/4 (**L** 130 X 130 X 20)	345	45	650	74	3.8	38
L 5 X 5 X 7/16 (**L** 130 X 130 X 11)	210	27	416	46	3.9	36
L 4 X 4 X 5/8 (**L** 100 X 100 X 16)	230	30	280	40	3.0	31
L 4 X 4 X 7/16 (**L** 100 X 100 X 11)	165	21	207	29	3.1	29
L 4 X 4 X 5/16 (**L** 100 X 100 X 8)	120	15	154	21	3.1	28
L 3 1/2 X 3 1/2 X 7/16 (**L** 90 X 90 X 11)	140	18	136	21	2.7	26
L 3 1/2 X 3 1/2 X 5/16 (**L** 90 X 90 X 8)	105	13	102	16	2.7	25
L 3 X 3 X 7/16 (**L** 75 X 75 X 11)	120	15	83	15	2.3	23
L 3 X 3 X 5/16 (**L** 75 X 75 X 8)	90	11	63	12	2.3	22
L 2 1/2 X 2 1/2 X 5/16 (**L** 60 X 60 X 8)	73	9	35	8	1.9	19
L 2 1/2 X 2 1/2 X 3/16 (**L** 60 X 60 X 5)	45	6	23	5	2.0	18
L 2 X 2 X 3/8 (**L** 50 X 50 X 10)	68	9	20	6	1.5	16
L 2 X 2 X 3/16 (**L** 50 X 50 X 5)	36	4.6	11	3	1.5	14
L 1 3/4 X 1 3/4 X 3/16 (**L** 45 X 45 X 5)	31	4.0	7.4	2.4	1.36	13
L 1 1/2 X 1 1/2 X 3/16 (**L** 40 X 40 X 5)	26	3.4	4.6	1.7	1.16	11

TABLE A-10 Area, Location of the Centroid, Moment of Inertia, and Section Modulus of Basic Figures

FIGURE	AREA	LOCATION OF CENTROID	MOMENT OF INERTIA I_x	SECTION MODULUS S_x
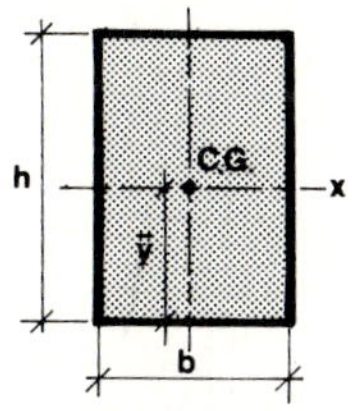	bh	$\bar{y} = \frac{h}{2}$	$\frac{bh^3}{12}$	$\frac{bh^2}{6}$
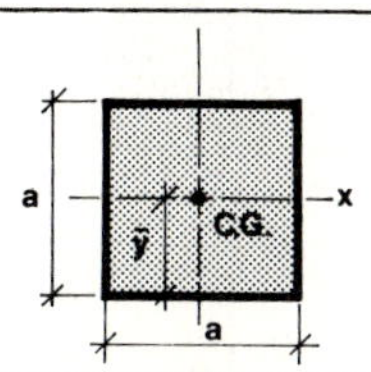	a^2	$\bar{y} = \frac{a}{2}$	$\frac{a^4}{12}$	$\frac{a^3}{6}$
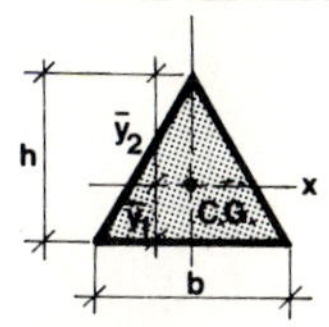	$\frac{bh}{2}$	$\bar{y}_1 = \frac{h}{3}$ $\bar{y}_2 = \frac{2h}{3}$	$\frac{bh^3}{36}$	$S_1 = \frac{I_x}{\bar{y}_1} = \frac{bh^2}{12}$ $S_2 = \frac{I_x}{\bar{y}_2} = \frac{bh^2}{24}$
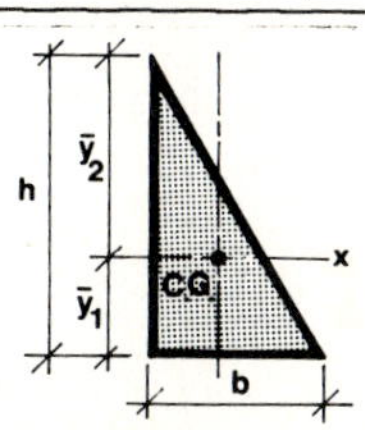	$\frac{bh}{2}$	$\bar{y}_1 = \frac{h}{3}$ $\bar{y} = \frac{2h}{3}$	$\frac{bh^3}{36}$	$S_1 = \frac{bh^2}{12}$ $S_2 = \frac{bh^2}{24}$
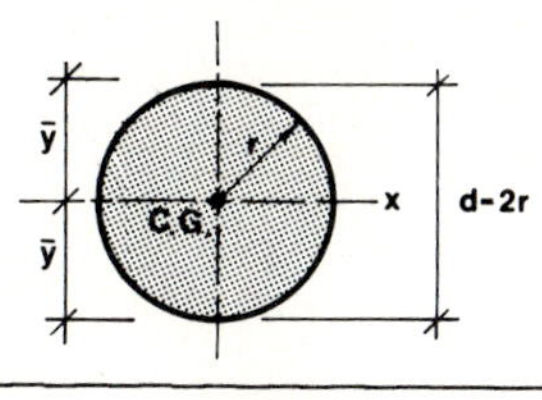	πr^2 or $\frac{\pi d^2}{4}$	$\bar{y} = r$ or $\bar{y} = \frac{d}{2}$	$\frac{\pi r^4}{4}$ or $\frac{\pi d^4}{64}$	$\frac{\pi r^3}{4}$ or $\frac{\pi d^3}{32}$
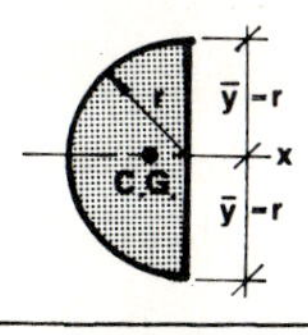	$\frac{\pi r^2}{2}$	$\bar{y} = r$	$\frac{\pi r^4}{8}$	$\frac{\pi r^3}{8}$
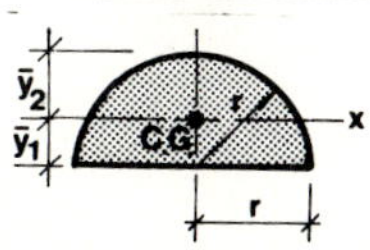	$\frac{\pi r^2}{2}$	$\bar{y}_1 = 0.5756r$ $\bar{y}_2 = 0.4244r$	$0.1098r^4$	$S_1 = 0.1907r^3$ $S_2 = 0.25886r^3$

TABLE A-10 Continued

FIGURE	AREA	LOCATION OF CENTROID	MOMENT OF INERTIA I_x	SECTION MODULUS S_x
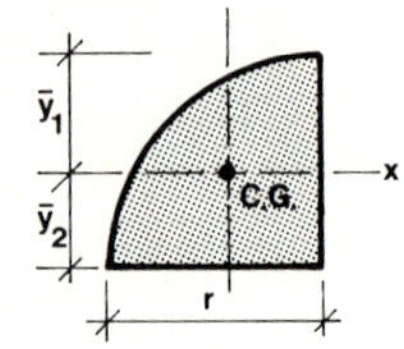	$\frac{\pi r^2}{4}$	$\bar{y}_1 = 0.5756r$ $\bar{y}_2 = 0.4244r$	$0.0549r^4$	$S_1 = 0.0953r^3$ $S_2 = 0.1293r^3$
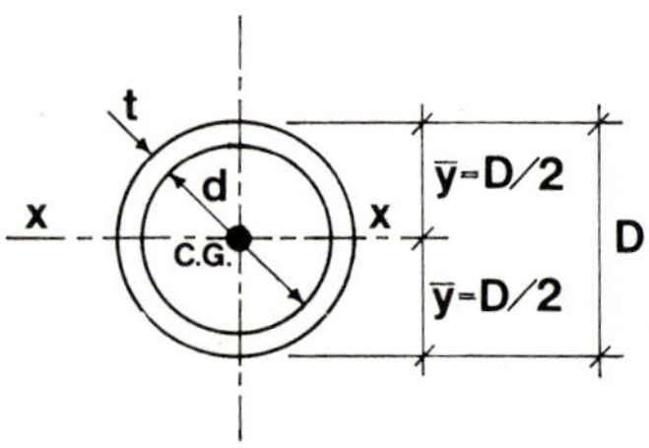	$\frac{\pi}{4}(D^2 - d^2)$	$\bar{y} = \frac{D}{2}$	$\frac{\pi D^3 t}{8}$ †	$\frac{\pi D^2 t}{4}$ †

†This formula holds for t much smaller than D.

TABLE A-11 Approximate Values of the Modulus of Elasticity E of Typical Structural Materials

MATERIAL	MODULUS OF ELASTICITY E kips/in^2	kN/cm^2
Steel	30 000	20 700
Wrought iron	28 000	19 300
Brass	15 000	10 300
Cast iron	11 000	7 500
Aluminum	10 000	7 000
Concrete (in compression)	3 000–5 000	2 000–3 400
Timber	1 760	1 200
Granite	1 280	880
Limestone	900	600
Brick	400	280
Plexiglass	400	280
Rubber	1.0	0.7

TABLE A-12 Typical Mechanical Properties of Some Metals

MATERIAL	ULTIMATE TENSILE STRESS σ_u kips/in^2	ULTIMATE TENSILE STRESS σ_u kN/cm^2	YIELD TENSILE STRESS σ_y kips/in^2	YIELD TENSILE STRESS σ_y kN/cm^2
Structural steel A-36	60	41.4	36	25
Aluminum alloy	40	27.6	35	24
Wrought iron	47	32.4	28	19.3
Cast iron	—	—	20 in tension	14
			80 in compression	55
Brass	60	41.4	50	34.5

TABLE A-13 Typical Values of Allowable Stresses σ_{all}

MATERIAL	TENSION kips/in^2	TENSION kN/cm^2	COMPRESSION ON SHORT MEMBERS kips/in^2	COMPRESSION ON SHORT MEMBERS kN/cm^2	BENDING kips/in^2	BENDING kN/cm^2	SHEAR kips/in^2	SHEAR kN/cm^2
Structural steel A-36	22	15	22	15	24†	16.5	14.5	10
Aluminum alloy	19	13	19	13	20	13.8	12	8.3
Concrete: From	0.085	0.06	1.35	0.93	1.35	0.93	0.060	0.04
Concrete: To	0.115	0.08	2.25	1.55	2.25	1.55	0.085	0.06
Douglas fir								
Perpendicular to grain	—	—	0.40	0.28	—	—	—	—
Parallel to grain	1.2–2.0	0.83–1.4	1.0–1.5	0.69–1.03	1.2–2.0	0.83–1.4	0.10	0.069

†Allowable stress in bending for adequately braced (laterally supported) beams which do not have very thin, wide flanges.

TABLE A-14 Maximum Deflection of Beams

TYPE OF BEAM LOADING	y_{max}	LOCATION OF POINT WITH y_{max}
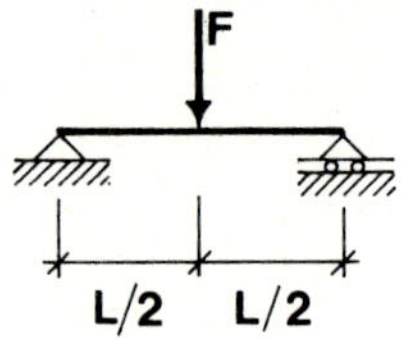	$\frac{FL^3}{48EI}$	Midspan
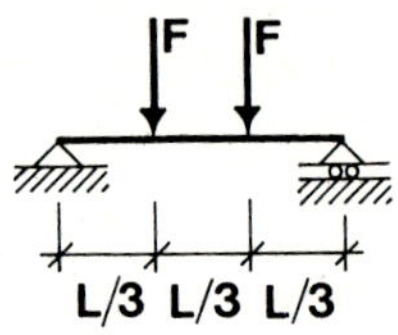	$\frac{23FL^3}{648EI}$	Midspan
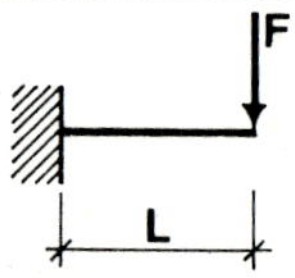	$\frac{FL^3}{3EI}$	At free end
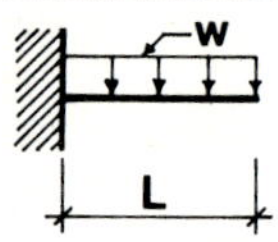	$\frac{wL^4}{8EI}$	At free end

TABLE A-15 Allowable Stresses σ_{all} (kips/in^2) for Compression Members of A-36 Steel

$\frac{KL}{r}$	σ_{all}	$\frac{KL}{r}$	σ_{all}	$\frac{KL}{r}$	σ_{all}	$\frac{KL}{r}$	σ_{all}	$\frac{KL}{r}$	σ_{all}
1	21.56	41	19.11	81	15.24	121	10.14	161	5.76
2	21.52	42	19.03	82	15.13	122	9.99	162	5.69
3	21.48	43	18.95	83	15.02	123	9.85	163	5.62
4	21.44	44	18.86	84	14.90	124	9.70	164	5.55
5	21.39	45	18.78	85	14.79	125	9.55	165	5.49
6	21.35	46	18.70	86	14.67	126	9.41	166	5.42
7	21.30	47	18.61	87	14.56	127	9.26	167	5.35
8	21.25	48	18.53	88	14.44	128	9.11	168	5.29
9	21.21	49	18.44	89	14.32	129	8.97	169	5.23
10	21.16	50	18.35	90	14.20	130	8.84	170	5.17
11	21.10	51	18.26	91	14.09	131	8.70	171	5.11
12	21.05	52	18.17	92	13.97	132	8.57	172	5.05
13	21.00	53	18.08	93	13.84	133	8.44	173	4.99
14	20.95	54	17.99	94	13.72	134	8.32	174	4.93
15	20.89	55	17.90	95	13.60	135	8.19	175	4.88
16	20.83	56	17.81	96	13.48	136	8.07	176	4.82
17	20.78	57	17.71	97	13.35	137	7.96	177	4.77
18	20.72	58	17.62	98	13.23	138	7.84	178	4.71
19	20.66	59	17.53	99	13.10	139	7.73	179	4.66
20	20.60	60	17.43	100	12.98	140	7.62	180	4.61
21	20.54	61	17.33	101	12.85	141	7.51	181	4.56
22	20.48	62	17.24	102	12.72	142	7.41	182	4.51
23	20.41	63	17.14	103	12.59	143	7.30	183	4.46
24	20.35	64	17.04	104	12.47	144	7.20	184	4.41
25	20.28	65	16.94	105	12.33	145	7.10	185	4.36
26	20.22	66	16.84	106	12.20	146	7.01	186	4.32
27	20.15	67	16.74	107	12.07	147	6.91	187	4.27
28	20.08	68	16.64	108	11.94	148	6.82	188	4.23
29	20.01	69	16.53	109	11.81	149	6.73	189	4.18
30	19.94	70	16.43	110	11.67	150	6.64	190	4.14
31	19.87	71	16.33	111	11.54	151	6.55	191	4.09
32	19.80	72	16.22	112	11.40	152	6.46	192	4.05
33	19.73	73	16.12	113	11.26	153	6.38	193	4.01
34	19.65	74	16.01	114	11.13	154	6.30	194	3.97
35	19.58	75	15.90	115	10.99	155	6.22	195	3.93
36	19.50	76	15.79	116	10.85	156	6.14	196	3.89
37	19.42	77	15.69	117	10.71	157	6.06	197	3.85
38	19.35	78	15.58	118	10.57	158	5.98	198	3.81
39	19.27	79	15.47	119	10.43	159	5.91	199	3.77
40	19.19	80	15.36	120	10.28	160	5.83	200	3.73

Index